Information Technology Plant Pathology & Biodiversity

Published By

Oxford Book Company

Regd. Off. "Bony Residency", Gate No. 2
Opp. Tilak Public School, Vishweswariya Nagar,
Gopalpura Road, Jaipur - 302018, India
Ph: 0141 - 2761280, Telefax: 2761381
email - oxfordbookjpr@gmail.com
website - oxfordbookcompany.com

Delhi Office: 102, 1st Floor, Satyam House,
4327/3, Ansari Road, Darya Ganj,
New Delhi - 110002, India
Ph: 011-45652440

Distributed by
KEYA INTERNATIONAL PRESS LLP
E-mail: keyainternationalpressllp@gmail.com

Published in India, 2025

ISBN : 9789380179551

Typeset by Shivangi Computers, Jaipur

Printed and bound in India by Replika Press Pvt. Ltd.

Preface

With the existing new technologies and innovations emerging every other day, the information technology is an increasingly interesting platform not just for developers, but also for all sorts of non-technical common people. With the need for making things easier in the daily life and with the technological advancements, more and more daily activities are shifting online. Computer based technology and information systems are actually quite large and vast spread in their utility, have broader spectrum and details. Internet plays a role of a large knowledge base and a crucial place for the latest news, trends and information.

Information technology has dramatically reduced the costs, increased the speed, and improved the productivity of life sciences research and development. Life sciences, in turn, has opened up new challenges and opportunities for IT applications. This virtuous cycle has contributed to a whole new frontier for knowledge generation. For example, the confluence of IT and biological advances made possible the mapping of the entire human genome and genomes of many other organisms in just over a decade. These discoveries, along with current efforts to determine gene and protein functions, have improved our ability to understand the root causes of human, animal and plant diseases and find new cures.

In recent years, innovations where IT and the life sciences converge have created vast quantities of data. The development of automated DNA sequencing and other innovative methods have reduced the costs and time needed to discover the genetic makeup of various organisms. The expanding complexity and diversity of information also pose new challenges. For example, data produced from studies of protein and gene function and interaction can require the consolidation of information in various formats and from diverse sources. The ability of researchers to manage and analyse these diverse and extensive data is critical to the future success of life sciences R&D.

This book is unique because it presents a broad overview of the applications of IT to plant pathology and biodiversity. Chapters cover subjects as diverse as plant diseases, biological databases, geographic information systems, and biodiversity informatics. The book will be an excellent introductory resource for all those concerned.

Dr. Parthav Narayana Murthy

Contents

	Preface	*v*
1.	Role of IT in Life Sciences	1-22
2.	Using IT to Improve Plants	23-46
3.	Information Technology and Precision Farming	47-92
4.	Plant Disease Epidemiology	93-118
5.	Pests and Diseases in Plantations	119-130
6.	Biological Control of Plant Pathogens	131-150
7.	Genetic Technology and Disease-resistant Plants	151-178
8.	Forests and Biodiversity	179-204
9.	Computerised Forest Management	205-238
10.	Medicinal Plants and Biodiversity Issues	239-254
11.	Biodiversity Conservation Services	255-286
12.	Biodiversity Informatics	287-300
	Bibliography	301-302
	Index	303-304

1

Role of IT in Life Sciences

Just as the Industrial Revolution led to changes in existing laws, regulations, management practices, and patterns of social interactions, so too is the Information Age reshaping today's economic and social environment. Narrowly viewed, the Industrial Revolution resulted from manufacturers applying newly available power systems to their production processes that, in turn, enabled the emergence of mass production. This innovation increased productivity and led to the production of completely new products at relatively low costs. However, the Industrial Revolution also drove change in the economic, social, and legal landscapes: towns grew, workers began to organize and, in time, concerns about issues such as plant safety and air quality arose. Similarly, the information technology (IT) that underlies the digital revolution is creating new economic, social, and legal challenges, even as it increases productivity. One of the most obvious shifts resulting from the digital revolution is the change in our relationship to information itself. We now expect that any information we need is easily and almost instantaneously accessible. However, that expectation is based on the assumption that information has been stored so it can be easily found and retrieved. Further, it requires that new, updated information continues to be produced and made available.

Digital capabilities are also reshaping interactions among individuals and organizations. Communication devices and channels continue to proliferate, expanding opportunities for interaction. The power inherent in new IT applications is being harnessed to improve the performance of organizations of all types by automating key interaction points. Effective management in this digital environment presents special challenges—from dealing with "interaction overload" (from unwanted telephone calls and email) to a loss of control (due to the complexity introduced by IT). However, it is the issue of identity that may prove to be one of the most fundamental challenges we encounter in the shift to a digital economy. Challenges to the security of identity come in many forms. They range from identity theft to unauthorized access to a network (hacking) or a facility. One important area of IT research and development is in the evolution of reliable identity verification technologies. Such technologies are needed in both the physical and virtual worlds. Although such security tools are a necessity, they must be designed and implemented so that they do not ignite privacy concerns.

With the existing new technologies and innovations emerging every other day, the information technology is an increasingly interesting platform not just for developers, but also for all sorts of non-technical common people.

With the need for making things easier in the daily life and with the technological advancements, more and more daily activities are shifting online. Having said this, the web can be a very useful tool as well as an intimidating proposition at the same time.

Computer based technology and information systems are actually quite large and vast spread in their utility, have broader spectrum and details. For instance when anyone uses the web for browsing, sending or receiving e-mails, playing online games or even sharing multimedia files with others, all the data has to pass through a set of complicated networks and soft-wares. There are many processes involved that are responsible for management of such systems.

The prime concern of computer related technology is to provide effective and efficient environment, utility of information, softwares and even knowledge and solutions to the common man in a user-friendly manner.

Even the vast majority of people who are still unfamiliar with the technology of computers and the internet, surely find it handy in providing a great means of communication to the whole wide world.

Internet plays a role of a large knowledge base and a crucial place for the latest news, trends and information. Internet even is a big boon to the business or research persons.

Computer related technologies have a strong impact on the world. These have attracted many students and professionals to the field of information technology. There are thousands of web sites and web hosting opportunities available which are ever growing.

There was some problem in the information superhighway of computer technology in the early 90's because it was not envisioned at that time that the general public would be turning to it in such large numbers. It was supposed to be a walkway reserved for bespectacled physicists and university professionals. With the introduction of the World Wide Web which we know today as 'www' was once considered as a medium for sharing text files has gone a major facelift in a period of a decade or so. With the web browsers designed to quickly find and organize information, the internet seeped deeper into popular culture and has become an integral part of daily life and even office work.

The unique fusion of the user interface, animations, video and radio streaming are the ever developing capabilities and advancements of the browsers that have the developers intrigued and working tirelessly.

There are many researches already claiming that the effect of the computer technology, more specifically the online footprint on an internet user's life is at an average more than that is due to watching TV. What television is renowned for is its multimedia content. The internet with its rich applications and focused capacity of gaining more popular viewer ship is moving in the right direction.

INFORMATION

Improvements in technology continue to increase our ability to capture, store, manipulate, and display information. Declining costs and shrinking component size have accompanied these technological improvements. The combination of these trends has led to a sharp increase in the information component of many everyday devices—from disposable telephone cards that track message units to pocket-sized telephones that have contact databases and games. Businesses also benefit as IT enables the development of manufacturing equipment that not only produces the product, but also reports production-run quality and tracks its own maintenance schedule. Even the authors of this report have benefited from the improved functionality of our computers. Using sophisticated statistical software, we can now manipulate large datasets that, until recently, exceeded the storage capacity of most desktop computers. The Internet has become key in information expansion, by providing a common protocol for communication among devices. Although increased information content and functionality is beneficial for any single device, when devices are joined together in a network their potential expands dramatically. Using the Internet, people can locate everything from the mundane (looking up a pasta salad recipe, checking a bank account balance, verifying the movie schedule) to the important (finding information on evacuation routes following a disaster, or deodorizing a child who mistook a skunk for her puppy). The Internet also makes available information that was formerly beyond easy reach, such as: a list of ongoing clinical trials from the National Institutes of Health, the English language version of Al Jazeera, or historical photographs of the Wright brothers. In addition to providing benefits to private individuals, the Internet has also become an integral part of the landscape in which government, business, and organizations function. As individuals and organizations continue to weave the Internet into their activities and infrastructures, resolving issues of search, archiving, and protection of intellectual property rights is critical to realizing the full potential of this network of networks.

Search

Internet users depend on their Web browser's search engine, on navigational guides or portals (e.g., Yahoo!), or on stand-alone search engines (e.g., Google, Altavista, and NorthernLight) to locate information. Since search engines use different algorithms—algorithms that may or may not have advertising expenditures as a variable—to locate items on the Internet, results can vary substantially.

Nor is any search engine capable of searching everything. Two of the most prominent limiting factors are related to language and file format. During a recent Online Information conference it was noted that "the most obvious access problem is that 'all the world's information' will be provided in documented form in most of the world's languages, and while Google and other search engines have interfaces in the major languages, the information that is retrieved will not necessarily be in the language of the interface."Speakers at the conference went on to note that while there are almost 600 file formats according to one count (with at least half of these found on the Internet), search engines index only a small portion of these.

Archiving

Organizations charged with maintaining information collections, such as libraries, face significant opportunities and challenges in a digital environment. The opportunities arise because digital information is easy to replicate and transmit with no loss of quality, which makes sharing articles, books, movies, pictures, and audio recordings easy. The procession of improvements in information storage—from microfiche and microfilm to the Internet, CDs, and DVDs—has substantially increased the opportunities for information collection, storage, and sharing. This increase has many benefits. For example, teachers can bring historical photo- graphs from the Library of Congress collections directly to their classrooms, and voters can easily review statements made by candidates for office. However, from an archiving standpoint, content created digitally is of particular concern. Some historic documents that once would have circulated on paper may now only be circulated electronically—for example, an e-mailed note to a member of the President's staff containing comments on a proposed treaty.

Solving the problems of digital archiving requires both a strategy and improved IT tools. The National Archives and Records Administration (NARA), given its mission of preserving the "essential evidence that documents the rights of American citizens, the actions of federal officials, and the national experience," faces a particular challenge with regard to the growing volume of e-mail communications. Recognizing that its "current systems for archival preservation of electronic records are limited in capability and ad hoc in nature…NARA launched the Electronic Records Archives (ERA) initiative," with the San Diego Supercomputer Center to improve its capabilities.On a broader level, the National Digital Information Infrastructure and Preservation Program charges the Library of Congress to work with NARA, the Commerce Department, the White House Office of Science and Technology Policy, the National Library of Medicine, the National Agricultural Library, the National Institute of Standards and Technology and "other federal, research and private libraries and institutions with expertise in telecommunications technology and electronic commerce policy" with developing the "protocols and strategies for the longterm preservation of such materials, including the technological infrastructure required at the Library of Congress."Archiving is not only a concern for organizations, such as the National Archives, but also for other government entities, businesses, and individuals that collect, archive, and preserve digital communications and documents. Every individual and organization that deals with information must devise a system to organize what they want to, or must, keep—whether paper, disk, or e-mail attachment. Organizing and storing information that may be needed in the future is a growing challenge.

Control

It has always been difficult for creators of intellectual property to maintain control of their output and information technology has a long history of increasing that difficulty. Making copies of written text—whether the original was carved in stone or written on paper—has always been possible. However, technology continues to make it increasingly easier and cheap. The advent of the photocopier allowed any individual to make copies cheaply without regard to the wishes of the copyright owner and the spread of facsimile (fax) machines allowed for quick dissemination of written documents. Similarly, the creation of relatively low-cost audio and video recording devices presented challenges to those who owned music and film publishing rights. Maintaining control of intellectual property became significantly more difficult in the digital environment. Not only can one create copies and disseminate them at virtually no cost, digital copies are equal in quality to the original.

The audio recording industry is one interesting example of an industry trying to find its equilibrium in the digital world. Napster challenged the recording industry's business model by enabling Internet users to bypass its distribution systems. In doing so, Napster also made it more difficult for the industry to control, and thereby profit from, its intellectual property (the rights to music owned by record companies and recording artists). Although the recording industry succeeded in shutting down the free version of Napster in 2001, other services (ones that are more truly peer-to-peer and therefore more difficult to shut down) such as Morpheus and Kazaa have emerged. According to one estimate, Kazaa users number almost a quarter billion worldwide—triple the number of users that Napster had prior to its shutdown.The ability of users of these online music-sharing services to swap music over the Internet called into question the market for shrink-wrapped compact discs, as well as the licensing arrangements between recording artists and their publishers. Response to this continued challenge has been twofold—litigation and the development of legitimate online markets, such as iTunes.

INTERACTIONS

IT is also transforming information exchange. Not only do people and devices have the opportunity to interact over a growing number of channels, the differing attributes of these channels are altering the activities in which people engage. From high-profile shifts, such as the proliferation of e-mail and e-commerce sites, to more behind-the-scenes shifts in activities, such as supply chain management, these new options are having a profound effect on the economic environment. The increased potential for interaction brings benefits, but also raises the specter of interaction overload. Innovations in IT have resulted in a proliferation of communication devices operating over a variety of channels. These devices and channels are more than simple substitutes for each other. Their underlying technologies give them unique attributes that are redefining the terms under which communications and commerce occur.

COMMUNICATIONS

As anyone who has recently filled out a school form can attest, the number of communications channels available to the members of an average household is large—home telephone number(s), and home e-mail address(es), work telephone number(s), work e-mail address(es), mobile telephone number(s), beeper number(s), etc.

Communication devices and channels proliferate because each device or channel has a different set of attributes. With communication using basic landline telephony, you make a call to a specific physical location, and someone at the location does (or does not) answer. Landline telephony is extremely robust (system outages are rare, and dropped calls are not a problem) and relatively cheap. With more sophisticated telephone systems, people can forward calls, identify callers, and store messages. Mobile telephones bring another dimension to telephony by associating the telephone number with an individual, not a location. Channels based on Internet technologies enable the ability to send anything that can be stored in a digital form to multiple select recipients or to post it for the world-at-large. Text messaging, whether over a telephone or a personal digital assistant, is another new communications type.The variety of interactions conducted widens considerably as individuals and organizations use these communications tools to redefine how they conduct many common activities. One attribute of some of the channels listed above is to diminish the importance of a user's location. The development of landline telephony meant that the parties to a conversation could be located at great distances from each other, but at stationary locations. Mobile telephony means that, within certain geographic limits, one need not know where a person is physically to communicate with him. E-mails are also non-location dependent because one can send and receive messages anywhere in the world to any account accessible over the Internet.

Furthermore, Internet access is increasingly becoming a mobile communications channel due to the spread of broadband wireless Internet access (e.g., Wi-Fi).A barrage of unwanted messages—some legitimate marketing messages and other potentially fraudulent or dangerous—have accompanied this expansion in communications channels. Anyone with an e-mail account has experienced spam—unsolicited notifications spanning the range from sexual aids to illegal international money laundering schemes. Many individuals feel overwhelmed by the constant barrage. The volume of illegitimate messages also makes it more difficult for legitimate businesses to communicate with potential customers who would be interested in their offers. Attempts to provide appropriate remedies have had mixed success. Specific regulations are in place to protect children online, but, in general, individuals and organizations must rely on filtering protocols that are less than perfect. Efforts to reduce the number of telephone solicitations through the use of the "Do Not Call" list being administered by the Federal Trade Commission are currently under court challenge, even as anti-spam legislation was passed by Congress.

E-COMMERCE

A specific type of interaction that has received considerable attention over the past few years is electronic commerce—that is, buying and selling online. The evolution of online transactions has been both "less" and "more" than many analysts originally estimated. During the second quarter of 2003, the U.S. Bureau of the Census reported that retail e-commerce sales were $12.5 billion.Although this represents only 1.5 percent of total retail sales—far from replacing instore sales as some proponents promised—it does represent a 28 percent increase over the second quarter of 2002. E-commerce growth in the business-to-business space, has also fallen short of early expectations. Between 2000 and 2001 (the latest data available), manufacturing e-commerce (whether over the Internet or proprietary systems) increased from 18.0 to 18.3 percent of total shipments, and merchant wholesale e-commerce increased from 8.8 to 10.0 percent.However, if one considers only the dollar value of online transactions, the importance of e-commerce to the economy is underestimated. Even when the transaction does not take place online, the terms and conditions of the commercial interaction are altered by the availability of e-commerce options. A car buyer can go to the nearest dealership armed with detailed research obtained online. The local bookstore must now consider the pricing and service policies of online competitors.

The availability of IT products and services also impacts the processes that underlie the interactions. The term "e-business processes" refers to business activities that use information and communications technologies. E-commerce is a specific type of e-business process, as are human resource information systems, and enterprise resources planning systems. One particularly interesting area of e-business applications is that of supply chain management. Use of IT products and services has enabled interactions between contract participants that closely rival (if not match) the quality of interaction that occurs within a firm. This gives businesses considerable leeway in determining which functions to conduct in-house and which to outsource.

While the ability to automate interactions has been available to large organizations for some time, the development of low cost, "off-the-shelf" tools has greatly expanded the use of e-business applications among smaller organizations. IT-driven changes in the ways that businesses are managing their supply chains has been so great that the statistical agencies have had to reexamine their data collection in a number of areas.

When a resource becomes essential to competition but inconsequential to strategy, the risks it creates become more important than the advantages it provides. Increasingly, IT is a prerequisite rather than an option for governments, businesses, volunteer groups, and households. Possessing information technology may not put you ahead, it may just keep you from falling behind. In such an environment, effective management of IT resources is critical.

The supply chain is one area of business process where IT is providing the means for businesses to streamline and reduce costs. While new supply chain efficiencies benefit the companies undertaking such investments, the new business structures pose challenges for economic data collection. Existing statistical programs classify business locations into industries based on their underlying production function—businesses doing similar activities are grouped together. This classification system assumes that manufacturers, wholesalers, retailers, and service businesses each perform a distinct set of functions that fit neatly into separate boxes.

As electronic information management changes the way businesses interact with each other, boundaries between these formerly distinct sectors are blurring. Manufacturers, wholesalers, and retailers all may be selling "services." Service firms in transportation and logistics may be leveraging their expertise to take on new functions such as inventory management, or other functions that traditionally have been associated with manufacturers and distributors.

An unstructured approach to IT investment can cause management problems. The actual interactions occurring between people, processes, data, and the technology become obscured under a wild tangle. Many organizations facing this problem have turned to broad-scale approaches, such as enterprise architecture (EA), to gain an understanding of how the organization is operating and, therefore, how it can be improved. Further, since IT increases the amount of information available to every member of the organization, decision-makers in these organizations are thinking about how technology can be used to empower and enable each individual involved.

An enterprise's leaders must have an accurate understanding of how the enterprise operates to manage it effectively. As an enterprise's complexity increases, this task becomes more difficult. One technique for accomplishing this goal is to construct a systematic description of the enterprise—an enterprise architecture—that relates the outcome of the enterprise's activities to the contributions of its people, business processes, data, and technology. An EA provides the holistic, "outside looking in," view that enables decision makers to understand where incompatibilities, redundancies, and overlaps exist within the enterprise. It also provides insight into the extent to which individual activities are contributing to (or detracting from) the enterprise's mission.

An enterprise's EA typically reveals a very different organization than the enterprise's organizational chart, and it forces enterprise management to confront how it actually conducts business. When all the points of interaction are described, management then has the information needed to begin a review of the effectiveness of these interactions in supporting the mission and goals of the enterprise.

COLLABORATIVE TOOLS

IT offers opportunities for communities of interest to form, share information, and work toward common goals. Some of these collaborative spaces are casual, such as online chat rooms. Others are formal, such as a team working on a design project. Businesses, for example, are increasingly using collaborative software programs to facilitate group discussion and decisionmaking. Such software allows participants to interact in dedicated online spaces, engage in discussions, and share and track information. Collaborative software, like most IT tools, cannot create efficiency in a vacuum. Organizers and participants must establish rules for interaction (i.e., who can participate, how action items are distinguished from extended conversations, who is in charge of deleting obsolete items, etc.).

IDENTITY

Ten years ago Peter Steiner succinctly captured one of the key issues in Internet interaction in a New Yorker cartoon showing one dog sitting at a computer talking to another dog with the caption "On the Internet, nobody knows you're a dog."And the identity issue remains one of critical importance today. The shift online of many work and personal activities requires that users have some level of assurance about the identity of the people or businesses from whom they receive information or with whom they conduct business. Effective use of networks also requires that participants are confident that information and transactions are not altered during transmission or storage. They must also be confident that access to sensitive or proprietary information is limited to users entitled to access that information. Without adequate safeguards,businesses and individuals will bear the brunt of increased cost due to fraud and theft and, as a society, we will not fully realize the potential benefits of online activity. Obviously, identity concerns exist beyond the online environment. The need to verify identity is of critical importance in many contexts. Port security requires that seafarers entering a harbor on a cargo ship are correctly identified and adequately screened. Financial market stability requires that investors are confident that their financial transactions occurred in the manner requested and that no one has tampered with their account balances. Residents living near a chemical or power plant need assurance that only actual employees can access the plant's control facilities.

Challenges to the security of identity include such disparate activities as identity theft, unauthorized access to a network (hacking), and unauthorized access to a facility. One important area of IT research and development is the development of technologies capable of verifying identity with a high degree of certainty. However, the use of such security tools can ignite privacy concerns.

BIOINFORMATICS

Bioinformatics is the application of information technology to the field of molecular biology. The term bioinformatics was coined by Paulien Hogeweg in 1979 for the study of informatic processes in biotic systems. Bioinformatics now entails the creation and advancement of databases, algorithms, computational and statistical techniques, and theory to solve formal and practical problems arising from the management and analysis of biological data. Over the past few decades rapid developments in genomic and other molecular research technologies and developments in information technologies have combined to produce a tremendous amount of information related to molecular biology. It is the name given to these mathematical and computing approaches used to glean understanding of biological processes. Common activities in bioinformatics include mapping and analyzing DNA and protein sequences, aligning different DNA and protein sequences to compare them and creating and viewing 3-D models of protein structures.

The primary goal of bioinformatics is to increase our understanding of biological processes. What sets it apart from other approaches, however, is its focus on developing and applying computationally intensive techniques (e.g., data mining, machine learning algorithms, and visualization) to achieve this goal. Major research efforts in the field include sequence alignment, gene finding, genome assembly, protein structure alignment, protein structure prediction, prediction of gene expression and protein-protein interactions, genome-wide association studies and the modeling of evolution.

Bioinformatics was applied in the creation and maintenance of a database to store biological information at the beginning of the "genomic revolution", such as nucleotide and amino acid sequences. Development of this type of database involved not only design issues but the development of complex interfaces whereby researchers could both access existing data as well as submit new or revised data. In order to study how normal cellular activities are altered in different disease states, the biological data must be combined to form a comprehensive picture of these activities. Therefore, the field of bioinformatics has evolved such that the most pressing task now involves the analysis and interpretation of various types of data, including nucleotide and amino acid sequences, protein domains, and protein structures. The actual process of analyzing and interpreting data is referred to as computational biology. Important sub-disciplines within bioinformatics and computational biology include:

a) the development and implementation of tools that enable efficient access to, and use and management of, various types of information.

b) the development of new algorithms (mathematical formulas) and statistics with which to assess relationships among members of large data sets, such as methods to locate a gene within a sequence, predict protein structure and/or function, and cluster protein sequences into families of related sequences.

Major Research Areas

Sequence analysis

Since the Phage F-X174 was sequenced in 1977, the DNA sequences of hundreds of organisms have been decoded and stored in databases. The information is analyzed to determine genes that encode polypeptides, as well as regulatory sequences. A comparison of genes within a species or between different species can show similarities between protein functions, or relations between species (the use of molecular systematics to construct phylogenetic trees). With the growing amount of data, it long ago became impractical to analyze DNA sequences manually. Today, computer programs are used to search the genome of thousands of organisms, containing billions of nucleotides. These programs would compensate for mutations (exchanged, deleted or inserted bases) in the DNA sequence, in order to identify sequences that are related, but not identical. A variant of this sequence alignment is used in the sequencing process itself. The so-called shotgun sequencing technique does not give a sequential list of nucleotides, but instead the sequences of thousands of small DNA fragments (each about 600-800 nucleotides long). The ends of these fragments overlap and, when aligned in the right way, make up the complete genome. Shotgun sequencing yields sequence data quickly, but the task of assembling the fragments can be quite complicated for larger genomes. In the case of the Human Genome Project, it took several days of CPU time (on one hundred Pentium III desktop machines clustered specifically for the purpose) to assemble the fragments. Shotgun sequencing is the method of choice for virtually all genomes sequenced today, and genome assembly algorithms are a critical area of bioinformatics research.

Genome annotation

In the context of genomics, annotation is the process of marking the genes and other biological features in a DNA sequence. The first genome annotation software system was designed in 1995 by Dr. Owen White, who was part of the team that sequenced and analyzed the first genome of a free-living organism to be decoded, the bacterium *Haemophilus influenzae*. Dr. White built a software system to find the genes (places in the DNA sequence that encode a protein), the transfer RNA, and other features, and to make initial assignments of function to those genes. Most current genome annotation systems work similarly, but the programs available for analysis of genomic DNA are constantly changing and improving.

Computational evolutionary biology

Evolutionary biology is the study of the origin and descent of species, as well as their change over time. Informatics has assisted evolutionary biologists in several key ways; it has enabled researchers to:

- trace the evolution of a large number of organisms by measuring changes in their DNA, rather than through physical taxonomy or physiological observations alone,
- more recently, compare entire genomes, which permits the study of more complex evolutionary events, such as gene duplication, horizontal gene transfer, and the prediction of factors important in bacterial speciation,
- build complex computational models of populations to predict the outcome of the system over time
- track and share information on an increasingly large number of species and organisms

The area of research within computer science that uses genetic algorithms is sometimes confused with computational evolutionary biology, but the two areas are unrelated.

Biodiversity

Biodiversity of an ecosystem might be defined as the total genomic complement of a particular environment, from all of the species present, whether it is a biofilm in an abandoned mine, a drop of sea water, a scoop of soil, or the entire biosphere of the planet Earth. Databases are used to collect the species names, descriptions, distributions, genetic information, status and size of populations, habitat needs, and how each organism interacts with other species. Specialized software programs are used to find, visualize, and analyze the information, and most importantly, communicate it to other people. Computer simulations model such things as population dynamics, or calculate the cumulative genetic health of a breeding pool (in agriculture) or endangered population (in conservation). One very exciting potential of this field is that entire DNA sequences, or genomes of endangered species can be preserved, allowing the results of Nature's genetic experiment to be remembered *in silico*, and possibly reused in the future, even if that species is eventually lost.

Analysis of gene expression

The expression of many genes can be determined by measuring mRNA levels with multiple techniques including microarrays, expressed cDNA sequence tag (EST) sequencing, serial analysis of gene expression (SAGE) tag sequencing, massively parallel signature sequencing (MPSS), or various applications of multiplexed in-situ hybridization. All of these techniques are extremely noise-prone and/or subject to bias in the biological measurement, and a major research area in computational biology involves developing statistical tools to separate signal from noise in high-throughput gene expression studies. Such studies are often used to determine the genes implicated in a disorder: one might compare microarray data from cancerous epithelial cells to data from non-cancerous cells to determine the transcripts that are up-regulated and down-regulated in a particular population of cancer cells.

Analysis of regulation

Regulation is the complex orchestration of events starting with an extracellular signal such as a hormone and leading to an increase or decrease in the activity of one or more proteins. Bioinformatics techniques have been applied to explore various steps in this process. For example, promoter analysis involves the identification and study of sequence motifs in the DNA surrounding the coding region of a gene. These motifs influence the extent to which that region is transcribed into mRNA. Expression data can be used to infer gene regulation: one might compare microarray data from a wide variety of states of an organism to form hypotheses about the genes involved in each state. In a single-cell organism, one might compare stages of the cell cycle, along with various stress conditions (heat shock, starvation, etc.). One can then apply clustering algorithms to that expression data to determine which genes are co-expressed. For example, the upstream regions (promoters) of co-expressed genes can be searched for over-represented regulatory elements.

Analysis of protein expression

Protein microarrays and high throughput (HT) mass spectrometry (MS) can provide a snapshot of the proteins present in a biological sample. Bioinformatics is very much involved in making sense of protein microarray and HT MS data; the former approach faces similar problems as with microarrays targeted at mRNA, the latter involves the problem of matching large amounts of mass data against predicted masses from protein sequence databases, and the complicated statistical analysis of samples where multiple, but incomplete peptides from each protein are detected.

Analysis of mutations in cancer

In cancer, the genomes of affected cells are rearranged in complex or even unpredictable ways. Massive sequencing efforts are used to identify previously unknown point mutations in a variety of genes in cancer. Bioinformaticians continue to produce specialized automated systems to manage the sheer volume of sequence data produced, and they create new algorithms and software to compare the sequencing results to the growing collection of human genome sequences and germline polymorphisms. New physical detection technology are employed, such as oligonucleotide microarrays to identify chromosomal gains and losses (called comparative genomic hybridization), and single nucleotide polymorphism arrays to detect known *point mutations*. These detection methods simultaneously measure several hundred thousand sites throughout the genome, and when used in high-throughput to measure thousands of samples, generate terabytes of data per experiment. Again the massive amounts and new types of data generate new opportunities for bioinformaticians. The data is often found to contain considerable variability, or noise, and thus Hidden Markov model and change-point analysis methods are being developed to infer real copy number changes. Another type of data that requires novel informatics development is the analysis of lesions found to be recurrent among many tumors .

Prediction of protein structure

Protein structure prediction is another important application of bioinformatics. The amino acid sequence of a protein, the so-called primary structure, can be easily determined from the sequence on the gene that codes for it. In the vast majority of cases, this primary structure uniquely determines a structure in its native environment. (Of course, there are exceptions, such as the bovine spongiform encephalopathy - aka Mad Cow Disease - prion.) Knowledge of this structure is vital in understanding the function of the protein. For lack of better terms, structural information is usually classified as one of *secondary, tertiary* and *quaternary* structure. A viable general solution to such predictions remains an open problem. As of now, most efforts have been directed towards heuristics that work most of the time.

One of the key ideas in bioinformatics is the notion of homology. In the genomic branch of bioinformatics, homology is used to predict the function of a gene: if the sequence of gene *A*, whose function is known, is homologous to the sequence of gene *B*, whose function is unknown, one could infer that B may share A's function. In the structural branch of bioinformatics, homology is used to determine which parts of a protein are important in structure formation and interaction with other proteins. In a technique called homology modeling, this information is used to predict the structure of a protein once the structure of a homologous protein is known. This currently remains the only way to predict protein structures reliably.

One example of this is the similar protein homology between hemoglobin in humans and the hemoglobin in legumes (leghemoglobin). Both serve the same purpose of transporting oxygen in the organism. Though both of these proteins have completely different amino acid sequences, their protein structures are virtually identical, which reflects their near identical purposes.

Comparative genomics

The core of comparative genome analysis is the establishment of the correspondence between genes (orthology analysis) or other genomic features in different organisms. It is these intergenomic maps that make it possible to trace the evolutionary processes responsible for the divergence of two genomes. A multitude of evolutionary events acting at various organizational levels shape genome evolution. At the lowest level, point mutations affect individual nucleotides. At a higher level, large chromosomal segments undergo duplication, lateral transfer, inversion, transposition, deletion and insertion. Ultimately, whole genomes are involved in processes of hybridization, polyploidization and endosymbiosis, often leading to rapid speciation. The complexity of genome evolution poses many exciting challenges to developers of mathematical models and algorithms, who have recourse to a spectra of algorithmic, statistical and mathematical techniques, ranging from ,exact, heuristics, fixed parameter and approximation algorithms for problems based on parsimony models to Markov Chain Monte Carlo algorithms for Bayesian analysis of problems based on probabilistic models.

Modeling biological systems

Systems biology involves the use of computer simulations of cellular subsystems (such as the networks of metabolites and enzymes which comprise metabolism, signal transduction pathways and gene regulatory networks) to both analyze and visualize the complex connections of these cellular processes. Artificial life or virtual evolution attempts to understand evolutionary processes via the computer simulation of simple (artificial) life forms.

High-throughput image analysis

Computational technologies are used to accelerate or fully automate the processing, quantification and analysis of large amounts of high-information-content biomedical imagery. Modern image analysis systems augment an observer's ability to make measurements from a large or complex set of images, by improving accuracy, objectivity, or speed. A fully developed analysis system may completely replace the observer. Although these systems are not unique to biomedical imagery, biomedical imaging is becoming more important for both diagnostics and research. Some examples are:

- high-throughput and high-fidelity quantification and sub-cellular localization
- morphometrics
- clinical image analysis and visualization
- determining the real-time air-flow patterns in breathing lungs of living animals
- quantifying occlusion size in real-time imagery from the development of and recovery during arterial injury
- making behavioral observations from extended video recordings of laboratory animals
- infrared measurements for metabolic activity determination
- inferring clone overlaps in DNA mapping, e.g. the Sulston score

Protein-protein docking

In the last two decades, tens of thousands of protein three-dimensional structures have been determined by X-ray crystallography and Protein nuclear magnetic resonance spectroscopy (protein NMR). One central question for the biological scientist is whether it is practical to predict possible protein-protein interactions only based on these 3D shapes, without doing protein-protein interaction experiments. A variety of methods have been developed to tackle the Protein-protein docking problem, though it seems that there is still much work to be done in this field.

Software and tools

Software tools for bioinformatics range from simple command-line tools, to more complex graphical programs and standalone web-services available from various bioinformatics companies or public institutions. The computational biology tool best-known among biologists is probably BLAST, an algorithm for determining the similarity of arbitrary sequences against other sequences, possibly from curated databases of protein or DNA sequences. BLAST is one of a number of generally available programs for doing sequence alignment. The NCBI provides a popular web-based implementation that searches their databases.

Web services in bioinformatics

SOAP and REST-based interfaces have been developed for a wide variety of bioinformatics applications allowing an application running on one computer in one part of the world to use algorithms, data and computing resources on servers in other parts of the world. The main advantages lay in the end user not having to deal with software and database maintenance overheads. Basic bioinformatics services are classified by the EBI into three categories: SSS (Sequence Search Services), MSA (Multiple Sequence Alignment) and BSA (Biological Sequence Analysis). The availability of these service-oriented bioinformatics resources demonstrate the applicability of web based bioinformatics solutions, and range from a collection of standalone tools with a common data format under a single, standalone or web-based interface, to integrative, distributed and extensible bioinformatics workflow management systems.

COMPUTATIONAL BIOLOGY

Computational biology is an interdisciplinary field that applies the techniques of computer science, applied mathematics and statistics to address biological problems. The main focus lies on developing mathematical modeling and computational simulation techniques. By these means it addresses scientific research topics with their theoretical and experimental questions without a laboratory. It encompasses the fields of:

Bioinformatics, which applies algorithms and statistical techniques to the interpretation, classification and understanding of biological datasets. These typically consist of large numbers of DNA, RNA, or protein sequences. Sequence alignment is used to assemble the datasets for analysis. Comparisons of homologous sequences, gene finding, and prediction of gene expression are the most common techniques used on assembled datasets; however, analysis of such datasets have many applications throughout all fields of biology.

Computational biomodeling, a field within biocybernetics concerned with building computational models of biological systems.Computational genomics, a field within genomics which studies the genomes of cells and organisms. High-throughput genome sequencing produces lots of data, which requires extensive post-processing (genome assembly) and uses DNA microarray technologies to perform statistical analyses on the genes expressed in individual cell types. This can help find genes of interests for certain diseases or conditions. This field also studies the mathematical foundations of sequencing.

Molecular modeling, which consists of modelling the behaviour of molecules of biological importance. Protein structure prediction and structural genomics, which attempt to systematically produce accurate structural models for three-dimensional protein structures that have not been determined experimentally. Computational biochemistry and biophysics, which make extensive use of structural modeling and simulation methods such as molecular dynamics and Monte Carlo method-inspired Boltzmann sampling methods in an attempt to elucidate the kinetics and thermodynamics of protein functions.

IT IN LIFE SCIENCE R&D

Information technology (IT) has dramatically reduced the costs, increased the speed, and improved the productivity of life sciences research and development (R&D). Life sciences R&D, in turn, has opened up new challenges and opportunities for IT applications. This virtuous cycle has contributed to a whole new frontier for knowledge generation. For example, the confluence of IT and biological advances made possible the mapping of the entire human genome and genomes of many other organisms in just over a decade. These discoveries, along with current efforts to determine gene and protein functions, have improved our ability to understand the root causes of human, animal and plant diseases and find new cures. Furthermore, many future IT innovations will likely be spurred by the data and analysis demands of the life sciences.

In recent years, innovations where IT and the life sciences converge have created vast quantities of data. The development of automated DNA sequencing and other innovative methods have reduced the costs and time needed to discover the genetic makeup of various organisms.Additionally, innovations such as high-throughput screening and microarrays are enabling studies of gene and protein functions (i.e., genomics and proteomics). The expanding complexity and diversity of information also pose new challenges. For example, data produced from studies of protein and gene function and interaction can require the consolidation of information in various formats and from diverse sources. Additionally, data may not be in a standard or readily usable format, particularly if they are not available electronically or are text-based information from scientific literature, patents and clinical trials. Critical biological information also is found throughout various scientific disciplines, such as the chemical sciences.The ability of researchers to manage and analyze these diverse and extensive data is critical to the future success of life sciences R&D.

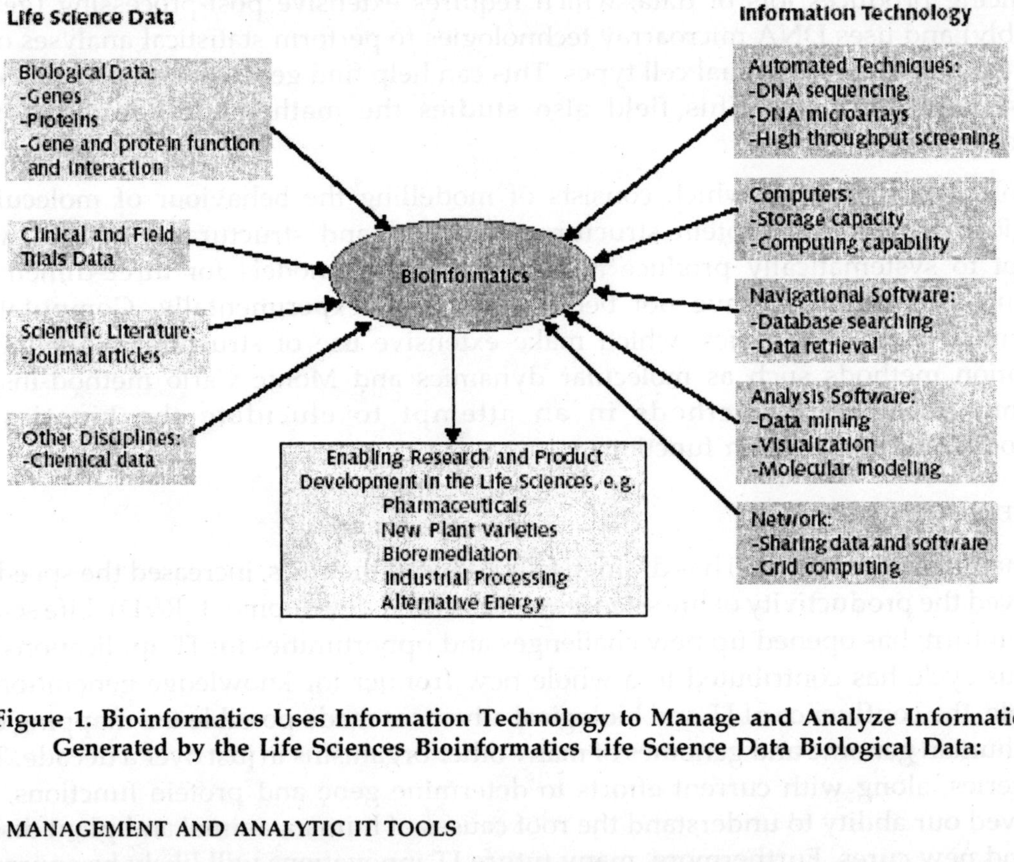

Figure 1. Bioinformatics Uses Information Technology to Manage and Analyze Information Generated by the Life Sciences Bioinformatics Life Science Data Biological Data:

DATA MANAGEMENT AND ANALYTIC IT TOOLS

The needs described above have spawned the research field "bioinformatics,"which focuses on the use of IT to collect, organize, store, interpret, share, and analyze biological data. Developments in bioinformatics will be critical for facilitating R&D in areas such as human and animal health, agriculture, industrial processing, natural resource recovery, and environmental remediation. Figure 1 illustrates the IT and life sciences components of bioinformatics.

COMPUTING POWER AND NETWORKS

Data storage, management, and analysis requirements in the life sciences are outpacing current computing capabilities, even though computing power continues to increase. Computing power has expanded at roughly the same rate as public DNA sequences located at the National Institutes of Health (NIH)—both have been doubling every 18 to 24 months for several years (Figure 2). In 1971, only 2,250 transistors were on an integrated circuit; by 2002 there were 42 million. Similarly, only 606 DNA sequences were housed at NIH's Genbank in 1982. The number of sequences climbed to 22.3 million in just 20 years.

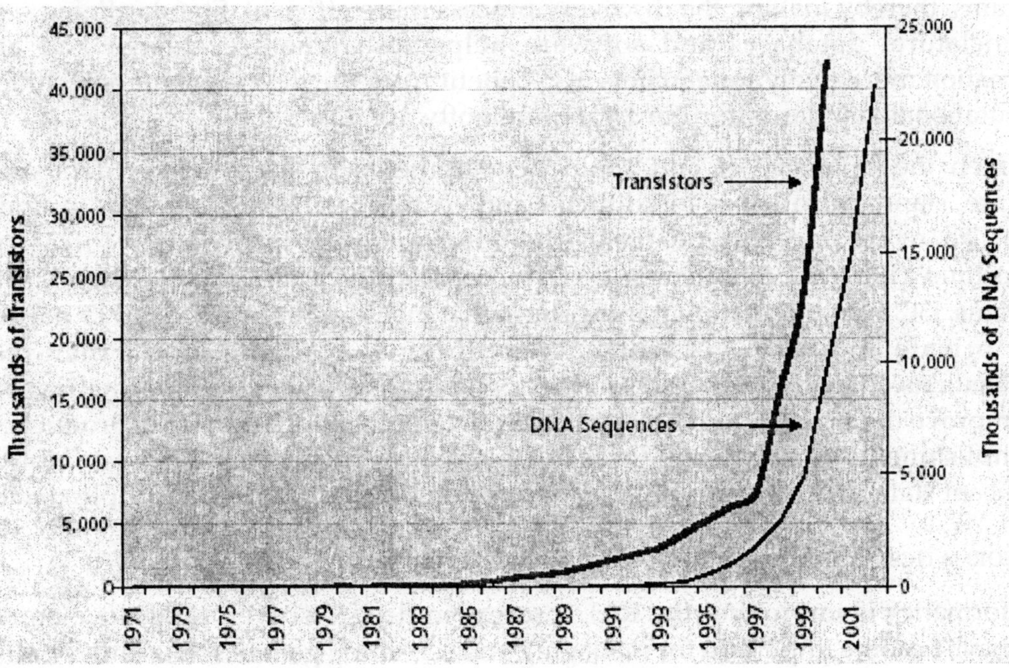

Figure 2. Number of Transistors per Integrated Circuit vs. DNA Sequences in Genbank

Researchers are exploring strategies to increase computing capabilities for diverse life sciences purposes, such as computer-based drug design. One strategy is to expand the potential of supercomputers. Another is to unify computing resources through development of global, or grid, computing networks. These networks link the power of individual PCs or supercomputers.Researchers have achieved teraflop speeds (a trillion floating point operations per second) using these grid networks. In the future they hope to achieve petaflop speeds (a thousand trillion floating point operations per second).The Internet is a particularly valuable tool for life sciences researchers, especially for those with limited resources. It enables researchers to tap into data, software, and computing power.

New software developments also promise life sciences researchers' the ability to acquire, format, search, and analyze disparate data sources and types. Data mining and visualization softwareare improving scientists' ability to screen data and identify important relationships. One software analysis tool, 3-D molecular modeling, facilitates studies of the function and relationships between biological molecules. These analysis tools are enabling scientists to conduct some research at the computer instead of at the lab bench, reducing the time and costs of life sciences R&D.

Standardization enables researchers to link different databases and software programs, thereby making the discovery process more efficient. Standardizing scientific nomenclature, databases and software helps researchers exchange and analyze information more easily.Researchers also benefit from the development of technical and computational standards for hardware and software.

Given the diversity of and rapid developments in life sciences databases and software, standardization is too difficult and costly for any organization to undertake alone. To overcome this obstacle, the Interoperable Informatics Infrastructure Consortium (I3C)was founded in 2001 to collectively address some standardization problems. I3C is an international consortium that includes life science and IT participants from private industry, government institutions, academia and other research organizations. They develop and promote "global, vendor-neutral informatics solutions that improve data quality and accelerate the development of life science products." I3C's accomplishments include standards developed to identify and access biologically significant data and a method that simplifies data retrieval from multiple databases.

BIOINFORMATICS ASSISTS DRUG DISCOVERY AND DEVELOPMENT

Bioinformatics is improving the R&D process in drug discovery and development. IT tools have become important for managing and screening genetics data and for modeling outcomes in drug development. New developments in bioinformatics and genetics, such as pharmacogenetics, will enable researchers to identify quickly a patient's genetic predisposition to contract certain diseases as well as their potential drug response. Costs of drug discovery have been escalating, nearly tripling since 1991.It now costs $900 million and takes 15 years on average to develop a new drug.Clinical trials constitute most of these costs. Additionally, about 75 percent of drug development costs can be attributed to failures. Analysts anticipate that advances in IT and the biological sciences—such as the computer-enabled ability to quickly screen drug candidates and predict drug responses—could lower failure rates substantially, reducing costs by as much as one-third and time by as much as two years.

BIOINFORMATICS IS A KEY LIFE SCIENCES R&D ACTIVITY

IT tools and bioinformatics R&D are key to remaining competitive for biotechnology and pharmaceutical companies. These companies are expanding IT capabilities by developing inhouse R&D programs in bioinformatics, acquiring bioinformatics companies, and partnering with IT companies, bioinformatics firms, and the public sector (e.g., the federal government and universities).

PUBLIC INSTITUTIONS SUPPORT BIOINFORMATICS R&D

Public institutions have made significant investments in research areas where the life sciences and IT converge. They supported and organized endeavors to map the genomes of several organisms, created databases, developed data retrieval and analysis software, conducted IT and other bioinformatics-related R&D, and funded bioinformatics training programs. These activities help drive the demand for IT goods and services in the life sciences. The National Institutes of Health (NIH) represents a majority of Federal spending for life sciences R&D, which is focused on human health. A five-year campaign to double NIH R&D funding to $27.3 billion was completed in 2003.

Several NIH programs and initiatives that are closely linked with bioinformatics R&D, such as the Human Genome Project,will get funding boosts in 2004.Also in 2004, NIH plans to devote $35 million for developing new life science approaches and technologies, such as bioinformatics. The National Institute of Allergy and Infectious Diseases, which is the lead institute in bioterrorism research, will receive increased funding. Bioinformatics is a key element in developing drug and vaccine candidates as countermeasures to potential bioterrorism pathogens. NIH also makes sizable investments in IT R&D. In 2003, NIH allocated $336 million for information technology R&D, representing about 1.3 percent of their $23.6 billion R&D budget.

Other federal agencies, such as the National Science Foundation (NSF) and the Department of Energy (DOE), also support life sciences and IT R&D, especially in disciplines other than human health (e.g., environmental biology, the plant sciences and alternative energy sources).A little over 10 percent of NSF's $5 billion budget and 2 percent of DOE's $22 billion budget are devoted to biological R&D. NSF's biological sciences R&D budget in 2004 shows a 20 percent increase (to $82 million) in emerging frontiers research, which includes bioinformatics. Of this, information technology research for the biological sciences would rise 10.3 percent to $7.5 million. DOE is a major supporter of advanced scientific computing research. R&D spending for this program is expected to increase 4.2 percent, to $173.5 million.

IT Market and Job Opportunities in the Life Sciences

The life sciences are opening up a new frontier for profitable IT innovations and applications. Although the life sciences represent only a small fraction of the entire IT market, that portion is growing at a substantial rate. A number of market research reports describe the life science market for IT in the United States and globally as dynamic with vast growth potential.Global bioinformatics market estimates vary widely, ranging from $0.8 billion to $14.6 billion in 2002.The life sciences market for IT is small relative to the whole IT market. The most optimistic estimate of the U.S. life science market for IT was $7.4 billionin 2002. This was less than 0.9 percent of the value added of the IT-producing industries (which was $829 billion in 2002).However, the life sciences market segment appears to have strong growth potential. During the economic downturn of 2001, growth in value added of IT producing industries declined by 5.6 percent. Market research firms reported that during this same period growth rates in the IT life science market were high overall. They also expect growth rates to continue to rise, ranging from 19 to 25 percent annuallyuntil at least 2006.

JOB OPPORTUNITIES IN BIOINFORMATICS

Increasingly companies and research organizations are seeking workers with more formalized training that have the skills of both computer and life scientists. The high starting salaries of bioinformaticians, $65,000 to $90,000 per year, reflect the strong demand for bioinformatics employees.Computer specialistsin the life sciences are among the high-technology employment categories currently experiencing job growth. Occupational projections by the Bureau of Labor Statistics(BLS) suggest that employment of computer specialists in the Drug and Research & Testing Services industrieswill continue to grow rapidly for several years. BLS estimates that Drug industries will increase computer specialist jobs by 60 percent, from 5,545 in 2000 to 8,859 in 2010. BLS expects computer specialist jobs in Research & Testing Services industries to increase 68 percent, from 42,567 in 2000 to 71,549 in 2010.

A survey of biotechnology use in U.S. industries further demonstrates the demand for computer specialists in the life sciences.While computer specialists only accounted for 6.2 percent of the respondents' biotechnology R&D workforce, growth in this occupational category was the fastest among the biotechnology R&D job categories. Between 2000 and 2002, responding companies added 1,236 computer specialist jobs, a 20 percent increase.Researchers have become increasingly reliant on IT tools to reduce the costs and boost the productivity of life sciences R&D. At the moment, the life sciences market for IT goods and services is small. However, the life sciences market for IT remained strong even during the 2001 recession and is expected to grow rapidly. Bioinformatics employment opportunities also have been expanding.

REFERENCES

Stephan, Paula E. and Grant Black, "Bioinformatics: Emerging Opportunities and Emerging Gaps," *Capitalizing on New Needs and New Opportunities in Biotechnology and Information Technologies: Government-Industry Partnerships in Biotechnology and Information Technologies,* ed. Charles W. Wessner., Board on Science, Technology and Economic Policy, National Research Council (Washington, DC: National Academy Press, 2001).

The Boston Consulting Group, *A Revolution in R&D: How Genomics and Genetics Are Transforming the BioPharmaceutical Industry,* (2001) at http://www.bcg.com and http://www.bcg.com/publications/files/eng_genomicsgenetics_ rep_11_01.pdf (accessed Oct., 2003).

2

Using IT to Improve Plants

The greatest improvement of efficiency in northern European cropping systems, can be found in the sensible use of Information Technology, to help manage three forms of arable variability: Spatial variability, temporal variability and a third class called predictive variability. The many crop, soil and yield maps that now exist across farms throughout Europe have quantified spatial variability. Temporal variability are the changes over time and predictive variability is where a manager must try to predict what the following year will bring and make plans to deal with it. The main factor that affects the temporal and predictive variabilities is the weather. Many existing tools and techniques help the manager to make sense of all this extra, complex data, but there is little awareness of how to incorporate weather data into the management process. As this is the largest perturbation, it should have a higher priority than it does at present.

There is a general consistency in the methods of arable farming in the British Isles, especially regarding the production of major crops such as cereals, oil seeds, pulses and root crops. Most farmers tend to follow recommended practices regarding field operations and input levels, with relatively minor modifications to suit local conditions especially soils and climate. The conventional production system is a relatively high input: high output system, whereby mechanized field operations, hybrid varieties, and chemical additives and controls are used in combinations which, subject to the uncertainty of weather conditions, achieve reasonably consistent yield performance year to year.

Relative to other business activities, the manager of a farm contends not only with the risks of the market place, but also with the uncertainties of the production system, given its reliance on natural and biological processes. There are a large number of external, non-controllable factors, which influence the technical and economic performance of the farm system, such as prevailing weather conditions or the incidence of pests and diseases. Decisions on cropping practices are required "upfront" without full knowledge of future events and resultant outcomes. Furthermore, the response of the system to actions taken cannot be predicted for sure, and decisions made at one point in the production cycle have considerable, albeit uncertain, influences on later outcomes and decisions. In this context farmers tend to be risk averters, adopting risk minimizing or satisfying strategies which have a reasonable chance of reliable performance across the range of possible eventualities. This is evident in practice. Farmers tend to concentrate on crops for which markets and prices are relatively stable or guaranteed. They may negotiate contracts to supply against agreed forward prices. They may also adopt methods of production such as chemical application levels, which are likely to achieve minimum acceptable, rather than maximum value-added, especially when the latter runs the risk of failure.

Conventional farming practices have evolved from a combination of experience and scientific knowledge. They are modified according to local circumstances such as soils and climate, but for the most part represent prescriptions for achieving the financial sustainability of the arable farm business. High yields and output are perceived to be the key to success. This is best achieved by relatively high variable inputs such as chemicals, which give relatively average returns per unit applied (even though in many years or in many parts of the farm, marginal returns may be low or negligible). In this respect, an element of the expenditure in the conventional package is an insurance premium, which accommodates the lack of knowledge about the future or the inherent variability of the production system. For instance, in many years the capacity of the machinery inventory or the level of pesticide application may be greater than necessary. In any one year the variation in crop condition or soil type across the farm may not warrant uniform treatment. In an uncertain business environment, however, over capacity or blanket prescriptions are often preferred to taking risks, even though, over a period of years, lower cost, risky strategies may give better average financial returns. Furthermore, the conventional package reduces the demands on farm management expertise. Risky systems are much more difficult to manage. Conventional methods economize on management skill. Alternative farming strategies, such as low input: low output, extensification, and integrated crop management systems have been tried and indeed promoted as financially attractive options. The results from such trials show that outputs can be similar to, and inputs significantly lower, than conventional systems.

However, such analysis misses the point. For the majority of farmers, these options are not attractive for widespread adoption because they are inherently much more risky and troublesome than the conventional package. The conventional system is a surer bet.

POTENTIAL CONTRIBUTION OF IT TO FARM MANAGEMENT

In the context of farm management, IT involves the use of digital information and control systems which acquire and use information to reduce risk, improve the efficiency of decision making, and assist with the implementation and evaluation of improved business management strategies. IT improves the knowledge base and increases the ability to control the production system in order to reduce risk and uncertainty, better identify the extent of variation in the system and thereby formulate strategies which explicitly address this variation, and supports the processes to implement these strategies. In this respect IT is essentially concerned with risk management.

From a management viewpoint, IT can be classified into two broad categories:

1. Information management and decision support:
   - Methods for the collection, assembly, logging and retrieval of data
   - Methods for data analysis, interpretation and decision making
2. Implementation: monitoring and control systems
   - Current trends
   - Since the early 1980s, there has been increasing concern that agricultural practice has caused damage to 'the environment', in a range of ways. These include:
   - Possible deleterious effects on the farmland ecosystem itself, which may reduce the sustainability of agricultural practice
   - Reductions or changes in desirable non-farmed habitats within rural areas,
   - Direct effects on human health
   - Social effects, such as the non-renewable-resource costs of current agricultural practice, unpopular changes in landscape appearance, in rural employment and social structure.

Predicting the effects of changes in management which are likely to arise from the adoption of a 'high information' strategy is only possible in very general terms, although some attempts at quantification of new strategies have been made in the USA. Most of the research to date has concentrated on quantifying the effects of historic changes in practice, which, until at least the mid 1980s, usually involved increasing use of fertilizers, biologically active inputs, and capital. It would not be appropriate simply to assume that the results of a reduction in inputs would be the exact reverse of the effects of increasing inputs.

There are now limited data from experiments which deliberately look at such reductions but these are insufficient to provide quantitative estimates of effects on a wide scale. Data from organic farming systems are more widely available, but represent an extreme form of management that is unlikely to be reflected in majority practice. Similarly, data from the initial years of set aside may not provide a reliable basis for estimating the effects of selective reductions in use of inputs within active husbandry systems. A basic comparison will be made between 'conventional' and a 'high information' strategy, which assumes that much more detailed management information is available, so that 'what-if' investigations can be made readily. Within such a system, the final decisions made will vary, depending on the values of the decision maker, but the opportunity will exist to take into account environmental as well as economic management variables.

The weighting put on environmental variables will depend on the decision maker, and may well range from zero to relatively high rankings, so the actual implementation arising from a 'high information' strategy will also vary. However, it is assumed that in general, the total use of biologically active inputs will be reduced in the 'high information' strategy. In the arable and grassland context, traditional managers have tried to treat fields uniformly. All the field operations have been managed to produce an even distribution, even though it is known that the conditions may vary across the field. High or low yielding areas may have been identified but only on the small traditional farms was there the capability to treat selectively. On the bigger farms the technology for varying the treatments has not been available.

In recent years the problem has become worse as arable fields have increased in size due to economies of scale that were gained from the mechanization sector. Mechanization is no longer the limiting factor to efficient production but has now become the arable inputs such as seed, fertilizer, sprays, etc. From initial studies in Europe and the USA it has been found that there are significant potential benefits from selective treatment, in both the economic and environmental senses. If the spatial variability of the land and crops are investigated, then the potential savings become apparent. The higher the variability, the higher the potential savings.

THE FUTURE SITUATION

The long term trend facing farming in Northern Europe is that of a continuing deterioration in the terms of trade whereby the prices of agricultural outputs fall relative to those of inputs. The response by farmers in the past has been a mixture of intensification and amalgamation, protecting incomes by yield improvement through high use of variable inputs such as fertilizer, and economizing on fixed inputs such as labour and machinery by enlarging the scale of operation. Given the financial pressures on farmers, improved farm business management, supported by IT, becomes a more critical factor in ensuring financial survival.

More recently, farmers have found it beneficial to target their output to the changing needs of the food market, especially regarding quality and supply assurance. The liberalization of agricultural production and marketing will hasten this process. These changes will further encourage the use of IT as part of a sophisticated production and marketing system.

Pressures to protect incomes is likely to encourage the use of IT as a means of achieving more economic use of inputs as well as enabling production and marketing activities which increase the "added-value" at farm level. The use of IT, particularly for business management, is likely to become an integral part of commercial farming systems and a pre-requisite for successful management. To this extent, IT will encourage its own diffusion amongst farmers. Those who do not use it will be disadvantaged.

In addition to economic pressure, there are likely to be greater constraints on land use and farming practice in order to meet environmental criteria. These constraints are emerging in the form of policy initiatives such as the Nitrogen and Environmental Sensitive Area Schemes and the EU Nitrification Directive. For the most part, these schemes are voluntary with incentives to participate. It is likely that they may become compulsory, with, and in some cases without, compensation for income lost. In such circumstances, conventional management practices no longer apply. The environmental constraints may be imposed in terms of the use of inputs and practices such as nitrogen use, or on limits on environmental impacts such as leachate in ground or surface water. In either case, farmers will have to record and report that they are meeting the set criteria. IT has a role to play for monitoring and confirming compliance.

Constraints on farming for environmental purposes initially are likely to come in technological packages, similar in concept to the blanket prescriptions of the conventional approach. Environmentally sensitive low input - low output systems are unlikely to sustain a viable farm sector without significant levels of income support. However, IT could help to formulate and implement locally relevant farming strategies, which, through better-targeted input use, achieve financially viable farming systems, which simultaneously comply with environmental quality criteria. In its most sophisticated form, IT would involve instrumentation and control for spatially variable field operations. Thus, IT by reducing the business risk inherent in environmentally sensitive farming systems, and providing the means of protecting environmental quality without compromising financial performance, could provide the key for sustainable agriculture in European arable farming.

INFORMATION TECHNOLOGY AND MANAGEMENT

Managers are increasingly using information technology to help them improve their decision making capabilities, particularly when trying to manage the inherent variability found within the crop environment.

One example is called Precision Farming, which is a systems approach to managing crops and land selectively, according to their needs. It utilizes expertise from many disciplines and integrates the latest information technology tools and techniques to enable farm managers to get a better understanding and control of their fields. Management is the essential factor to achieve a stated outcome for the farm. Managers should identify their own strategies and practices that allow them to deal effectively with the variability found on their farm.

Three types of variability have been identified. The first type is spatial variability, which can be seen as changes across the field. An example would be where one side of the field yields higher than the other side. Temporal variability is identified when parameters change over time. This can be seen when a crop starts by growing well but results in a poor yield.

Predictive variability, is the difference between what the manager predicted would happen and what actually happened. The classical example of predictive variability is where the manager predicts that a certain yield will be achieved if a certain amount of fertilizer is applied, but the crop does not achieve it because the weather changes. Each type of variability must be measured, assessed and possibly influenced, according to how significant it is.

MEASURING VARIABILITY

The first stage in the process is to measure important factors that indicate or affect the efficiency of the growing crop. The two main approaches are to create yield maps through instrumenting the harvesting system or assessing soil parameters by sampling. Both techniques give information about different parts of the cropping system. Yield maps are historic and cannot be used while the crop is growing, but record the actual yield during harvest. Soil sampling can be expensive but many soil parameters such as texture and horizon depths do not change over time, so are a good investment. Measuring soil nutrient status must be treated with care as repeatability, let alone accuracy, is difficult to achieve. Sampling strategies based on a simple grid tend to be expensive and better-targeted sampling techniques are being developed [Thomas et.al. 1999]. Asset surveys can also be carried out to record physical features, such as field and crop boundaries, high trees that may cause shading, compaction in gateways, etc. Other high-density measurement techniques are rapidly becoming more important such as Remote Sensing and ADP (Aerial Digital Photography) or non-contact sensing (e.g. Electro-magnetic induction). ADP can give real-time information of the crop canopy and allow management to be modified while the crop is growing.

ASSESSING THE SIGNIFICANCE OF VARIABILITY

The spatial variability must be assessed for significance to the manager. Normally this is done by looking at the spread of the data or seeing if the extreme values lie outside acceptable thresholds, such as indices for soil nutrients. One technique is to reclassify yield data into 'gross margin' maps. (Fig. 1). This technique deducts the variable costs from the income, which varies spatially with the yield, resulting in a gross margin map that shows which areas generated more income than others did. Some gross margin maps have shown areas that actually loose money. Given enough detail, a similar map based on inputs could be produced to show environmental impact such as nitrogen fertilizers in a nitrate sensitive area. The temporal effect of weather and prices are shown in Fig. 2.

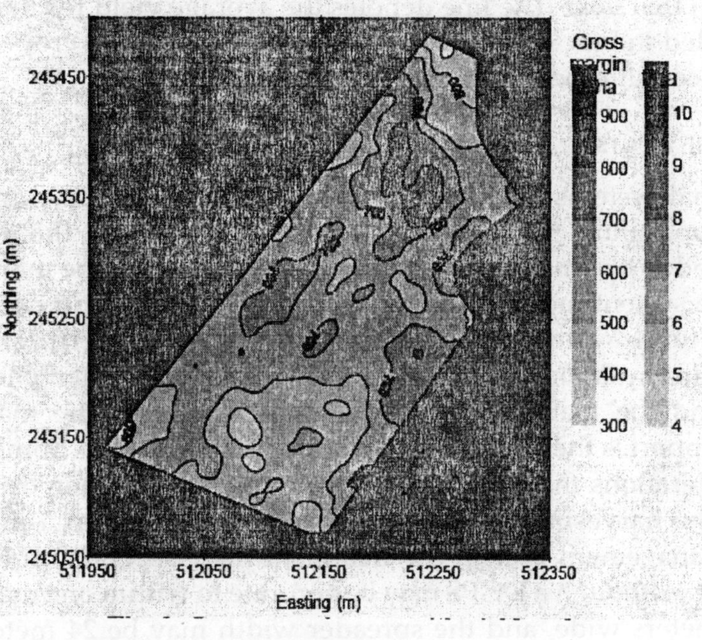

Fig. 1. Gross margin map with yield scale

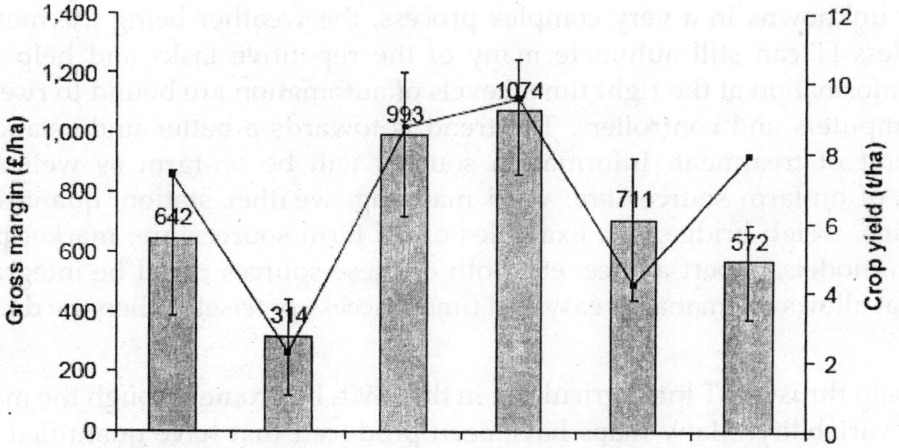

Figure 2. Gross margins and the in-field variability between 1993 and 1998

The bars in Fig. 2 show the change in average gross margin over 6 years. Superimposed are the whiskers denoting maximum and minimum gross margins experienced in that year. The line denotes the average yield for that year and the selling price is included in brackets. Note that in 1994 and 1997 the crop was oil seed rape and the wheat prices slumped in 1998.

MANAGEMENT OF INPUTS

Most traditional systems over-apply inputs such as seed, spray and fertilizer to reduce the risk of crop failure. With better assessment techniques, the inputs can be reduced or redistributed to optimal levels and the risk of failure can be managed. This will result in making the system more efficient. Regardless of the country and crop, management of an agricultural system is complex. To improve the efficiency, computer based Management Information Systems (MIS) must be sophisticated enough to deal with this complexity and the manager's strategies and practices. The management input and computing support is the same in each country and each crop. Some crops may well have special considerations that should be taken into account when designing the MIS, such as planning the harvesting logistics when supplying crop to a processing factory. The size of the management unit in the treatment map depends on the ability to measure and manage it. Although a DGPS may be accurate to within one metre, a combine header may be ten meters wide, and the spreader width may be 24 meters. The management unit will be limited by the spreader width. A draft methodology for dealing with in-field variability has been proposed.

The ultimate goal is a full understanding of the whole process and rational decisions being made following the management philosophy. This will never happen, as there are too many unknowns in a very complex process, the weather being the most obvious. Nevertheless IT can still automate many of the repetitive tasks and help to provide accurate information at the right time. Levels of automation are bound to rise, as are the use of computers and controllers. The trend is towards a better understanding and a smaller unit of treatment. Information sources will be on-farm as well as offfarm. Examples of on-farm sources are; yield mapping, weather station, quantities held in storage bins, weigh bridge, etc. Examples of off farm sources are; market prices, pest prediction models, expert advice, etc. Both of these sources could be integrated into a system that allows the manager easy and timely access precisely when the decision must be made.

The main thrust of IT into agriculture in the 1990s has come through the management of spatial variability. Many maps have been produced that have quantified this effect, but only a few management guidelines of how to deal with it are currently available. As time progresses better understanding of temporal variability is becoming apparent. A seasonal change in the weather is the main factor that gives rise to temporal variability, which leads to gross predictive errors. Some method of classifying the current year to better manage weather trends is needed.

ROLE OF IT IN AGRICULTURAL DEVELOPMENT

There is broad consensus in the international community that the exchange of information and knowledge - both local and global - by individuals and communities using new information and communications technologies (ICTs) will have a critical role in achieving sustainable development and food security in the 21st century. Many examples exist to illustrate this observation, such as the ability of governments to predict areas of food insecurity and vulnerability using appropriate information and software tools, so that action can be taken to prevent or reduce the likelihood of an emergency. Put bluntly, information that is properly applied can save lives and improve livelihoods.

Information can only empower when the user has access to it. Leaving aside mass media such as radio and television, information used to be disseminated in paper-based form, and it had to be brought physically to the user. More recently digital media such as diskettes and CD-ROMs began to be used for particular dissemination functions. These early digital technologies quickly led to the emergence of wide-area connectivity through the Internet, overcoming the limitations of paper-based dissemination programmes. The exploitation of the technology revolution has made information on agriculture available world-wide and on-demand. In the context of FAO, approximately 1,000 users per year requested information from the FAOSTAT database in the 1980s when it was held on a mainframe computer, whereas the Internet version available through WAICENT currently has about 60,000 user sessions per month.

Current estimates indicate that there are about 5 million web-sites and 1.6 billion web-pages. Demand from users has increased in a similar way to the amount of information on the Internet, and statistics clearly demonstrate the global reach of internet usage only 10 years after it was invented. In addition, this growth is accelerating faster in developing countries than anywhere else. The total number of Internet users in Africa, Latin America and the Caribbean and Eastern and Central Europe is expected to quadruple by the end of 2001, and Internet usage should grow even more in the Asia-Pacific region. Accessibility in remote rural areas will also improve, as satellite communications make it possible to connect rural telephones at appropriate bandwidth and at reasonable cost.

Problems associated with Access to Information

A new set of problems is emerging now that we have the ability to transmit and visualize information in digital form which has to be solved if the full potential of the medium is to be realized.

Digital Divide

Internet connectivity alone will clearly not resolve development issues. It is active participation in and contribution to the world's growing digital knowledge bank that has the potential to educate and reform in many sectors, such as commerce, education, agriculture and health. The Internet has become a vast and growing global network that people use to converse, debate, meet, teach, learn, buy, sell and share virtually every type of information imaginable. According to the Gartner Group, business to business (B2B) e-commerce transactions will skyrocket in value from $US 145 billion in 1999 to $US 7.29 trillion by 2004.

Like many communication technologies before it, the Internet potentially enables rural communities to receive outside information and knowledge that can spur development. However, the observation has also been made that unlike other media (such as radio and TV), the Internet allows every user to potentially be a sender or receiver. It offers new opportunities for two-way, interactive and horizontal communication. Some observers also believe that the Internet can be an effective medium for the articulation of development needs and perceptions.

However, Internet access is likely to be available only to a small proportion of the people in the poorest countries for the immediate future; within these countries, the rural areas, and specific groups within rural areas (e.g., women), will be left even further behind. This phenomenon has been dubbed the "digital divide". It has been widely recognized that developing countries in particular are pursuing a moving target, as the high-income industrial countries constantly push the knowledge frontier outward. There is a new generation of digital applications in agriculture, and the emergence of such innovations is serving to widen the technology gap between rich and poor countries. Indeed, even greater than the knowledge gap is the deficit in the capacity to capture knowledge in digital form. Differences in some important measures of knowledge creation are far greater between rich and poor countries than the difference in income.

Closing this digital divide will not be easy, especially in the wider context of the rural environment. Regarding women's access to the Internet: "Estimates suggest that the global Internet gender ratio has remained static for a number of years, with about 63% of users being men and 37% women. Less optimistic is the claim, made by the Association for Progressive Communication, that 'male domination of computer networks' is as high as 95%". A recent FAO publication stresses not only the importance of improving the quantity and accessibility of ICT infrastructure for rural women, but also increasing the relevance of information (content) to the needs of rural women and training them in computer skills. The global information revolution could paradoxically become a tool of even greater inequality and worsening poverty for developing countries.

Human Resource

The goal of increased access is impeded by the relatively poor abilities of certain key stakeholders to contribute to the knowledge community. It is clear that this is an area which will require major development, and capacity building is one of the most important processes for promotion of access for which a policy framework is needed. A separate key-note Working Document COAIM 1/3 addresses this important area.

Need for Standards and Procedures

The Internet has grown and continues to grow exponentially. A huge variety of organizations have embraced the medium, adding information in diverse formats for every conceivable audience, and the path is clear for the Internet to become the most important platform for publishing and dissemination. However, the Internet has in some ways become a victim of its own success, given the issues of navigation and retrieval in the increasingly unstructured and chaotic environment. Users are having to search for information using wholly inadequate tools, as all the major so-called "search engines" index only a very small fraction of the relevant Internet sites. According to various recent independent surveys, such facilities cover only about 2 to 16 per cent of the searchable part of the Internet. A second related issue is that there is no quality control for material on the Internet and the user has no way of assessing material that is indexed by the major search engines. Accurate retrieval has become the top priority issue of the digital world, and novel approaches are required. All stakeholders in knowledge exchange for development recognize that the Internet environment has to change considerably if it is to fulfil its potential.

FAO feels that there are two major approaches to addressing the complex problem of information retrieval. The first is the development of standards and procedures role, and the second focuses on the development of innovative software tools and applications in the public domain that can underpin the access revolution.

Areas for FAO's Action

Information was recognized as being a fundamental part of FAO's activities that it is the first article of the organization's Constitution

- It is important to put FAO's role in information and knowledge exchange into the context of what the organization can and is doing given trends in information and communication technologies (ICTs). It is clear that there are some areas where FAO has strengths and other areas where other agencies have key roles to play.
- FAO has a major capital resource, namely its information on key issues related to agricultural development food security spanning over 50 years, now being made accessible through WAICENT. This is a valuable and publicly accessible component of the global agricultural knowledge base.
- FAO has intergovernmental status, and thereby has an important part to play in a normative role by providing a neutral authority/facilitator/advocate in the process of development and in information/knowledge systems in particular.

— On the other hand, infrastructure development, especially in telecommunications, is the responsibility of governments and increasingly the private sector. FAO's approach is to partner with those organizations active in the telecoms hardware sector, including the International Telecommunication Union, that are working to bridge the digital divide.

Much of the organization's focus is on working with the intermediaries who are in a position most effectively to communicate and to share information with farmers and other rural people. This has included primarily public sector information specialists, researchers, extension agents, educators and others who provide support services to rural areas.

STRATEGY AND POLICY

In order to foster universal access to information and ICTs, the international community needs to pay attention to the broader policy areas of helping developing countries to create and implement strategies for joining the information economy. The shortcomings of older strategies of central control of information must be recognized. This involves giving clear and unambiguous messages to decision-makers about the cost-benefits for dissemination of information in embracing the new technologies, the danger of being left behind, and the inevitability of global flows of information. It also involves promulgating the mechanisms and methods to underpin policy reform that (a) promotes the growth of the information economy, (b) enables the free, multidirectional flow of information and knowledge, both within and among countries, and (c) coordinates more effectively the training and capacity-building efforts of the Member Countries in policy areas. More attention must be given to building the capacity of non-governmental groups in developing countries, including the private sector, to play an active role in creating the enabling environment for the information economy and also to assert the rights and principles of universal access to information and knowledge. Part of the role of the non-governmental sector should be to foster an environment of greater accountability and transparency in policy for information access.

Community-based Initiatives

A major set of contributions that FAO and its member countries can make to increasing access for rural and disadvantaged communities is by helping to find creative solutions to address demand in rural and underserved communities. This can be achieved through acting as a catalyst to bring together investment partners to support those opportunities, including the public sector, the private sector and the international community.

Technology is simply a means of accessing information, and yet many development initiatives have focused too heavily on the provision of connectivity and computers. Clearly, low computer literacy is an enormous barrier, but significant effort must be put into generation of relevant content. Recent experience has shown that when rural communities are provided with access to reliable information sources using new technologies, most user groups rapidly become accustomed to the medium. The agent of change is the provision of previously unavailable content, in particular where that material relates to financial security in the form of market information (inputs and outputs), networking between peers (small farmers' interest groups), and information on technical aspects of primary production. In addition, solving conflicts related to land and water utilization is easier when the stakeholders have relevant and reliable information.

With the costs of ICTs plummeting, exchanging knowledge can be more efficient and effective than before, especially when combined with other more traditional technologies such as mass media in the form of rural radio etc. and printed materials. Given these advances, the stage appears to be set for a rapid narrowing of the knowledge gap and a surge in economic growth and human well-being. The task of providing information access to the disadvantaged, as well as building the social infrastructure required to provide sustainable purchasing power, will require a partnership of the public and private sectors to put in place the necessary legislative and regulatory framework to stimulate change, the ICTs, and the content. The development assistance community is set to play a critical role in assisting the process of change in favour of the world's poor, and in providing relevant systems and content.

A technical workshop organized as a satellite event for the Consultation will consider the best ways to establish such initiatives. It will consider evidence that already exists to show that new ICT-based approaches can help developing countries to move quickly from agrarian societies to rural knowledge societies. There is a growing realization that knowledge can be exchanged in digital form through facilities at the community level in poorer countries. Experience has shown that some essential steps if such ventures are to be sustainable, and these are also considered in some depth in the Working Document COAIM 1/3 on "Strengthening information and knowledge management capacities". Finally, it is important to recognize that private enterprise has a key role to play in transforming local conditions, given the critical role of telecommunications providers, sources of finance (banks and credit unions), agribusiness (fertilizer, seed and agrochemical manufacturers), and the food marketing and retail sectors. Partnerships between the local community, the government sector, the private sector and international development assistance agencies will be most suited to building sustainable technology-based systems of knowledge access and exchange in the rural environment.

DIVERSITY OF GLOBAL INFORMATION/KNOWLEDGE FLOWS

Even a highly successful, coordinated international effort to foster universal access to information will not advance the underlying goal of sustainable, equitable development if it only assures global access to the information produced by advanced economies and international agencies. FAO, its Member Countries and other partners need to work together to create opportunities for the development and dissemination of information and knowledge from a variety of sources, in a variety of languages, representing a variety of views and traditions. Here, selectivity and focus are critical success factors. In addition, the aim should be to support indigenous capacity and create the conditions for more effective generation and dissemination of content.

It is clear that the continued generation of new agricultural technologies and exploitation of currently available ones will be essential if the world's growing population is not to suffer ever-increasing incidence of malnutrition. Yet, it is a generally recognized fact that many agricultural scientists in developing countries are unable to publish their research output so that their per capita publication rate is substantially lower than in developed countries, leading to a corresponding under-representation in the world's leading scientific literature. This phenomenon is stifling the dissemination of important technologies and lessons learnt. The underlying reasons for this are in some ways linked to the gradual reduction in research output as governmental services in some parts of the world have seen their operating budgets reduced. A much more fundamental reason relates to the acceptance criteria of most mainstream primary journals, which rule out publications based on adaptive research and even negative results. This impasse can only be broken by a new approach based on an open system accepting a wider range of content and formats that maintains some essential editorial standards, and yet which makes such material easily retrievable. This type of approach will be considered as part of a new plan for the AGRIS system, with information management methodologies that are more suited to local needs, and AGRIS will be considered by a technical workshop convened especially during the Consultation.

APPLICATIONS AND SYSTEMS DEVELOPMENT

FAO is strategically positioned at the global level to implement, and provide support in the use of, specialized tools and applications for information delivery and communication in the areas of food security and agricultural development, especially in the context of the organization's normative role. In addition, the organization should contribute to the emergence of applications relevant to the needs of Member countries and the poor by supporting initiatives which foster an environment locally for innovation and investment, and also by helping the establishment of indigenous innovators and developers.

Information systems will be developed and implemented in selected areas within the WAICENT programme, often in partnership with Members. Outreach services to countries are discussed in the Working Document COAIM 1/3 on "Strengthening information and knowledge management capacities". The WAICENT strategy in the area of systems development is that outputs should have the following characteristics:

a. The ability to operate in various information technology and system environments.

b. The adherence to major non-proprietary standards

c. In order to grow with and adapt to technological changes and to be able to accommodate other products, based on similar standards, where appropriate.

In addition, the system should ideally be:

d. Decentralized, with data remaining under the stewardship of Member Countries and other partners, to ensure that property rights are respected, resources are up to date and that quality can be maintained.

e. Public goods, permitting unlimited distribution of the applications to partners and users with no licensing or royalty costs.

Most of the issues to be addressed in sustainable development and food security require information from many disciplines, making it necessary to access many types of information resources. More recent developments in information management at the global level have enabled the establishment of distributed networks of databases and content, which lend themselves to the requirements stated above and improving the access to the wealth of information available. The new technologies can also improve the effectiveness of searching and retrieving relevant data and information across distributed networks.

WAICENT is already tackling the development of a new approach to accessing information that could have major implications internationally. A web-based information/knowledge management system is being developed with the aim of creating a specialized gateway. The system will provide a framework with a comprehensive set of web-based modular functions geared to aiding the administration, organization, indexing, cross-referencing, uploading and retrieval of information in a decentralized and participatory world-wide network. A modular approach will maximize scalability and portability, and will permit the incorporation of additional features into the system or the re-use of modules such as mapping systems developed in FAO for other purposes.

THE WAY FORWARD FOR FAO AND ITS MEMBERS

FAO has identified key areas where the greatest need and opportunity exist for effective collaboration with its Members and the international community to facilitate the advance towards the goal of universal, sustainable, and economically viable access to information and communication technologies. These areas compare well with the major issues associated with "access" identified at the Global Knowledge II Action Summit. The Consultation is invited to consider these issues and opportunities and to provide country and regional perspectives to guide FAO's further activities. In particular, guidance is sought on the issues and proposals that have direct implications on FAO's current and future activities in this field.

The Consultation may wish to raise to the attention of Members that access to knowledge and information is critical for sustainable development, and for its components that are relevant to FAO, namely agricultural productivity, food security, and environmental protection.

The Consultation may wish to consider the priority of improving access to information in FAO's work through WAICENT with Members and other stakeholders, taking full advantage of the opportunities provided by the new generation of information and communications technologies. The Consultation may also wish to guide FAO on the importance of empowering rural communities and individuals with and through information and communication, focusing especially on vulnerable groups such as women and youth.

To reflect the inherent cross-sectoral nature of information management, the Consultation may wish to consider its future role in coordinating a FAO-wide programme covering key aspects of access to information and knowledge.

OPEN ARCHIVE NETWORK IN AGRICULTURE

AGRIS, operational since 1975, was a very early international initiative based on a collaborative network of institutions, whose aim was to build a common and freely accessible information system for science and technology in agriculture and related subjects. Until the late 1990s, AGRIS-related outputs mainly comprised a centralized bibliographical database and associated products. Since 2000, however, the AGRIS network has directed its efforts increasingly towards building up decentralized capacities in its participating resource centres, and empowering those centres to improve agricultural information management in their own institutions. One of the main undertakings of AGRIS is to address issues related to accessing scientific and technological publications, and scholarly papers, but also and especially grey literature. Grey literature contributes significantly to agricultural research, especially in developing countries, but it is not yet widely accessible to the agricultural community.

Rapid technological changes have opened up new opportunities for sharing data, information and knowledge. Over the past few years, the communications world has witnessed several new developments, two of which are the Open Access (OA) and Open Archives Initiatives (OAI). These are complementary initiatives whose key objectives are to improve visibility and access to research publications, thus maximizing their impact. Open Access is described as "a free availability on the public Internet, permitting users to read, download, copy, distribute, print, search - or link to - the full texts of scholarly and scientific articles; crawl them for indexing, pass them as data to software, or use them for any other lawful purpose, without any financial, legal, or technical barriers other than those inseparable from gaining access to the internet itself". The Open Access model allows for the widest dissemination of research output, and maximum visibility, while, at the same time, removing the constraints common to traditional diffusion methods of scientific literature. It provides the means for researchers to avail themselves of full text content, using two paths: Open access publishing journals, which make articles openly accessible immediately upon publication, and Open Access archiving which allows authors "to deposit a digital document in a publicly accessible Web site, preferably an OAI-compliant open archive". The Open Archives Initiative (OAI) develops and promotes interoperability solutions that aim to facilitate the efficient dissemination of content. The initiative is built on the principle that interoperability will federate the distributed open archives, thus encouraging the development of value-added services such as portals, subject gateways, and specialized search engines, with the overall benefit of increasing the visibility of the Open Archives. The ultimate objective is to overcome existing barriers to interoperability by using open archives for all digital materials.

The Open Access publishing model and the Open Archives initiative relate strongly to the AGRIS principles. By applying the former's concepts, the AGRIS community will not only build up the capacity of its resource centres to disseminate and share their research outputs, but will also bring accessibility and visibility to scholarly publications in agriculture. It goes without saying that the AGRIS network will also need to comply with the requirements and specific conditions outlined for the Open Archives.

OBJECTIVES

The AGRIS network will promote the Open Access publishing paradigm and apply OAI standards and methodologies since these latter significantly contribute to the realization of the long standing goal of AGRIS, namely to improve access to and exchange of information and knowledge in agricultural science and technology between developing countries. The specific objectives include increasing the visibility and accessibility of content including that which was not previously accessible and promoting common subject-oriented standards thus underpins the main principles of the AGRIS network. The implementation of a new architecture using tools related to Open Access will greatly benefit the agricultural research community: both the visibility and readership of researchers will increase.

Furthermore, the Open Archives offer easy access to documents for potential readers who would otherwise have limited access due to the cost of mainstream journals. The initiative enables linking of local and international research as well as providing a better picture of a country's agricultural research output. The new model provides an architecture in which communication between two tiers of partners can be defined: data providers, those who have metadata to expose, and service providers, who harvest the metadata. Communication between the two is enabled through an interoperability solution known as "metadata harvesting", which allows data providers to expose their metadata via an open interface to service providers, who use the metadata as a basis for value-added services.

The quality of description of agricultural information will be based on the use of common subject- and network specific standards. Some of these standards include the AGRIS Application Profile (AGRIS AP) and Knowledge Organization Systems (KOS) such as the AGROVOC Thesaurus, elaborated further in Section 5. These will give the subject service providers in agricultural sciences better possibilities to build enhanced services from the metadata collected. Standards and tools will be promoted within the AGRIS network in order to guarantee the success of the new architecture.

ELEMENTS OF THE ARCHITECTURE

The proposed AGRIS network architecture, provides the main elements and the foreseen dynamics of information flow in the network. Conceptually, the workflow has three principal activities: content management, exposing metadata and the provision of value-added services.

The content management activity describes the current process of capturing, describing and storing information using appropriate tools and methodologies. A list of appropriate standards, methods and software is available on the AGRIS and the AIMS1 (Agricultural Information Management Standards) Web sites.

The second part of the workflow is to expose the metadata. This involves AGRIS partners becoming data providers and interoperating with service providers via the application of OAI standards and procedures. To guarantee visibility, the data providers will have to register and describe their collections in a registry available via the AGRIS web site. The AGRIS network intends to go beyond the exchange of simple Dublin Core metadata (a pre-requisite to participating in the OAI community), by recommending the use of the AGRIS AP - a richer exchange format for harvesting and exchanging agricultural content.

The AGRIS AP and KOS are principal factors in the description of content. These allow the subject service providers to select and harvest quality data which ultimately improves the quality of their services. The last action of the workflow is that of subject services providers, who give added value to the metadata harvested, by offering platforms over which the researchers will be able to interact and share information. The AGRIS network has established discipline-based communities which share common activities and services; these groups already form a basis on which the tasks and responsibilities of the service providers can be defined.

The service providers will also have the possibility to create specific subject gateways for different disciplines in the area of agricultural sciences. Furthermore, material held by the data providers will be also harvested by other service providers outside the AGRIS network. The overall picture of the new AGRIS network is that of a space where all the above described elements play an active role in the sharing of metadata and knowledge. The implementation of networks of open archives is the main element for the new architecture; it will allow the agricultural research community to both access and share its outputs.

DATA PROVIDERS

Embracing the OA model, at the institutional level will mandate that the data providers to put mechanisms in place to support the implementation of digital repositories. These are open access archives that may include content such as already-published articles (postprints), pre-published articles (pre-prints), theses, manuals, teaching materials or other documents that the authors or their institutions wish to make publicly available without financial or other access barriers. The technical infrastructure for setting up institutional repositories is readily available as seen from the variety of free open source software that adhere to the required standards and the limited costs associated to the set up. Set up of institutional ICT/M strategies and policies that include sound publishing and documentation workflow will also enhance the institutionalization of the Open Access model. Other issues to consider are the rights associated with author self-archiving of published articles.

The ROMEO database (Rights Metadata for Open Archiving) available via the Securing a Hybrid Environment for Research Preservation and Access (SHERPA) Web site provides information on the publisher copyright policy and self-archiving. To facilitate this process, the benefits of OA should be clearly advocated and demonstrated amongst researchers as well as policy makers who can institute policies that favour and/ or mandate OA. Libraries should champion this course within the institute by playing the important role of facilitating the adoption of OA through awareness campaigns to educate its researchers and policy makers on the role and benefits of the model in promoting visibility and impact to the institution's research output.

EXPOSING METADATA

Data providers in the new architecture are not comprised only of the existing AGRIS resource centres: any agricultural research and technology institution that wants to publish its output in the AGRIS network can register as a partner and exploit the common standards and technologies of the network. The data provider must be made aware that a repository or an open archive is not only a database with full text documents and their metadata.

Three different types of data providing mechanisms are defined within the AGRIS network:

— *Dynamic repository:* a software layer which enables the repositories to be OAI-compliant and acts as an intermediary between open archives and the OAI-based service providers. The layer accepts OAI requests and generates dynamic responses;

- *Static repository*: used with small metadata collections. A static repository can be generated periodically by a script that extracts information from an existing database and
- *Hosted repository:* for repositories that cannot expose their own metadata on the Internet; other institutions can host and expose it for them. A data provider can expose different metadata formats, allowing them to be harvested by different service providers, thus giving them more outlet channels, and greater visibility.

Unqualified Dublin Core is the basic metadata format for participation in the OAI community; however, it is not sufficient for the requirements of the agricultural scientific community. The AGRIS AP has been created specifically to improve the quality and interoperability of metadata in this domain. Data providers can offer both unqualified Dublin Core and the AGRIS AP in order to be visible to more search engines and subject service providers in the AGRIS network and beyond. The AGRIS Initiative will encourage the exchange of metadata using the AGRIS AP standard so as to guarantee the high quality description of agricultural information resources.

METADATA HARVESTING

Metadata harvesting is an interoperability solution to allow service providers to communicate with data providers. The interoperability solution in this case comprises uniform naming (standardized metadata), a common syntax, and a metadata harvesting protocol. The Open Archives Initiative Protocol for Metadata Harvesting (OAI-PMH) is a protocol developed with the objective of collecting the metadata of records found in open archives. OAI-PMH is based on a client-server architecture (service providers request metadata from data providersopen archives). Consequently, the service providers can build up their services using metadata from many different archives. During the past few years the OAI-PMH has become widely known as a way to establish a connection with distributed open archives. The main reason for its acceptance is the simplicity of its implementation. Also, several tools have been developed to allow for the creation, management, and interoperability of OAI compliance of the open archives. Uniform naming is achieved through the standardization of metadata semantics.. A common syntax, XML, is used for representing and transporting archive-specific metadata sets. Most current software is able to generate XML records.

SERVICE PROVIDERS: ENHANCING THE ACCESS

The AGRIS network currently has several service providers, which include: the AGRIS database at FAO, the commercial AGRIS database at Ovid Technologies, the AGRI2000 service from Sistema de Informacion y Documentacion Agropecuaria de las Americas (SIDALC) , the Consultative Group on International Agricultural Research (CGIAR) virtual information centre, and many other smaller regional service providers. The new architecture delineates a much easier and more efficient environment to set up specific services. AGRIS Services Providers harvest, from more than one open archive depending on the scope and nature of their collections, records that contain only metadata, not full text documents. The functionality of a service provider to successfully harvest data is based on the provider's metadata schemas, metadata collection and open archive update frequency, the extent of the availability of the open archives, conditions of use of records, mention of source, and link from service provider to data provider. In 2003, Foulonneau and Dawson outlined the key points to consider in the implementation of a service provider, most of which are already fulfilled in the AGRIS network as follows:

— Data collection & analysis: most of the AGRIS members are already collaborating in the AGRIS database; Common metadata schema: the AGRIS AP is the established metadata standard for exchange of the agricultural information;
— Granularity of descriptions: the AGRIS AP provides a qualified metadata set that enables richer description of agricultural documents. It has agricultural specific schemes; Metadata crosswalks: there are crosswalks from the AGRIS AP to other metadata schemas;
— Terminology issues: several KOS in agricultural science have been developed and used; Multilingual environment: one of the principal characteristics of the AGRIS network.

The harvested metadata can offer associated full texts, to which the service providers enable access. Different service providers are envisaged such as subject portals, digital libraries, gateways or web pages exposing only the metadata. Services that can be built using the harvested data include the following: data indexing; advanced searching and browsing; usage statistics; citation parsing; the possibility to make comments on documents; multi-language support; user authentication; aggregated news feeds; e-forums; automated email notification alerts, i.e. Selective Dissemination of Information services (SDI); RSS news feeds on conferences, events and courses in different disciplines; tailored services (customization by the user).

NETWORKING FOR INTEROPERABILITY

The AGRIS Web site is a communication channel between the data and service providers and the specific user. It will be a gateway to various information resources including a repository for the outputs of the Task Forces, the AGRIS Open Archives directory, data management standards and tools, as well as other relevant information. The Web site plays an important role in the proposed AGRIS network architecture, and is consistent with the priority placed on advocacy. The Web site is a key tool in promoting the adoption of the broad vision of coherence in agricultural information in the context of the new AGRIS initiative. The AGRIS Open Archives Directory is a platform for the registration of all the AGRIS open archives. It will be searchable on the AGRIS Web site, and used to identify all the data providers in the community. The directory will categorize and list the wide range of open archives using a set of metadata that will provide relevant information. It will be searchable by country, location, subject, metadata format, software, type, material included and other criteria.

ASSURING QUALITY IN METADATA CREATION

Content description through quality metadata creation and use of standard terminologies is the basis of efficient content management as well as the development of value added services. Metadata standards such as Dublin Core (DC) and the AGRIS Application Profile (AGRIS AP) provide mechanisms for sharing information in a standardized manner by recommending the use of common semantics and interoperable syntaxes. The OAI standards mandates the use of unqualified Dublin Core metadata schema for exposing metadata through the OAI-Protocol for Metadata Harvesting, however this is not sufficient for the requirements of the Agricultural Scientific Community. Use of unqualified Dublin Core inhibits the development of efficient value added services due to the poor quality and loss of data amongst other things.

As mentioned before, the AGRIS AP, a richly defined standard, was created to improve the quality, and better exchange of metadata within the agricultural science community. The qualification of the metadata element facilitates better aggregation and subsequent retrieval of content. Standardized metadata, in terms of syntax and semantics used, facilitates interoperability., however, poor quality metadata can mean that a resource is essentially invisible within a repository and remains unused. The more specialized subject service providers will use AGRIS AP as a basic metadata format for metadata harvesting. To ensure a minimum level of quality, certain metadata fields and the use of relevant controlled vocabularies need to be agreed upon. The AGRIS AP achieves this by: mandating five required elements and promoting the use of agriculture-specific controlled vocabularies such as AGROVOC, CABI codes etc.

AGRIS Application Profile

Dublin Core is a metadata format whose structure is the result of an international consensus. The Dublin Core Metadata Element Set (DCMES) defines fifteen elements for simple resource description and discovery, all of which are recommended and none of which are mandatory. The DC has been extended with further optional elements, element qualifiers and vocabulary terms. Unqualified Dublin Core is the metadata format required for basic interoperability for OAI Protocol Metadata Harvesting. However, within subject areas and communities other metadata specifications may be required; for example, it may be necessary to describe resources with complex structures and in a specialized way, as in the case of the AGRIS network. The AGRIS AP was designed for the agricultural domain with its specific characteristics and requirements. It is a standard created specifically to enhance the description, exchange and subsequent retrieval of agricultural Document-Like Information Objects (DLIOs). The AGRIPS AP is a format that allows sharing of information across dispersed bibliographic systems. It provides further qualifications to the simple Dublin Core element set; local extensions which were considered necessary for the comprehensive description of agricultural information resources. The standard, developed in light of the new AGRIS vision, enhances the quality of the description of agricultural information resources, enabling greater processing possibilities by service providers.

The use of AGRIS AP by the data providers within the AGRIS Community offers certain advantages to the service providers: defines a richer interoperability layer to aid secondary and tertiary resource discovery through the use of a qualified element set; gives flexibility to provide more relevant query results through targeted searching; provides rich, qualified metadata; gives adequate information about the content of the resource through the use of agriculture specific qualifiers such as classifications schemes, thesauri, etc.; and provides the capability to browse resources by subject, by country, by year etc. Although to be "officially" OAI-compliant, data providers will have to be simple Dublin Core compliant, as far as the AGRIS network is concerned, where "subject services" are provided, it is recommended that data providers be also AGRIS AP-compliant.

KNOWLEDGE ORGANIZATION SYSTEMS

Knowledge Organization Systems (KOSs) are organized knowledge structures, examples of which include: Authority files, Classification system/Schemes, Concept maps, Controlled lists, Dictionaries, Fact sheets, Glossaries, Ontologies, Subject headings, and Gazetteers. These resources fall along a continuum, according to the explicitness of their semantics and their amenability to machine interpretation. KOS are words or phrases taken from a standardized set that help to resolve two main issues: the problem of two or more words that can be used to mean the same concept, like Fishing vessels/Fishing boats or Health risks/Health hazards; and the problem of two or more words that have the same spelling but represent different concepts, e.g. vessel (blood)/ vessel (fishing) or Ling (a heath plant)/Ling (fish of the cod family).

The use of KOSs allows for the creation of more specialized services such as: browsing by keywords, country of coverage, searching by type of Document (patent, books, etc.), and limiting search results to one or more specific years or language or semantic navigation within the result set based on keywords identified in controlled vocabularies. These services are enabled through the use of controlled vocabularies and their explicit mention in the data exchanged, a possibility provided by the AGRIS AP. The use of machine readable formats like XML and standard vocabularies to exchange metadata are the building blocks for the future semantic web in the area of agriculture. They are the building blocks on which applications, that exploit the benefits of the semantic expressivity in bibliographic metadata, can be created to improve access to agricultural information.

REFERENCES

Abdul-Hadi, Z. et. al. 2006. Rapid Generation of Plant Protection Expert Systems.Computers in Agriculture and Natural Resources, 4th World Congress Conference, Proceedings of the 24-26 July 2006 (Orlando, Florida USA) Publication Date 24 July 2006.

Abdulsamadi M. , et.al., "Role of Information Technology in Agriculture" East Azarbaijan Research Center for Agriculture and Natural Resources, Tabriz-Iran

CLAES. 2007. National agricultural research management information system. Available at http://www.arc.sci.eg . CLAES, Egypt.

Mark S. Boguski, "Bioinformatics—A New Era," *Trends Guide to Bioinformatics* (Elsevier Science, 1998).

Swortzel, Kirk A, "Role of Information Technology in Agriculture". *Agricultural Education Magazine,* v73 n6 p26-27 May-Jun 2001.

Rafea, A. 1998. Agriculture. In J. Liebwitz (ed) *Handbook of applied expert systems.* CRC Press, Boca Raton.

3

Information Technology and Precision Farming

Precision farming or precision agriculture is an agricultural concept relying on the existence of in-field variability. It requires the use of new technologies, such as global positioning (GPS), sensors, satellites or aerial images, and information management tools (GIS) to assess and understand variations. Collected information may be used to more precisely evaluate optimum sowing density, estimate fertilizers and other inputs needs, and to more accurately predict crop yields. It seeks to avoid applying inflexible practices to a crop, regardless of local soil/climate conditions, and may help to better assess local situations of disease or lodging.

To many, *precision farming* might appear an oxymoron. With mud up to axles and 400 acres left to plow, precision seems worlds away. Yet *site-specific management* makes sense to a rapidly growing number of farmers as years of experience confirm the variability in field conditions and yield. Mapping and analyzing this variability and linking the spatial relationships to management action places production agriculture at the cutting edge of *GIS* applications. Its use down on the farm is both down to earth and downright ambitious.

UNDERLYING PRINCIPLES

Until the 1990s maps played a minor role in production agriculture. Soil maps and topographic sheets, for the most part, were too generalized for application at the farm level. Acquisition of spatial data with the detail and information farmers needed for operations were beyond reach. The principle of *whole-field management*, based on broad averages of field data, dominated management actions. Weigh-wagon, or grain elevator measurements, established a field's yield performance. Soil sampling determined the typical nutrient levels within a field. From these and other data the best overall seed variety was chosen and a constant rate of fertilizer applied, as well as a bushel of other decisions—all treating the entire field as uniform within its boundaries.

Site-specific management, on the other hand, recognizes the variability within a field and is about doing the right thing, in the right way, at the right place and time. It involves assessing and reacting to field variability by tailoring management actions, such as fertilization levels, seeding rates and variety selection, to match changing field conditions. It assumes that managing field variability leads to both cost savings and production increases, as well as improved stewardship and environmental benefits.

Site-specific farming isn't just a bunch of pretty maps, but a set of new technologies and procedures linking mapped variables to appropriate management actions. It requires the integration of several key elements: the *global positioning system* (GPS), *on-the-fly data collection devices, geographic information systems* (GIS) and *variable-rate implements*. Modern GPS receivers are able to establish positions within a field to about a meter. When connected to a *data collection device*, such as a yield/moisture meter, these data can be "stamped" with geographic coordinates. Several portable "heads-up" digitizing devices allow farmers to sketch conditions, such as weed infestations, on a map or aerial photo backdrop. A *GIS* is used to map the field data so a farmer can see the conditions throughout a field. The GIS also can be used to extend map visualization of yield to analysis of the relationships among yield variability and field conditions. Once established these relationships are used to derive a "prescription" map of management actions required for each location in a field. The final element, *variable rate implements*, notes a tractor's position through GPS, continuously locates it on the prescription map, then varies the application rate of field inputs, such as fertilizer blend or seed spacing, in accordance with the instructions on the prescription map. Combining the technologies of GPS, GIS and *intelligent devices and implements* (IDI) provides the mechanisms for managing field variability. The maturation and commercialization of these technologies have made the concept practical.

PROCESSING PRECISION FARMING DATA

To date, most analysis has been visual interpretations of yield maps. By viewing a map, all sorts of potential relationships between yield variability and field conditions spring to mind. These "visceral visions " and explanations can be drawn through the viewer's knowledge of the field. More recently, data visualization is being extended through map analysis at three levels: cognitive, analysis and synthesis.

The foundation of precision farming occurs at several levels:

— cognitive

— analysis

— synthesis

The *cognitive level* manages and stores mapped data on the desktop. The *analysis level* discovers relationships among the mapped variables, such as yield and soil nutrient levels. This step is analogous to a farmer's visceral visions of relationships but uses the computer to establish more detailed mathematical and statistical connections. Although this step is somewhat an uncomfortable "leap of scientific faith," it extends data visualization by investigating the coincidence of the patterns of variation among a set of maps. The results relate yield goals to specific levels of farm inputs—traditional agricultural research, but tailored to a farmer's "backyard."

The *synthesis level* evaluates newly derived relationships to formulate management actions for a new location (change in space) or another year at the same place (change in time). The result is a prescription map used to guide the intelligent implements as they "variable rate control" the application of field inputs. Or, the analysis might discover an area of abnormally low yield as aligning with a section of old drainage tile in need of repair. Further analysis might locate areas whose simulated yield increases under drier conditions justify installation of additional drainage tiles.

TECHNICAL ISSUES

The precision farming process can be viewed as four steps: data logging, point sampling, data analysis and spatial modeling. *Data logging* continuously monitors measurements, such as crop yield, as a tractor moves through a field. *Point sampling*, on the other hand, uses a set of dispersed samples to characterize field conditions, such as phosphorous, potassium, and nitrogen levels. The nature of the data derived by the two approaches are radically different— a "direct census" of yield consisting of thousands of on-the fly samples versus a "statistical estimate" of the geographic distribution of soil nutrients based on a handful of soil samples.

In data logging, issues of accurate measurement, such as GPS positioning and material flow adjustments, are major concerns. Most systems query the GPS and yield monitor every second, which at 4 mph translates into about 6 feet. With differential positioning the coordinates are accurate to about a meter. However the paired yield measurement is for a location well behind the harvester, as it takes several seconds for material to pass from the point of harvest to the yield monitor. To complicate matters, the mass flow and speed of the harvester are constantly changing when different terrain and crop conditions are encountered. The precise placement of GPS/Yield records are not reflected as much in the accuracy of the GPS receiver as in "smart" yield mapping software.

In point sampling, issues of surface modeling (estimating between sample points) are of concern, such as sampling frequency/pattern and interpolation technique. The cost of soil lab analysis dictates "smart sampling" techniques based on terrain and previous data be used to balance spatial variability with a farmer's budget. In addition, techniques for evaluating alternative interpolation techniques and selecting the "best" map using residual analysis are available in some of the soil mapping systems.

In both data logging and point sampling, the resolution of the analysis grid used to geographically summarize the data is a critical concern. Like a stockbroker's analysis of financial markets, the fluctuations of individual trades must be "smoothed" to produce useful trends. If the analysis grid is too course, information is lost in the aggregation over large grid spaces; if too small, spurious measurement and positioning errors dominate the information.

The technical issues surrounding *mapped data analysis* involve the validity of applying traditional statistical techniques to spatial data. For example, regression analysis of field plots has been used for years to derive crop production functions, such as corn yield (dependent variable) versus potassium levels (independent variable). In a GIS, you can use regression to derive a production function relating mapped variables, such as the links among a map of corn yield and maps of soil nutrients—like analyzing thousands of sample plots. However, technical concerns, such as variable independence and autocorrelation, have yet to be thoroughly investigated. Statistical measures assessing results of the analysis, such as a spatially responsive correlation coefficient, await discovery and acceptance by the statistical community, let alone the farm community.

In theory, *spatial modeling* moves the derived relationships in space or time to determine the "optimal" actions, such as the blend of phosphorous, potassium and nitrogen to be applied at each location in the field. In current practice, these translations are based on existing science and experience without a direct link to data analysis of on-farm data. For example, a prescription map for fertilization is constructed by noting the existing nutrient levels (condition) then assigning a blend of additional nutrients (action) tailored to each location forming a if-(condition)-then-(action) set of rules. The issues surrounding spatial modeling are similar to data analysis and involve the validity of using traditional "goal seeking" techniques, such as linear programming or genetic modeling, to calculate maps of the optimal actions.

The application of GIS within production agriculture has been rapid. Since its inception in the early 90s, precision farming has moved from a fledgling idea to operational reality on millions of acres. Its current expression emphasizes the generation of yield maps by linking GPS with on-the-fly yield monitors. Valuable insight is gained by visualizing field variability, particularly when yield maps for several years are considered. More advanced applications include analysis of soil nutrient maps to derive a prescription map used in variable rate control of fertilizer, terrain analysis for variable seeding rates and spatial modeling for timing and spot application of herbicides/pesticides.

The infrastructure for precision farming is coming online. Most manufacturers offer precision farming options with their farm vehicles and implements. A growing number of service providers offer advice to farmers in their adoption of the new technology. At present, however, a full implementation of precision farming is in the hands of the developers and researchers. Advancements in the data analysis and spatial modeling phases await contributions from the GIS community. The considerable knowledge and methodologies of the agricultural science community need to be reviewed for their spatial inferences. Opportunities abound in one of GIS's more important applications and we all benefit from precision farming's fruits—check it out at your local super market.

OPERATIONALIZATION OF PRECISION FARMING

Agriculture is the backbone of the Indian Economy"- said Mahatma Gandhi five decades ago. Even today, as we enter the new millennium, the situation is still the same, with almost the entire economy being sustained by agriculture, which is the mainstay of the villages. Not only the economy, but also every one of us looks up to agriculture for our sustenance too. Therefore, it is no surprise if agriculture gets the celebrity status in the name of Precision farming (PF). In recent times, the researchers in the field have been busy formulating methodologies and fabricating new implements for precision farming. It is here the challenge arises considering the implementation of the technology at various levels in the Global community. The need of the hour is not application of the technology but the adoption of appropriate technology, which would suit the particular level of the global community. In India, the farming practices are too haphazard and non-scientific and hence need some forethought before implementing any new technology.

Applications of agricultural inputs at uniform rates across the field without due regard to in-field variations in soil fertility and crop conditions does not yield desirable results in terms of crop yield. The management of in-field variability in soil fertility and crop conditions for improving the crop production and minimizing the environmental impact is the crux of precision farming. Thus, the information on spatial variability in soil fertility status and crop conditions is a pre-requisite for adoption of precision farming. Space technology including global positioning system (GPS) and GIS holds good promise in deriving information on soil attributes and crop yield, and allows monitoring seasonally- variable soil and crop characteristics, namely soil moisture, crop phenology, growth, evapotranspiration, nutrient deficiency, crop disease, and weed and insect infestation, which, in turn, help in optimizing inputs and maximizing crop yield and income.

Though widely adopted in developed countries, the adoption of precision farming in India is yet to take a firm ground primarily due to its unique pattern of land holdings, poor infrastructure, lack of farmers' inclination to take risk, socio-economic and demographic conditions.

The conventional agronomic practices follow a standard management option for a large area irrespective of the variability occurring within and among the field. For decades now, the farmers have been applying fertilizers based on recommendations emanating from research and field trials under specific agro-climatic conditions. Since soil-nutrient, characteristics vary not only from one region to another, but also from field to field. Even within a field, there is a need to take into account such variability while applying fertilizers to a particular crop. Consideration of in-field variations in soil fertility and crop conditions and matching the agricultural inputs like seed, fertilizer, irrigation, insecticide, pesticide, etc. in order to optimize the input or maximizing the crop yield from a given quantum of input, is referred to as precision farming or precision agriculture or precision crop management.

The term "precision farming" means carefully tailoring the soil and crop management to fit the different conditions found in each field. It is defined as the application of technologies and principles to manage spatial and temporal variability associated with all aspects of agricultural production. It is also referred to as "prescription farming", "site specific farming" or "variable rate technology."

By catering to this variability, called precision farming, one can improve the productivity or reduce the cost of production and diminish the chance of environmental degradation caused by excess use of inputs. Thus, mapping and analysis of within field variability is an essential input for precision crop management. Thus, PF involves acquiring the variations in crop or soil properties, mapping, and analyzing the variations, adopting suitable management techniques to maximize the yield. Farmers have been applying fertilizers based on recommendations emanating from research and field trials under specific agro-climatic conditions, which have been extrapolated to a regional level. Since soil nutrient characteristics vary not only between regions and between farms but also from plot to plot, and within a field or plot, there is a need to take into account such variability while applying fertilizers to a particular crop. Consideration of in-field/plot variations in soil fertility and crop conditions and matching the agricultural inputs like seed, fertilizer, irrigation, insecticide, pesticide, etc. in order to optimize the input or maximizing the crop yield from a given quantum of input, is referred to as precision farming or precision agriculture or precision crop management.

The information for variability map can be obtained from soil tests for nutrient availability, yield monitors for crop yield, soil samples for organic matter content, information in soil maps, or ground conductivity meters for soil moisture. Generally, the fields are manually sampled along a regular grid and the analyzed results of the samples are interpolated using geostatistical techniques. These techniques are time consuming, labour intensive and in many cases destructive especially, for agricultural situation in India. With small size of landholdings and low income of farmers, the adoption of this methodology in its present form is not feasible. Various workers have shown the advantages of using remote sensing technology to obtain spatially and temporally variable information for precision farming. In an earlier work, Ray et al. have shown the usefulness of IRS merged data in mapping the variability.

COMPONENTS OF PRECISION FARMING

- Remote Sensing
- GIS
- DGPS
- Variable rate Applicator

Remote Sensing

This is for Data acquisition of the farms to find the soil, vegetation and other parameters that are amenable for remote sensing. Remote sensing techniques play an important role in precision farming by providing continuous acquired data of agricultural crops. Remote sensors image vegetation, which is growing on different soil types with different water availability, substrate, impact of cultivation, and relief. These differences influence the state of the plants and cause heterogeneous regions within single fields. Hence, the heterogeneous vegetation acts as an interface between soil and remote sensing information, because vegetation parameters describing the state of the plants can be deduced from remote sensing imagery. The analysis of the variability occurring within the field was carried out by measuring soil and plant parameters through conventional methods as well as through spectral techniques using ground truth spectroradiometer (350-1800 nm) and satellite data. Ray et al. (2001), in general, have reviewed by Moran et al. (1997) and for Indian conditions the potentials of remote sensing in providing information required for precision farming.

Geographical information system

The Geographic Information System (GIS) contributes significantly to precision farming by allowing presentation of spatial data in the form of a map. In addition, GIS forms an ideal platform for the storage and management of model input data and the presentation of model results, which the process model provides.

DGPS

'Do the right thing, in the right place and in right time'. This is where GPS comes into picture. In addition, the accuracy, which is the important factor in PF, demands for DGPS (Differential Global Positioning System). GPS makes use of a series of military satellites that identify the location of farm equipment within a meter of an actual site in the field. The value of knowing a precise location within inches is that:

- Locations of soil samples and the laboratory results can be compared to a soil map,
- Fertilizer and pesticides can be prescribed to fit soil properties (clay and organic matter content) and soil conditions (relief and drainage),

- Tillage adjustments can be made as one finds various conditions across the field, and
- One can monitor and record yield data as one goes across the field.

The Global Positioning System (GPS) technology provides accurate positioning system necessary for field implementation of variable rate technology (VRT). The Internet makes possible the development of a mechanism for effective farm management using remote sensing.

Variable Rate Applicator:

The variable rate applicator has three components:

- Control computer
- Locator and
- Actuator

The control computer coordinates the field operation. It has a map of desired activity as a function of geographic location. It receives the equipment's current location from the locator, which has a GPS in it, and decides what to do based upon the map in its memory or data storage. It then issues the command to the actuator, which does the input application.

Practical problems in Indian agriculture

Precision Agriculture has been mostly confined to developed countries. Reasons of limitations of its implementation in developing countries like India are:

- Small land holdings,
- Heterogeneity of cropping systems and market imperfections,
- Lack of technical expertise knowledge and technology (India spends only 0.3% of its agricultural Gross Domestic Product in Research and Development)
- High cost.

In India, major problem is the small field size. More than 58 percent of operational holdings in the country have size less than 1ha. Only in the states of Punjab, Rajasthan, Haryana and Gujarat more than 20 per cent of agricultural lands have operational holding size of more than four hectare. There is a scope of implementing precision Agriculture for crops like, rice and wheat especially in the states of Punjab and Haryana. Commercial as well as Horticultural crops shows a wider scope for precision Agriculture.

In India, broadly two types of agriculture viz., high input agriculture characterized by the provision of assured irrigation and other agricultural inputs, and subsistence farming, which is confined mostly to rain-fed, or dry land regions, are prevalent.

Nearly two-third arable land in India is rain-fed. The crop yields are very low (˜1 t ha-1) and very good potential exists for increasing productivity of rain-fed Cropping systems.

Methodology

Using a GPS along with a digital drainage map, the farmer is able to apply these pesticides in a safer manner. The spraying equipment can be preprogrammed automatically to turn off when it reaches the distance limitation or zone of the drainage feature. Additionally, farmers can preprogram the rate of pesticide or fertilizer to be applied so that only the amount needed determined by the soil condition is applied varying this rate from one area of the field to another. This saves money and allows for safer use of these materials.

Data collected by the GPS operations can be automatically recorded with the GIS program. Remotely sensed data can be analyzed and added to the GIS using soil maps, digital terrain and field operations information as ground truth. Additionally the farmers are able to record observations through the growing season such as weed growth, unusual plant stress or coloring and growth conditions. This can be used to guide further field operations like spraying, fertilizing and irrigating and it is part of the permanent record. Precision farming will make a strong impact on the way farmers manage their farm operations in the future.

Seed spacing

The yield from the seeds is maximum when adequate care is taken to adequately space them to enable each plant the optimum requisites for its growth, namely the soil nutrients, water, sunlight and protection against pest infestation. This is best accomplished by varying the seeding rate even within a single piece of land according to the soil conditions such as texture, organic matter and available soil moisture. One would plant fewer seeds in sandy soil as compared to silt loam soils because of less available moisture. The lower seed population usually has larger heads (ears) of harvested seeds providing for a maximum yield. Since soils vary even across an individual farm field, the ability to change seeding rates as one goes across the field allows the farmer to maximize this seeding rate according to the soil conditions. A computerized soil map of a specific field on a computer fitted on the tractor along with a GPS can tell farmers where they are in the field allowing the opportunity to adjust this seeding rate as they go across their fields.

Steps to be taken for implementing PF in India:

In the present existent situation, the potential of precision agriculture in India is limited by the lack of appropriate measurement and analysis techniques for agronomically important factors (National Research Council, 1997). High accuracy sensing and data management tools must be developed and validated to support both research and production. The limitation in data quality/availability has become a major obstacle in the demonstration and adoption of the precision technologies. The adoption of precision agriculture needs combined efforts on behalf of scientists, farmers and the government. The following methodology could be adopted in order to operationalise precision farming in the country.

— Creation of multidisciplinary teams involving agricultural scientists in various fields, engineers, manufacturers and economists to study the overall scope of precision agriculture.

— Formation of farmer's co-operatives since many of the precision agriculture tools are costly (GIS, GPS, RS, etc.).

— Government legislation restraining farmers using indiscriminate farm inputs and thereby causing ecological/environmental imbalance would induce the farmer to go for alternative approach.

— Pilot study should be conducted on farmer's field to show the results of precision agriculture implementation.

— Creating awareness amongst farmers about consequences of applying imbalanced doses of farm inputs like irrigation, fertilizers, insecticides and pesticides.

Realizing the potential of space technology in precision farming, the Department of Space, Government of India has initiated eight pilot studies in well-managed agricultural farms of the ICRISAT, the Indian Council of Agricultural Research and the Agricultural Universities, as well as in farmers' fields. The pilot studies aim at delineating homogeneous zones with respect to soil fertility and crop yield, estimation of potential yield, yield gap analysis, monitoring seasonally-variable soil and crop conditions using optical and microwave sensor data, and matching the farm inputs to bridge the gap between potential and actual yield through Spatial Decision Support Systems (SDSS). The test sites are spread over a fairly large area across a cross section of agro-climatic zones of the Indian sub-continent, and cover some of the important crops like wheat, rice, sorghum, pigeon pea, chickpea, soybean and groundnut.

The next step would be to generate detailed-level information on soil resources addressing potentials and limitations of individual fields since except for states like Punjab, Haryana, Madhya Pradesh and Maharashtra where fields size is quite large, practically individual field could be treated as a homogenous management unit for the purpose of precision farming.

DGPS Network:

As a first step towards operationalization of PF, it is necessary to establish a DGPS infrastructure for the country. It would enable the farmers to get an accuracy of few centimeters in the various unit processes (sowing, fertilizer application, herbicide-pesticide application etc.,) involved in PF. A DGPS network would cater to the needs of multitude of applications (Meteorology, Transportation, Geodetic survey, Crustal Deformation studies, Disaster management and mitigation, etc.) of which PF is one. Each DGPS master reference is capable of providing the services within a radius of 100-kilometer radius. Hence, the number of master stations required to establish the DGPS infrastructure, which would cover the entire country, is calculated. This can be calculated as follows (only a random calculation)

Area of India	= 329 million hectares.
Area of GPS (circular area, PI = 3.14)	= PI * (100) ^2 sq. km.
	= 31400 sq. km.
	= 3.14 million hectares
Total no. of GPS reference stations required for the country	= 329/3.14
Total GPS reference stations required	= 105
Cost of a single DGPS set	= Rs. 3 lakhs
Total cost of the entire infrastructure	= Rs. 3.15 crores

This is affordable by any means for a country as a whole considering the innumerable applications it can cater. A simple assumed cost benefit analysis is carried out for precision farming, which is given below.

Cost Benefit Analysis

The real value for the farmer is that he can adjust seeding rates, plan more accurate crop protection programs, perform more timely tillage and know the yield variation within a field. These benefits will enhance the overall cost effectiveness of his crop production. Since the country is going to spend the amount, the GDP is considered rather than the farmer's personal income. Hence the cost-benefit analysis of the present project is

GDP of India for 1993	= $ 250.97 billion (FAO yearbook, 1994)
GDP in Indian Rupees	= Rs. 12,924.955 billion
Amount spent for R&D	= 0.3% of GDP = Rs. 38.775 billion
Cost-Benefit Analysis	= 38.775 billion/315 million = 12,309

Thus, it is proved that there is an enormous benefit in implementing the DGPS network, which would further pave the way for Precision Farming.

Discussion and Analysis

Precision farming has taken "agriculture into the space age". Farmers have services available that involve satellites collecting data, transmitting location information, or providing data from a variety of sources. Farmers can analyze this satellite information or they can rely on companies to do this service for them for a fee. Precision farming is undoubtedly relevant to Indian agriculture in the context of improving agricultural production and Stakeholders' income and minimizing environmental impact. Geoinformation technology offers immense potential for deriving information on soil fertility, crop conditions and crop yield, crop simulation models enable estimating potential crop yield and the decision support system to facilitate developing appropriate prescription for improving crop production while minimizing the cost of inputs. The major limitations of currently available high spatial resolution multispectral data are the fixed and broad spectral bands, very low repetivity, high cost which most of the Indian farmer could not afford. In order to match the technology with the requirement of precision farming, improvements in corresponding elements needs to be made to address the limitations to reap its benefits to fullest extent.

CONTINUOUS DATA LOGGING

Smart Farmers and Dumb Maps

We have been mapping for more than 8,000 years and farming even longer. So what makes the merger of maps and fields in precision farming so radically new and confusing? The precision farming revolution involves a new perspective of mapped data from one of description of the precise placement of physical features to one of prescription of appropriate actions based on spatial analysis. This transition involves extending the familiar paper map composed of inked lines and shadings to the somewhat intimidating digital map world of organized sets of numbers. In terms of mapped data it involves recognition of the differences between cartographic and geographic information systems (GIS) data quality. Cartographic quality uses the stored numbers simply as surrogates for pen colors of traditional map features. Computer-assisted design (CAD) drawings and graphical representations of maps use this approach and often are referred to as "dumb maps" because they merely automate descriptive mapping. The objective of these systems is drafting traditional map products, not map analysis for management action. Cartographic quality numbers also can serve as an index to feature attributes stored in a database enabling rapid access to database information and map displays of the data. The GIS quality numbers, on the other hand, respect all of the "map-ematical" rights, privileges and responsibilities needed for spatial statistics, analysis and modeling required for the prescriptive maps in precision farming.

GIS Quality Data

Cartographic quality data simply use the stored numbers as surrogates for pen colors in drafting traditional map features and "geo-query" of the data. The numbers in GIS quality data, on the other hand, respect all of the map-ematical rights, privileges and responsibilities you fondly recall from high school math. At the heart of precision farming is the recognition that maps are organized sets of numbers poised for analysis.

Consider the derivation of a map of soil phosphorous levels from a set of soil samples depicted in figure 1. Inset (a) shows the level of phosphorous (P) for 16 samples taken from a field. The normalized values range from almost 0 to 87, with the higher levels depicted as the taller points. Note the lowest levels occur on the left side of the field. A "first-order" estimate of the spatial distribution of the data in inset (b) assigns the P-level of the closest sample for each location throughout the field (discrete sample points to continuous map surface). In techy terms the blocks are called Thiessen polygons with sharp boundaries formed by the perpendicular bisectors among neighboring samples.

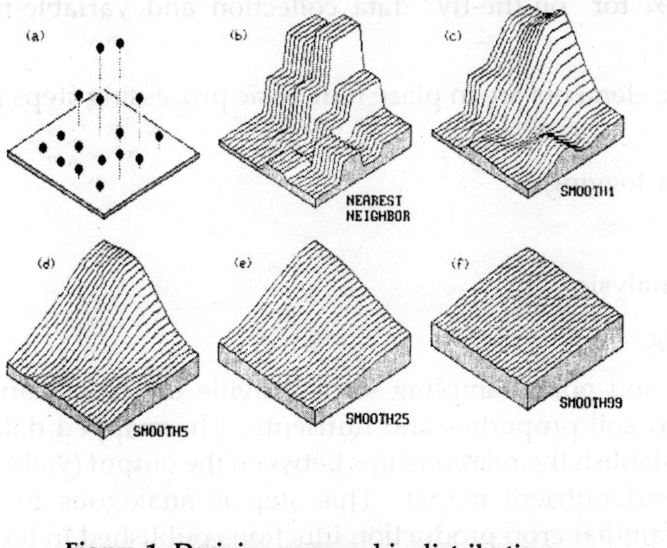

Figure 1. Deriving geographic distributions.

Inset (c) in the figure was derived by moving a smoothing window around the nearest-neighbor surface. When the window is centered near one of the sharp boundaries, it has a mixture of big and small values, resulting in an average somewhere in between—a whack off the top and a fill-in at the bottom. The remaining insets (d) through (f) continue to smooth the phosphorous surface. Eventually the surface will be completely eroded to a flat plane in inset (f) that approximates the average of the whole field—the ultimate in spatial aggregation.

That brings us to the crux of precision farming and variable rate technology (VRT). In the past you set the fertilizer rate for the average level of P in the field and let 'er rip—everywhere the same based on the field average. But a lot of information locked in the field samples was thrown away. In precision farming/VRT the onboard computer gets a reading from the GPS unit, positions itself on the phosphorous surface, checks the P value for that part of the field, then sets the rate so just the right amount is applied. As the tractor moves through the field, the rate is continuously varied based on the spatial calculations. How well the system works has a great deal to do with the data quality and the spatial stat/math used in the.

FOUR BASIC STEPS IN PRECISION FARMING

In the introduction we noted three basic technologies driving precision farming:

- global positioning systems (GPS) for positioning,
- geographic information systems (GIS) for analyses, and
- intelligent devices and implements (IDI)¾ such as yield monitors and variable-rate fertilizer rigs¾ for "on-the-fly" data collection and variable-rate control of field inputs.

Now that the basic elements are in place four basic processing steps in precision farming tie it all together:

—continuous data logging

—point sampling

—mapped data analysis

—spatial modeling

The data logging and point sampling steps provide the base maps of yield and field conditions, such as soil properties and nutrients. The mapped data analysis step uses the computer to establish the relationships between the output (yield map) and the inputs (soil properties and nutrient maps). This step is analogous to the ag researcher's derivation of the familiar crop production functions published in texts and reports. The main difference is that the analysis generates a production curve for your own backyard—not for a research field 70 miles away on a different soil and using different varieties. The spatial modeling step uses your tailored "backyard" relationships to determine appropriate management actions and, as Fran Pierce of Michigan State puts it, the "if <condition> then <action>" guidance a farmer needs. The difference from traditional research and guidance is that precision farming's recommendations respond to the unique conditions and variability within your field and is reported in the form of a map. Admittedly, this process might seem a bit overwhelming at first¾ and more appropriate for a "techno-farmer"¾ but it could be heralding the next revolution in agriculture.

Putting Yield in Its Place

The yield monitoring step in the above-mentioned process is rapidly becoming commonplace and seems to be fading as high tech talk at the coffee shop. The new buzz is about yield mapping, which hooks a yield monitor's output to a computer. The GPS "stamps" these data with earth coordinates, thereby enabling the computer to map them. There has been a fair amount of discussion about the accuracy of the GPS stamp, but less discussion about the process and accuracy involved in generating a yield map—the art and science of creating "pretty maps." The data logging of yield results in thousands of point measurements over a field. Some mapping systems simply place a small colored dot at each location, with the color indicating relative yield. The assembled mass of dots forms a point map with color patterns tracking the differences in yield throughout a field. Although point mapping generates acceptable map images, it doesn't provide the consistent data structure needed for map analysis. An alternative approach uses a rectangular grid to summarize yield for a continuous map by averaging the measurements falling in each grid cell. Many of these systems use smoothing windows that weight-average numerous points around a cell as a means to remove measurement noise and get a better generalization of the data's spatial distribution. Also, the summary for each cell provides statistics about localized yield variation, such as a standard deviation map, which directs attention to areas with highly variable yields—salt-and-pepper patterns of high and low yields. More importantly, the grid forms a consistent structure for relating different maps, such as the percentage of change in yield between two years or the correlation among yield and soil nutrient maps (mapped data analysis). However, there are several technical issues that, if ignored, turn the science of yield mapping into a large part art. These issues will be addressed throughout the book.

Putting Yield in Its Proper Place

Precision farming presumes you have good data to analyze. Before we forge ahead, it might be prudent to investigate the issues that can turn good data bad, specifically, the precise placement of yield measurements.

Your GPS may be accurate to a meter, but does that same precision hold for the yield data? Three factors beyond the GPS signal influence the precise placement of yield measurements: 1) antenna offset. 2) mass flow adjustment, and 3) synchronization of computer clocks.

Antenna offset is the easiest to deal with because it simply adjusts for the fixed distance from the harvesting point to where you anchored the antenna. Simply plug in the offset and correct for the point of harvest¼ sort of. That's the current point, but the material at the yield sensor at that moment was collected somewhere further back (inset T1). A simple fixed distance offset to the yield monitor won't do, as the path of the material doesn't conform to simple geography.

Several things contribute to the dynamics of mass flow through the harvester, and ultimately the calculation of the physical distance needed to put the yield measurement in its proper place. First, the mechanical design of the harvester follows a contorted internal path to the yield monitor. The twisted path coupled with the speed of the conveyor belts results in a processing time delay as the material moves through the system.

Under constant conditions the delay can be measured and the physical distance calculated by multiplying the standard delay by the harvester's speed. For example, a delay of 16 seconds at 4 mph repositions the current yield measurement about 70 feet behind—kind of like a trolling a lure behind a boat. However, things aren't that simple. What happens on turns? Harvesting pauses, speed changes, the line bends, the processing path clears out, then harvesting resumes; all of which makes repositioning the yield measurement a lot tougher.

Even on the straight-away, things are a lot more complex than simply trolling for data. There is a myriad of muddling factors. Consider moving down a slope in a dense part of the field¾ material accumulates so the processing delay is increased to 19 seconds and the speed is cut back to 3 mph to handle the flow. The physical distance under these conditions extends to nearly 84 feet (4.4 ft./sec. * 19 sec.).

The final issue involves synchronization of computer clocks. Like a metronome, the GPS and yield monitors are sampled at a given downbeat (e.g., each second). However, communications interference across the I/O bus in the hardware can stall things, particularly when recording conflicts with writing out data buffers or screen display updates. Like a lousy dancer, a dropped downbeat can really step on your toes.

All this seems to make an accurate yield map beyond reach and precision farming a pipe-dream. Not true; what it should do is alert you that "dumb maps" simply put a yield measurement at its initial GPS position. It should also arm you with some serious questions to look into about how "smart" your mapping software actually is. Many of these issues or questions are presented in appendix C, "Checklist for Yield Mapping Software."

Connecting the Dots

The sage advice of "what seems simple ain't" certainly holds for precision farming. The precise placement of GPS yield records isn't simply reflected in the accuracy of a GPS unit. A mass flow offset is required to place the measurement somewhere behind the instantaneous GPS location. The offset accounts for the time it takes for plant material to move from the harvest point to the yield monitor, and its calculation involves a complex interaction between mechanical design and amount of material flow. The delay multiplied by the intervening speed calculates the offset, which is used to reposition a yield record. The ability of yield mapping software to account for mass flow dynamics is the key to precise placement of yield measurements.

However, the geographic treatise of the precisely placed points determines, to a large extent, what you can do with the data. Figure 2.illustrates how yield options can be calculated. A point map, inset (a), is the simplest rendering of yield data. It uses a fixed symbol, such as a dot centered on a point's x,y coordinates, which is assigned a color corresponding to an interval of yield such as dark green for 150 to 200 bushels of corn. When viewed, groups of similarly colored dots form patterns of yield variation. The display is visually effective but lacks the geographic structuring needed for data analysis. In the computer, the coordinates and yield values simply form a long list of independent numbers—your eye and brain assembles the data into patterns of neighboring points.

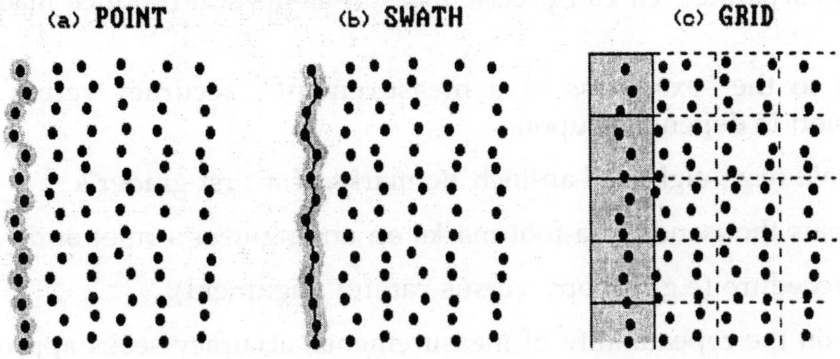

Figure 2 Basic types of yield options

A swath map, inset (b), provides a bit more data organization by linking points along a harvester's path. The swath can be sequential pairs of points, but often is represented as a mathematical vector (line segment), plotted as a rectangle with a width of the harvester's header and a length of a set number of consecutive points. The color fill of the rectangle (termed a swath element) is determined as the average of the yield points it contains. Swath elements can be plotted to align with the actual GPS coordinates forming a jagged path, or they can be "forced" to the straight lines characterizing adjoining harvester paths. The advantage of swaths over points is that the computer "sees" more of the relationships in the data—sequencing, length, alignment, and data smoothing. However, both data structures are too limited for data analysis as they are based on irregular geographic units (points and rectangles) of varying sizes and placement.

A grid map, inset (c), on the other hand, uses a regular grid to statistically summarize the data. Like the daily or weekly reports of commodity prices, averages are useful in determining trends from the volatility of moment-by-moment price fluctuations. In addition to a better representation of field trends, a regular grid provides the geographic consistency the computer needs to analyze data patterns within and among maps. Since the size and spacing of grid elements are consistent, the computer can easily locate neighboring cells for intramap analysis, such as average yield for different portions of a field. More importantly, the consistent partitioning into grid cells provides the structure for intermap analysis, such as relating a yield map to soil nutrient maps. Generally speaking, point and swath maps support graphical analysis (visualization), and grid maps support data analysis (map-ematical). Both perspectives are needed.

Precision and Accuracy

What's the difference between precision and accuracy? Is it merely semantics? Is a precise device always accurate? Or can precise measurements still produce inaccurate data?

Precision refers to the "exactness of a measurement"; accuracy refers to its "correctness." Precision is dependent upon

— measurement scale (e.g., eighth-of-an-inch tic marks of a first-grader's

— straightedge versus thousands-of-a-foot marks on an engineer's rule) and

— measurement procedure (e.g., sloppy versus careful alignment).

Precision is focused on the repeatability of measurements; accuracy seeks appropriate representation of a characteristic or condition.

One source of imprecision is termed random error. Sometimes measurements are more than and sometimes they're less than without any discernible pattern. In most instances, this "noise" is easily remedied—simply take a bunch of readings and average them. However it's impossible to repeatedly harvest and measure the same spot in a field so we assume random error is minimal in our yield monitors and choose to ignore it.

Another situation is termed biased error. Your bathroom scale might repeat time after time that you weight 201 pounds, but you know it's always 10 pounds over so your weight is actually only 191. In precision farming you might have a precise yield monitor that repeatedly records differences within a bushel, but always 10 over. That's easy to remedy—measure known amounts and calculate a calibration factor (subtract 10).

Things get a bit tougher if the bias changes with the level of measurements; say at least 10 more for high readings to 10 or less for low ones. Now you have to develop a calibration function whose adjustments change as the yield values change. Even more insidious is biased error that results from a complex interaction of several factors, such as the dynamic mass flow adjustments of yield discussed earlier.

So what sort of error does a weigh-wagon calibration address? If the actual weight is 10 percent more than the yield monitor estimate (spatial integral for you techy-types) most folks simply adjust each yield record by 10 percent. Sure, the mathematics works as both the weigh-wagon and monitor values for the area are forced to be the same. But does a spatially aggregated benchmark make sense as a spatial calibration factor? Actually it doesn't make any difference—spatially, that is. Every point simply is increased by 10 percent so the pattern (relative yield) hasn't changed. What's needed is research into the nature of yield map errors and effective calibration functions.

If "appropriate adjustment" of imprecise data is one way to make measurements more accurate (measurement accuracy), another way is to aggregate the data to characterize a larger unit (feature accuracy). For example, politicians do this all the time. They aggregate data on individuals into statistics about groups of people that best represent general characteristics of voters, then tailor their spiel to the nonexistent typical person representing each group. That's sort of what a grid map does to yield records—it summarizes them into spatial groups (grid cells). As in most statistical applications, the summary tends to enhance trends by dampening individual fluctuations. It recognizes that site-specific management doesn't operate at the individual plant level, but somewhere between that and whole-field management. What's needed is a better understanding about how best to define the aggregation unit (grid cell) and determine the best mix of management actions (fertilizer blend, seeding rate, variety selection, etc.) for each of them. Bet you never knew you had so much in common with a politician.

Resolving Yield Mapping Issues

There are two main advantages to this approach:

— It provides consistent geographic referencing among various mapped data layers.

— It smoothes out measurement noise while reporting statistics on localized yield variation.

However, there are a few of technical issues that need to be considered. First, the gridding resolution (cell size) can greatly affect the calculations. Insets (a) and (b) in Figure 3 depict two gridding resolutions. Note that the number of points falling in the cells is significantly different between the two resolutions; hence, the average yield reported will most certainly differ. If the grid pattern is too large, some of the information in the data will be lost (averaged-over). If it is too small, undue credence might be attached to differences arising simply from measurement and positioning errors.

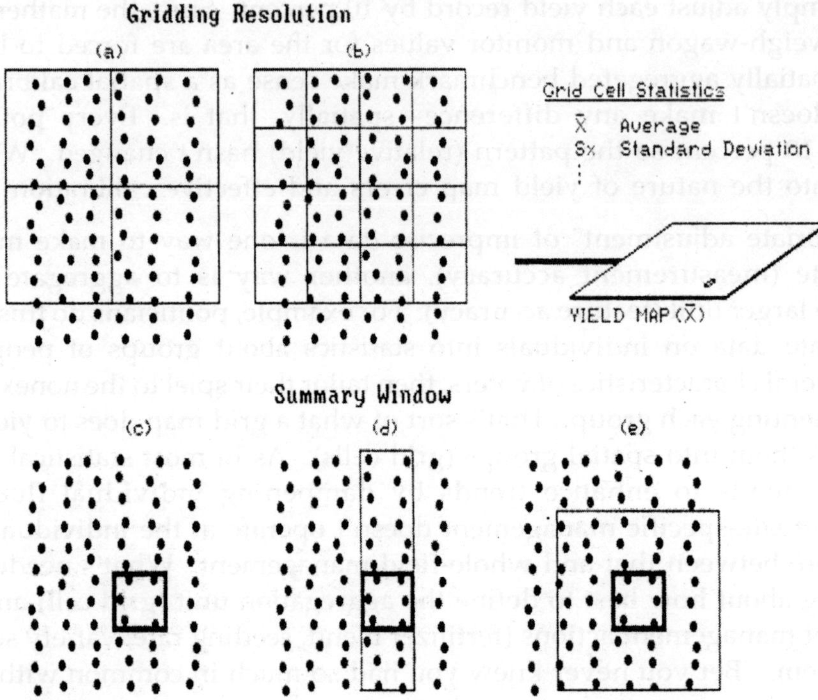

Figure 3. Summarizing yield measurements

Other technical issues involve the definition of the summary window and the summary procedure used. Insets (c), (d) and (e) show three different summary window designs—single-cell, inline filter and nearest-neighbors summary. The single-cell design uses only those measurements that actually fall within a cell for the summary calculations. The inline design uses the direction of travel to "filter" surrounding data. It can help in smoothing out measurement errors resulting from the "coughs and spits" due to uneven grain flows through a combine. The technique is analogous to moving averages routinely used in analyzing times series data, such as commodity prices or the stock market. Both the single and inline techniques can be used for "on-the-fly" data compression—simply keep the summary statistics and discard the pile of raw measurements.

The nearest-neighbors technique involves the post-processing of the raw data. It moves a window around the field sequentially centered on each grid cell. At each stop, it calculates a summary for the points falling within the window and assigns that value to that grid cell. In addition to window design, the summary procedure can vary—such as simple or weighted averaging. For example, a distance-weighted average is influenced more by nearby measurements than those that are farther away.

POINT SAMPLING

Dirty Stuff

While data logging continuously records measurements (e.g., crop yield) as a tractor moves through a field, point sampling uses a set of dispersed, discrete samples to characterize field conditions (e.g., phosphorous). The nature of the data derived by the two approaches is radically different—a "direct census" for a yield map versus a "statistical estimate" for a phosphorous map. In data logging, issues of accurate measurement, such as GPS positioning and material flow adjustments, are major concerns.

In point sampling, issues of spatial interpolation—estimating between sample points such as sampling frequency/pattern and interpolation technique—are the focus of concern.

Let's tackle the first area: the sampling design issues of frequency and pattern. There are four distinct considerations in sampling design:

— stratification,

— sample size,

— sampling grid and

— sampling pattern.

The first three considerations determine the appropriate groupings for sampling (stratification), the sampling intensity for each group (sample size) and an appropriate reference grid resolution (sampling grid). All are closely tied to the spatial variation of the data to be mapped. For example, if your field splits a couple of soil types, you might divide the field into two strata. If one is fairly consistent, you might allocate fewer samples than within the other, more variable soil unit (Figure 4). Also, you might decide to generate a third strata for even more intensive sampling around the soil boundary itself. Or, another stratification possibility is to use your yield map. If you feel the variations in yield are primarily "driven" by soil conditions, then that map could be used to divide the field into high, medium and low productivity areas.

This approach might respond to soil units that are below the resolution of the traditional soil type map or characterized by different nutrient levels. In the past, a single sampling frequency was used throughout an entire state or region. In part, a single frequency was chosen for ease and consistency of field implementation, as well as tradition. With the advent of GPS technology, however, variable sample frequencies is a modern alternative¾ simply navigate to each sample location.

as simple or weighted averaging. For example, a distance-weighted average is influenced more by nearby measurements that those that are farther away.

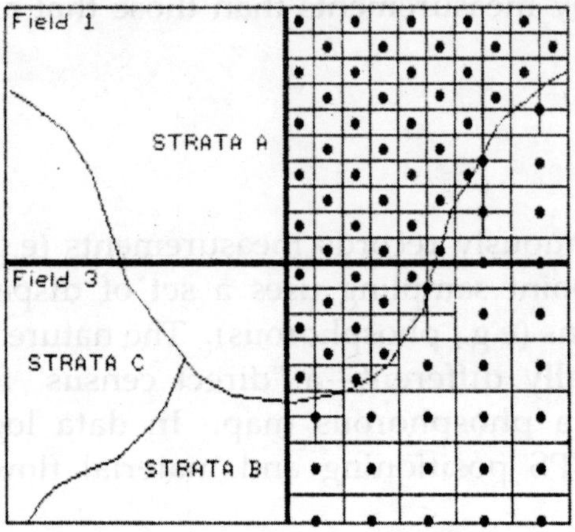

Figure 4. Field stratification

Sampling Patterns

The final step, sampling pattern, identifies individual sample locations. Traditional statistics tends to emphasize randomized patterns because they ensure maximum independence among samples, which is critical to calculating the central tendency of a data set (average for an entire field). Figure 5 identifies five systematic patterns, including a completely random one.

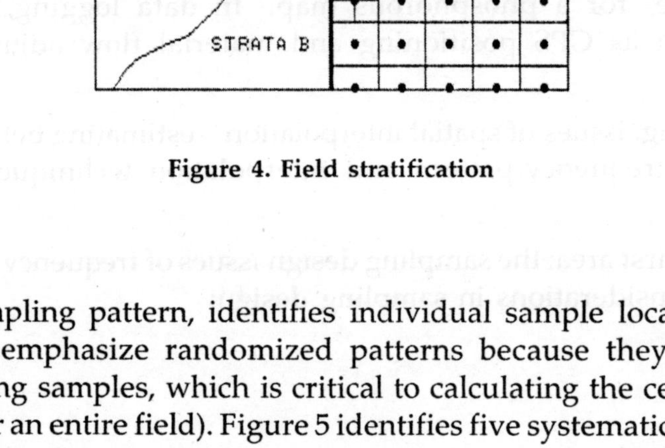

Figure 5. Systematic sampling patterns

Note that the regular pattern is uniformly distributed in geographic space. So is the staggered start one, except the equally spaced y-axis samples alternate the starting position at one-half the sampling grid spacing. The result is a "diamond" pattern rather than a "rectangular" one. This pattern is better suited for generating maps because it provides more intersample distances for spatial interpolation.

The random-start pattern begins each column transect at a randomly chosen y coordinate within the first grid cell, thereby creating even more intersample distances. The result is nearly a regularly spaced pattern, with "just a hint of randomness." The systematic unaligned pattern also results in a fairly regularly spaced pattern, but it exhibits even more randomness as it is not aligned in either the x or y direction. The pattern is formed by first placing a random point in the cell in the lower-left corner to establish a pair of x and y offsets. For rows in the grid, the x offset is held constant while y is randomly varied. For columns, the y offset is held constant while x is varied. The result is random starts in both the x and y axes.

The "dots" in the random cluster pattern establish an underlying uniform pattern (every other staggered start sample point). The "crosses" locate a set of related samples that are randomly chosen within the enlarged grid space surrounding each dot. Note that the patterned is not as regularly spaced as the previous techniques; however, it has other advantages. The random subset of points allows a degree of unbiased statistical inference, such as a t-test of significance differences among population means. The simple random pattern allows full use of statistical inference (whole field statistics), but the "clumping" of the samples makes it the worst for mapping (site-specific statistics). So which pattern should be used? Whichever works best with your data and objectives.

From Points to Maps

Spatial interpolation refers to the generation of a continuous map surface from discrete point samples. All spatial interpolation techniques establish a "roving window" that

— moves to a location in a field,

— calculates an estimate (guess) based on the point samples around it,

— assigns the estimate to the center of the window and

— moves to the next location.

The extent of the window (both size and shape) affects the result, regardless of the summary technique. In general, a large window capturing a bunch of values tends to "smooth" the data. A smaller window tends to result in a "rougher" surface with abrupt transitions.

Three factors affect the window's extent: its reach, the number of samples, balancing. The reach, or search radius, sets a limit on how far the computer will go in collecting data values. The number of samples establishes how many data values should be used. If there are more than enough values within a specified reach, the computer uses just the closest ones. If there aren't enough values, it uses all that it can find within the reach. Balancing attempts to eliminate directional bias by ensuring that the values are selected in all directions around window's center. Once a window is established, the summary technique comes into play. Inverse distance is an easy one to conceptualize. It estimates the value for a location as an average of the data values within its vicinity. The average is weighted in a manner that decreases the influence of the surrounding sample values as the distance increases.

Surf's Up

The inverse distance interpolation technique is fairly straight forward—distance weighted average of nearby sample points. Kriging and minimum curvature techniques give us two more spatial interpolation tools to explore before we can consider which is the best technique to determine residual analysis. Obliquely speaking, the Kriging interpolation technique uses "regional variable theory based on an underlying linear variogram." Without immersion into its theory, it is enough to know that there is a lot of math behind the technique. In effect, it develops a "custom" window based on the spacing and directional trends it detects in the sample data. The reach, or size of the window, is determined by intersample correlation. The summary technique weights the data values, aligning them along the trend's direction more than values opposing the trend. This trend-weighted moving average can result in estimated values that exceed the actual range of the sampled data. Also, there can be unexpected results in large areas without data values. The technique is most appropriate for systematically sampled data exhibiting discernible trends. The center portion of figure 6 depicts the Kriging surface of the sample data described above.

Another technique, minimum curvature (mincurve), first calculates a set of initial estimates for all map locations based on localized trends in the sampled data, then repeatedly applies a smoothing equation to the surface. The smoothing continues until successive changes at each map location are less than a specified "maximum absolute deviation," or a maximum number of iterations has been reached. In practice, the process is done on a coarse map grid and repeated for finer and finer grid spacing until the desired grid spacing and smoothness are reached. As with Kriging, the estimated values often exceed the range of the original data values and things can go berserk in areas without sample values.

The only thing certain is that there isn't a single best method that covers all situations. Some rules of thumb and general considerations have been discussed, but the only good way to determine the "best" map is to empirically verify the results. This involves generating several maps then testing the results against known measurements, aptly termed a test set. As in horseshoes and bocci, the closest one wins. Residual analysis summarizes the differences between map estimates and actual measurements.

Is there a significant difference between the various techniques in their estimates for the two different populations? (Get out your calculator.) I wonder if the conclusions change if the test and interpolation samples are swapped and another residual analysis performed? What if different sampling designs were tried? What would be the effect of different interpolation parameters (window reach, number of samples and data balancing)? Residual analysis gives us a place to start looking for answers.

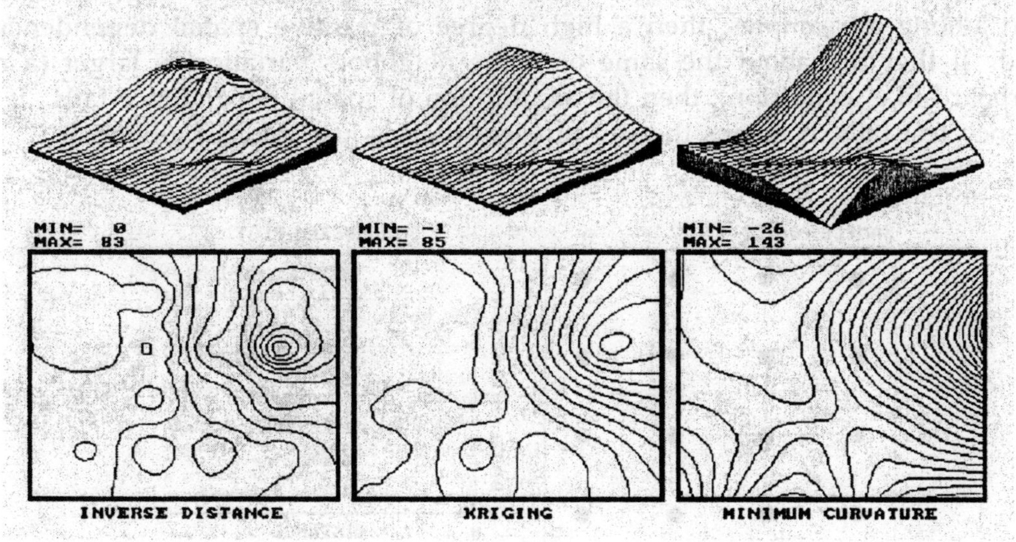

Fig 6. Results of Inverse Distance, Kriging vs. Minimum Curvature interpolation

To summarize, we have developed a procedure for checking the reliability of an interpolated map, such as a map of soil nitrogen derived from a set of soil samples. It suggests holding back some of the field samples to "empirically verify" the estimated values at the sampled locations. If you don't, then you are simply accepting the map as perfect. The difference between an interpolated "guess" (prediction) and "what is" (measurement) is a residual. A table of all the residuals for a test set is analyzed to determine the overall accuracy of an interpolated map. The residual analysis determines that the Kriging interpolation procedure has the closest predictions to the test set measurements. Its predictions, on average, are only 3.3 units off, while the other techniques were considerably more (5.1 and 10.5). The residual table identifies the best interpolated map and just how good it is, but it fails to identify where in the field the map is likely predicting too high or too low. That's the role of an error map.

If you have survived the previous discussions about the basics of "traditional" statistics, let's turn our attention to spatial statistics. Its basic concept involves spatial autocorrelation, referring to the degree of similarity among neighboring points (e.g., soil nutrient samples). If they exhibit a lot similarity, or spatial dependence, they ought to derive a good map. If they are spatially independent, then expect a map of pure, dense gibberish. So how can we measure whether "what happens at one location depends on what is happening around it?" Common sense leads us to believe more similarity exists among the neighboring soil samples (lines in the left side of Figure 7) than among sample points farther away. The Geary and Moran Indices consider the closest neighbors. They compute the differences in the values between each sample point and its closest neighbor. If the differences in neighboring values are a lot smaller than the overall variation among all samples, then a high degree of positive spatial dependency is indicated. If they are about the same or if the neighbors variation is larger (a rare "checkerboard-like" condition), then the assumption of spatial dependence fails. If the dependency test fails, it means the soil nutrient map is just colorful gibberish.

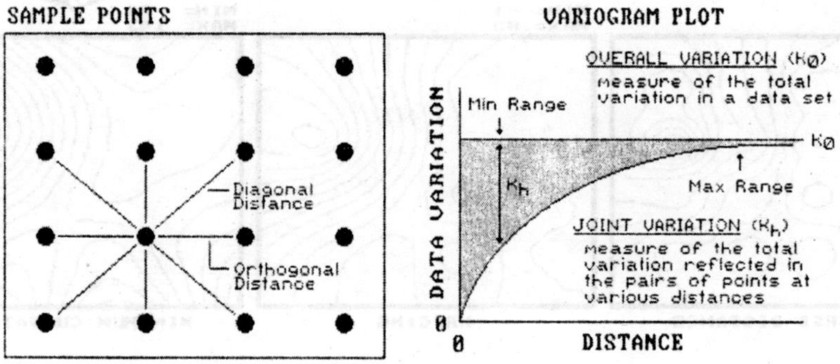

Figure 7. Variogram plot

These indices, however, are limited as they merely assess the closest neighbor, regardless of its distance. That's where a variogram comes in. It is a plot (neither devious nor spiteful) of the similarity among values based on the distance between them. Instead of simply testing whether close things are related, it shows how the degree of dependency relates to varying distances between locations. The origin of the plot at 0,0 is a unique case. The distance between samples is zero; therefore, the dissimilarity (data variation = 0) is exactly the same as itself.

As the distance between points increases, subsets of the data are scrutinized for their dependency. The shaded portion in the plot shows how quickly the spatial dependency among points deteriorates with distance. The maximum range position identifies the distance between points beyond which the data values are considered to be independent of one another. This tells us that using data values beyond this distance for interpolation actually can mess-up the interpolation. The minimum range position identifies the smallest distance actually contained in the data set. If most of the shaded area falls below this distance, it tells you there is insufficient spatial dependency in the data set to warrant interpolation. If you proceed with the interpolation, a nifty colorful map will be generated, but it'll be less than worthless.

Zones and Surfaces

Site-specific farming involves "carving" a field into smaller pieces that better represent the unique conditions and patterns occurring in the field. Two fundamental approaches are used: management zones and map surfaces. Management zones use a farmer's knowledge, air photos, terrain features, yield maps or other factors to identify discrete areas considered homogeneous. Sampling, analysis and management decisions are undertaken for each distinct zone—as if they were separate, mini-fields.

Map surfaces, on the other hand, treat a field as a continuous surface by partitioning it into thousands of grid cells that track gradual transitions throughout the field. The resulting grid spaces represent tiny snippets of the field, and information is assigned to each to track the pattern of variation.

Both approaches have their advantages, and disadvantages—management zones are intuitive, require minimal data collection and are less expensive to implement. Map surfaces, on the other hand, are not constrained by artificially abrupt boundaries, therefore they better describe field variability and have greater analysis capabilities. "Like all things GIS," an understanding of the nature of the data and the assumptions underlying the approaches provide insight into their differences.

UNDERSTANDING GIS

Sure, it's an acronym for *geographic information systems*, but it means different things to different people. To some, it's a tool to create and update maps. To others, it's a technology for combining and interpreting mapped data. However, all agree that in the broadest sense, it's a revolution in map structure, content and use.

In the real world, the landscape is composed of soil (never enough), rocks (too many), water (often too little), plants (more often weeds) and a vicissitude of biological life. In the paper world, words, numbers and graphics are used to represent these things. A traditional map is a graphical representation that accounts for geographic space (where) as well as the characteristic and condition (what).

A GIS takes the traditional map to new heights and may be described by its process, data, and analytical capabilities. The GIS process involves encoding, storage, processing and display of computerized (digital) maps.

Processing functions include:

- computer mapping;
- spatial database management; and
- spatial statistics, analysis and modeling.

These functions are descriptive and interpretive as well as prescriptive in nature. Computer mapping is a descriptive process that rapidly creates and updates map products. It provides tools for automating the cartographic process, but does not alter the inherent nature of a traditional map.

Spatial database management is used to combine and interpret sets of mapped data. It provides new procedures for *geo-query* of the data. For example, a database map can be searched for map compartments with certain requirements (i.e., low pH values in a certain soil type), then produce a map locating these areas. The spatial modeling capabilities extend the descriptive and interpretive uses of maps by deriving new information through the relationships among existing maps. For example, template maps can be "summarized" for typical characteristics (such as average slope for each mapping compartment), and the result can be added as a new column to the database.

Part of the revolution in GIS simply involves *digitizing* (converting ink to numbers) familiar maps. However, GIS goes beyond just a new form of maps. It is a powerful set of new tools for *map-ematical* processing of these data and modeling complex systems, such as crop production—describing, interpreting, predicting and prescribing the potential.

Data Formats

Traditional maps are graphical abstractions containing inked lines, shadings and symbols to locate landscape features. *Points* represent dimensionless features, such as wells. *Lines* depict features with length, such as streams or waterways. *Areas* (or *polygons*) have length and width, such as agricultural fields. Add depth and areas are extended to *volumes*, such as the entire body of water in a pond. Hypervolumes continue to extend the dimensions beyond simple geometry, such as considering time (mapping the rising and falling levels in a body of water throughout a year).

Most maps are composed of points, lines and areas. For example, a typical field map may identify an irrigation gate as a dot, an irrigation ditch as a squiggle, and a ponding area as a blob. A GIS can produce a similar graphic representation, but it doesn't store the data that way.

In a *vector*-based GIS world, map features are represented by *x*, *y* coordinates. Points are identified as a single coordinate pair. Lines are a connected set of points (like a connect-the-dot picture). Areas, such as soil type units, are identified by a list of coordinates defining the line segments forming their borders. The fundamental unit of a vector system is the point. A line is constructed by connecting individual points. An area is constructed by a line that connects to itself forming a border around the area of interest. The data structure and mathematics defining the implied line segments is extremely complex. The basic structure of points, lines and areas, however, is very familiar as it mimics the inked markings on traditional maps.

A less familiar format, *raster*, uses an imaginary grid of cells (similar to a spreadsheet layout) to represent the landscape. Points are individual column/row entities (*grid cells*), lines are a set of connected cells and areas are stored as all the cells within the interior of each map feature. Although this data structure has several advantages, it has a major disadvantage—lack of spatial precision. If a stream passes through an acre grid cell, the whole cell is identified as containing the stream. You don't know if it's at the top or bottom or wiggles several times through the center of the cell. At this juncture, it is important to note that vector systems "precisely locate" map features, whereas raster systems "statistically express the occurrence" of features. Generally speaking, vector is better for computer mapping and spatial database management processing functions, while raster is better for spatial modeling of the relationships within and among mapped data. In part, this distinction arises from the recognition of a fourth basic map feature—*surfaces*. While points, lines and areas form discrete spatial entities, a map surface depicts a continuously changing gradient. Elevation, for example, is constantly changing throughout geographic space. While representing elevation as a series of discrete contour lines may be familiar, it is an inaccurate representation (a hold-over from manual drafting). Similar arguments hold for the complex spatial distributions of crop yield and soil nutrient levels, as well as a multitude of other farm conditions affecting crop production.

Though less precise at depicting discrete features, raster systems are ideal for representing map surfaces. They partition geographic space into consistent spatial units (grid cells) that are analogous to sample plots—it's just that there is a sample plot everywhere. In techy-speak, raster's contiguous set of grid cells "sample" the continuous spatial distribution of each variable and maps its spatial autocorrelation. Also, analysis of the spatial dependency among maps is simplified by using a consistent "sampling" grid. Since the grid cells align in each map, the computer simply retrieves the necessary data stored for each cell location. A vector-based system, on the other hand, must calculate the coincidence for each irregularly bounded feature with every other feature on every map—a considerable task as most map's contain hundreds of irregular features defined by a multitude of mathematically implied line segments.

Data Links

How are maps linked to descriptive data? In the paper world, the human is the link, running back and forth between a map on a wall and data about map features stored in file cabinets. If a farm cooperative needs to know which fields have certain corn varieties growing on Cohasset soil, they flip through their files and note the field numbers of the ones they're after. Then they go to the map to locate them. If they wonder what the variety and soil type is for a neighboring field or forest stand, they run back to the files.

Dumb Maps

When you view a map, all sorts of things are apparent. If two blue lines come together you instantly recognize it as a fork in a stream. You easily comprehend which stream networks are connected, and which are not. Your interpretation of contour lines even tells you which way water flows.

A *GIS* isn't as lucky. The relationships among the features forming a map (*map topology*) has to be contained in the data's organization. The GIS may know that a coordinate pair (*x*, *y* location) is identified with a stream, but without topology it has no idea how that location relates to all of the other map locations.

You will recall that the fundamental element of map structure in a vector system is the point, represented by an *x*, *y* pair of values. These coordinates usually relate to a standard earth coordinate system such as latitude and longitude. Although the geo-referencing is the same, there are several types of *points*. *Tics* are geographic control points used in registering a map. *Discrete points* are used to represent information such as wells. These independent points also are used to associate data with areas and to position map labels. *Vertices* and *nodes* are points used to construct lines and polygons. This process is similar to "connect-the-dots" drawings. Vertices are merely passed through; nodes identify the special points where more than two lines meet.

The set of all line segments (termed *connected vertices*) between two nodes forms an *arc*. For linear features, such as a stream network, the GIS data structure keeps track of the connected arcs. When a stream flows into a lake, the node is tagged as an inlet. The stream node at the other end of the lake is identified as an outlet. You "see" this stuff; the computer has to be told.

Polygons are areas enclosed by arcs. Just as several points form an arc, a closed series of arcs form polygons. A *discrete point* inside each polygon (usually the *centroid*) serves as a quick link to information about it (i.e., soil, size and crop). When polygons are adjoining (i.e., soil types within a field), their shared arcs are tagged with a special code. In this manner, the GIS knows the adjoining polygons, their adjoining polygons, etc.

With minimal guidance from map annotations and labels, you see all these relationships in a paper map. But the GIS just "sees" a huge pile of numbers. It must incorporate all of the relationships you see into its data structure. When it jumps into the middle of a map (technically termed a *coverage*), it has to be able to sequentially construct all of the relationships among map features. Yep, digital maps are dumb without a data structure to keep track of all the details.

Raster Features

The vector *data structure* is constructed by connecting points, lines and arcs and explicitly storing their relationships (i.e., which polygons are touching). Raster systems store the "what" information with an implicit (rather than explicit) ID# that's based on the position of the information in a data file. The entire landscape is covered by an imaginary grid of cells, the basic unit of this data structure. Like a vector's point, there are different types of cells. A whole cell contains a single map characteristic throughout its interior (i.e., a particular soil type). A partial cell contains a mixture of characteristics (i.e., portions of two soil types) or an individual characteristic (i.e., a road or spring) that is too small to cover the entire cell.

It's the partial cells that account for the lack of precision in raster data. In this data structure, all of the spatial detail smaller than a cell gets lost. If a finer analysis grid is used, precision increases. In theory, the grid could be as fine as the *x*, *y* coordinates in a vector system, yielding identical spatial precision. However, the storage and processing demands at such a high resolution exceed the capacity of most modern computers.

Until a super computer sits on every desk, an oversized partial cell must be used to identify a single point in geographic space. Connected series of partial cells identify rasterized line features. A set of whole (*interior*) and partial (*border*) cells identify rasterized areas. To simplify things, the characteristic dominating a border cell can be assigned, thus making it a whole cell (and a whole lot easier to conceptualize and store).

The raster structure allows us to extend the basic map features from just discrete points, lines and areas to map surfaces. A *surface* describes the continuous distribution of gradient data. Elevation is a classic example, at least in hilly terrain. Each *grid cell* is assigned an elevation value that typifies the elevation within its boundary. The raster format of elevation data is termed a *digital elevation model* (DEM) and contains radically different information than the traditional contour lines on a topographic map.

Atmospheric pressure, temperature and cost surfaces are other examples of this new type of map feature. In fact, *site-specific management* implies a continuous treatment of geographic space. While the infrastructure of a farm (buildings, roads fences, ditches) are best represented as vector data forming traditional maps, most agronomic data (crop yield, soil nutrients, soil moisture, nematode concentration) are inherently continuous spatial gradients best represented as raster data forming map surfaces.

Converting between Lines and Squares

A common thread in most *precision farming* analyses is the data structure employed—

an imaginary grid of cells superimposed over an area of interest. The most important distinction affecting data analysis is that raster data characterizes a consistent "parceling" of geographic space while the vector structure contains sets of irregular parcels. The imaginary grid of uniform cells represents an area as a continuous surface that is identical throughout a single map and between map layers. The vector structure, on the other hand, represents discrete objects that provide little information about their spatial context and none about the consistency needed for efficient spatial analysis within or among maps.

Conversion between the two data structures is easily done, but at a cost in spatial precision and informational accuracy. Transforming a vector point simply identifies the raster grid cell containing the point. Reversing the process simply uses the centroid of a cell to represent a point feature. If the cells used the translation are small, the imprecision introduced isn't much. The positioning error is a function of cell size (worst case is one-half the diagonal distance of the cell) and dramatically increases with larger cells.

In practice, fat cells are used for blurry features with uncertain positioning and thin cells are used for well-defined features. Some linear features (i.e., an irrigation canal) can go either way—thin to best represent the canal's positioning or fat to statistically identify all cells that have direct access to water.

Polygonal features divide continuous space into cells best representing the condition within each cell. An interior cell is easily assigned when the same condition covers the entire cell¾ it's the edge cells that present the problems. Most procedures split the set of edge cells based on their dominant condition. Geometry is used to determine which side of a line contains the most area of a cell. Another approach uses fat cells to identify a separate "transition" category and identifies the mixed conditions within each edge cell (e.g., some soil A and some soil B).

Conversion from raster to vector is just as inexact and plagued with simplifying assumptions. With vector-to-raster conversion, precise locations are "blurred" and represented as occurring somewhere within a cell. Raster-to-vector conversion, on the other hand, takes a generalized location (somewhere within a cell) and infers precise positioning—an art form at best. The simplest approach uses the centers of the cells to extract a list of x,y coordinates that are used to position point features or a string of connected centroids to locate a line. Many systems use *spline functions* to mathematically smooth a line like a draftsman's french curve template.

An alternative for representing linear features "outlines" the sides of cells (usually fat cells) for a vector representation of "corridor" containing the line—in a sense, a more honest reporting than inferring precise placement from a smoothed set of connected centroids. In a similar vein, the edges of the border cells of a polygon can be outlined (and smoothed) to form a precise rendering of the implied boundary, or the sides of border cells can be identified to represent a transitional zone.

In precision farming, vector-to-raster or raster-to-vector conversions play an important role. Many of the input maps and popular map display systems (e.g., ArcView and MapInfo) utilize traditional vector data structure. However, most data analysis utilizes raster data. Conversion between the two is not an exact science and users should be aware of transformation approaches and sensitive to the interpretation of their results.

Workspaces

So far, we have identified the basic concepts and terminology in GIS data structures. However, there is a bigger picture that needs to be covered—*workspace organization*. A GIS workspace is like your desktop when you are working on a project. It has all of the information (books, maps, papers, calculator, etc.) that you need to tackle a problem. Within a computer, the data and programs you need are assembled into a workspace.

The proper term for a GIS map is a *coverage* (either vector or raster format). Three things make a GIS coverage different from a traditional paper map.

— It's digital (numbers instead of ink).

— It represents only one theme (the "what" of a map).

— It's seamless.

While a paper map composites several themes (i.e., political boundaries, roads, water and soils), a GIS stores each theme as a separate coverage. In addition, each of the categories defining a coverage must be mutually exclusive in geographic space—we can't have two soil types assigned to the same location. Seamless means a user can specify a set of corner coordinates and the GIS will automatically "cut and paste" data from the appropriate storage locations. This process is similar to your locating four adjoining topographic maps, identifying your project boundary on each, whacking away with your scissors, then taping the pieces together.

A layer is a set of adjoining sections having the same features and attributes. For example, a GIS database might contain separate layers for field boundaries, drainage tiles, soil type, nutrient levels, crop yield and terrain slope. Each layer is stored separately but can be printed as a "sandwich" of inked layers.

You must take great care in encoding each section to make sure that they *edge match*; the boundaries of features must continue from section to section. This requires precise digitizing (e.g., tracing map features into the computer) and consistent classification schemes for the mapped data.

While a *layer* describes the informational content of each GIS map, a *tile* describes the geographic area represented in each of the layers. It identifies the geographic extent covered in by the *workspace*. A extension agent's "tile" is likely to include an entire county or two, while a farmer's "tile" might cover just a few fields. In large part, the size of the tile determines the size of the database and, ultimately, the size of the computer system and complexity of the software needed to handle the data. Most farm-level workspaces easily fit in a standard personal computer environment. Larger "chunks" of land for co-op or government-level perspectives require larger computers and extensive data networks—like shoes, a "one-size" system does not fit all.

A *nested map* series contains maps at various *resolutions* (levels of detail) over the same geographic area (*tile*). This concept is particularly applicable to raster databases containing different *remote sensing* data. Frequently, a user will store two or more analysis grid resolutions of the same area—a coarse one for strategic planing and a fine one for tactical studies.The final workspace concept ties it all together—the *map library*. A map library lists all the GIS maps in a system. The listing is simultaneously organized by location, theme and detail. In a full-feature GIS, you can specify a project area, select the maps you need, then store them in your own workspace. With a healthy understanding of the errors involved, you can transform maps at various geographic scales and projections, resample maps of various levels of detail, as well as exchange vector and raster maps.All this has been an overload of important, but mundane and arcane terminology. Some of it makes common sense; some of it may make no sense at all. In many ways a digital map is as different as it is similar to traditional paper maps. Keep in mind that you're the intellectual superior of the GIS. You simply "see" things on paper while the computer has to organize everything in excruciating detail on disk. Although fundamentally different, you and your GIS need to be agreeable partners.

More on Spatial Topology

Your computer thinks of a map as simply an organized set of numbers. *Spatial topology* describes the linkage of numbers and map features. Think of it as the information that gets added to map coordinates so the computer can "see" their relative positioning. Different software allows the computer to see maps differently. A graphics package, or paint-brush type system, does a great job of coloring maps because the computer sees the map as "on-and-off dots," all assigned an appropriate color. Your eye assembles the dots into patterns, but the computer only sees a pile of independent dots. A graphics package can draw maps, but it doesn't supply the spatial topology needed for analysis.

A *mapping package* offers more than "connect-the-dots" topology. It divides up the set of coordinates into separate groups of map features. It uses the same approach as *computer-aided design* (similar to .DXF files), with points (a pair of coordinates), lines (a string of connected coordinates) and polygons (a line that closes on itself). A mapping package can draw maps and keep track of separate map features.

A *spatial database management system* (SDBMS) adds another ability: using a "connect-the-dots-to-records" structure for *thematic mapping* and *geo-query*. With this approach, a CAD-type structure of map features (*spatial table*) is linked to a database containing information about each feature (*thematic table*). It uses a common ID number for each feature, which is found in both the spatial and the thematic tables.

A *thematic map* is generated by assigning colors to a range of conditions identified with one of the map themes. For example, a farm co-op can generate a corn yield map for a county by assigning colors to low, medium and high yields for each field's aggregate yield. A farmer can generate a yield map of each field by assigning colors for low-, medium- and high-yield measurements recorded by a yield monitor with *GPS* positioning.

A *geo-query* selects sub-sets of the data. With the simplest geo-query, mouse-click on any map feature and the thematic data linked to it pops up. Conversely, the thematic table can be queried, which results in a map. For example, a farm co-op can select certain farmers' fields for thematic mapping of yield. Or a farmer might ask for a map of field locations that have both high yields and high moisture content. The power of an *SDBMS* is in the ability to visualize existing data.

A GIS fills in several gaps in the computer's understanding of the relationships in mapped data. A GIS provides a topology based on "connect-the-dots to records and to spatial relationships." This relational capability separates pretty maps from GIS maps—and provides software with the ability to perform map analysis. For a computer to find its way around a map, the connections you see must be stored in the spatial table, as well as the list of coordinates (vector *x,y* or raster column/row) positioning each feature. For example, a farm co-op might be interested in deriving a map of all high yielding fields within 20 miles of a grain elevator. Or a farmer might calculate a percent-change map for a field's yield over two years, then generate a map of locations that increased (or decreased) by a certain percentage.

Generally speaking, a topologically structured GIS allows the derivation of new spatial information based on existing data as well as on query and display. A SDBMS provides geo-query and thematic mapping of existing data but no spatial analysis. Mapping and graphics packages provide colorful map products but no query or analysis capabilities. The differences in the capabilities among these packages arise from their differences in topological structures. Topology differences also account for important differences in cost and complexity of precision farming software.

Data Pedigree

Metadata refers to "data about data," including specification of the spatial and thematic characteristics of mapped data. Five elements make up the metadata used in precision farming:

— data dictionary (measurement units, relationships),
— thematic data attributes (integer, floating-point),

— spatial data structure (vector, raster),

— geo-referencing (coordinate system, projection) and

— data quality (derivation procedures, level of detail).

The *data dictionary* might define "bushels per acre" as the common unit of measurement for wheat and "pounds per acre" or "kilograms per hectare" as interchangeable definitions for nitrogen applications. Once the units are identified, the thematic data attributes for each data element are defined and the structure of the thematic table is specified. For example, yield might be characterized as a positive, floating-point value while crop type might be set to a positive integer code and a large alphanumeric text string.

At this point the computer knows about the "what" information for each map variable. The spatial data structure identifies the composition and organization of the spatial table. Geo-referencing details the procedures used in registering the map. Now the computer knows "where" the data are on the face of the earth. Finally, data quality specifications complete the description of the data by noting "how good it is."

A benefit of adhering to common metadata definitions is the interchange of data among different software. Early systems were self-contained and relied on nonstandard (often proprietary) data formats. It was like the old days when a red hitch wouldn't accept a green implement. The details encapsulated in the metadata describe a data set in terms of its format and structure, which is particularly useful for software developers. But it also deals with interrelationships and common descriptions. It organizes data in meaningful terms for intended users who want to "surf" the conceptual framework of related data.

The details of metadata best are left to programmers, but the agriculture community needs to understand the importance of standards in the products they buy. The first four elements of metadata facilitate data exchange by detailing the "what" and "where" information while the data quality element describes its accuracy. Statistics on the level of detail and data confidence should be the focal point of end users (farmers)—"garbage in is garbage out," no matter how efficiently it is transferred.

We have identified the distinction between precision and accuracy and even suggested techniques for assessing and mapping uncertainty. Several areas of development in GIS technology extend these measures, thereby improving the data quality component of metadata. A map pedigree follows the lineage of a map like a hog's pedigree follows its genealogy. It is similar to the textual descriptions accompanying soil maps, with some significant enhancements.

A *map pedigree* logs in information such as who is "responsible" for the data, how base data are collected, how derived data are processed and what are the statistics on map precision and accuracy. For example, a GIS model of soil loss might use a base map of field elevation to calculate a derived layer of terrain slope. In turn, the slope map is interpreted for its effect on soil loss (increasing slope increases soil loss). The interpreted slope map is combined with base, derived and interpreted maps showing other factors (e.g., soil type, crop cover, tillage) to generate a modeled map of soil loss for a series of storm events. A map of crop production potential might be generated from similar set of mapped data. Or a prescription map for site-specific nutrient application might be modeled from a set of maps depicting the current levels of soil nutrients throughout a field.

What is common among these examples is that the modeled map is dependent upon logic and processing flow and not "the-precise-placement-of-physical-features-for-navigation" heritage of the traditional map. The GIS technology has extended the descriptive role of mapping to interpretation, prediction and prescription. As such, the concepts of map precision and accuracy must be extended to model logic and interactions among mapped data.

A *logic flowchart* is a diagram illustrating the interrelationships among maps leading to a modeled map. It is a graphical summary of the map pedigree for maps that interpret, predict or prescribe spatial conditions. It shows you each "thoughtful" step in the process with boxes representing maps and arrows connecting the boxes representing processing. It extends traditional data quality measures to the quality of the "conceptual" maps used in decision making.

Also, it extends the interaction with a GIS to the level of a "spatial spreadsheet." Users can ask "what if" questions by changing portions of the flowchart and rerunning the model. In the future, the logic flowchart will be the focus of discussions between a farmer and an agronomist when viewing a prescription map of management options (i.e., site-specific fertilizer application rates). "What would happen if phosphorous is increased over here, but reduced over there?" "If nitrogen levels were cut by 10 percent everywhere, would productivity plummet?" "If so, where would it plummet?" "Where wouldn't it matter much?"

It is fundamental to recognize that maps used in decision making are fundamentally different from maps used for navigation. Do you understand the sensitivity of the modeled map you're about to bet the farm on? Do you agree with the its assumptions and logic? After all, you'll be paying for them.

UNDERSTANDING GPS, REMOTE SENSING AND IDI WORK

All GIS databases require a common coordinate system to identify where in the world the data are and to spatially register various maps. A coordinate system is composed of two elements:

— a spheroid that mathematically describes the three-dimensional shape of the earth and

— a map projection that mathematically converts the spherical representation to two-dimensional coordinates for display on a flat screen or printed on a sheet of paper.

Mentioning the word "mathematically" twice in the same sentence is journalistic suicide, but it should reinforce the idea that maps within a GIS are numbers first, pictures later. As users of the technology you won't be tested on the intellectual elegance of a blackboard full of equations, but need to understand the basic concepts and the coordinate "got'chas" you might encounter.

The first is the choice of the equation of the *spheroid*. It is similar to enlarging and shrinking a giant balloon in an attempt to "best fit" the earth's surface. Keep in mind that the spinning earth is fatter at the equator so the equation of a simple sphere won't do. Nor is the spheroid with the best overall fit the best fit for all locations on the earth. Hence, there is a multitude of variations to the basic equation; blow up the balloon a bit and a little more squashing at the poles gives North America a better fit but messes up things for Bolivia.

There are two elements defining a GIS coordinate system—the *spheroid* and the *projection*. The spheroid describes locations in three-dimensional space (curved earth) while the projection converts them to a two-dimensional rendering for plotting on a screen or sheet of paper (flat map). Changes in either element change the relative positioning of mapped data. For example, locations in the United States "move" nearly 200 feet between the *North American Datum* established in 1927 (NAD27) and the revised earth surface equations of the *World Geodetic Spheroid* established in 1983 (WGS83). Major problems arise when your GPS is set to WGS83, but the maps you downloaded from the Internet are NAD27 based; your registration is off more than half a football field from the start. But that's nothing compared to the positioning errors that arise if you mix projections.

GPS: Basic Stuff

GIS technology allows you to view maps in the blink of an eye, visualize the spatial patterns in data sets and even model the complex interrelationships among mapped variables. But its abstract renderings (digital maps) require a real-world expression to make GIS a practical tool. For a long time farmers and other "field folk" have been breathing dust and swatting mosquitoes, while all the time lusting for a simple way to figure out where in the world they are and where they might be going. The celestial navigation, used by the early mariners as they gazed to the heavens, eventually gave way to the surveying and mapping sciences. But these solutions still seem beyond the grasp of the average bush-wacker. What was needed was a simple field unit that puts the power of GIS on the kitchen table, in a vehicle, or directly in our hands while standing in a field.

That's where the *global positioning system* (GPS) comes in. It allows us to link *GIS* maps and their related data sets to real-world positions and movements. The GPS is based on a constellation of 21 satellites, each of which circles the globe every 12 hours. The system can be thought of as a set of "man-made stars" for an electronic equivalent to celestial navigation. So, how does it work? And will it work for you?

Although the process involves complicated electronics, it uses the same calculations you used in geometry class involving that device with a pencil on one arm and a life-threatening sharp point on the other. Recall that you would stick the point at a known location (satellite position) on a piece of paper then extend the arms for a given distance (satellite to GPS receiver time lag * speed) and make a small arc. Repeat for a second point/distance and where the arcs cross determines where you are—in two-dimensional space.

In three-dimensional space, circles of a calculated distance are mathematically "drawn" about a set of satellites whose precise position are known at any instant in time through orbit mathematics. The intersection of the circles determines the position of the GPS receiver on the earth.

In trigonometric theory, only three channels (satellites) need to be monitored, but in practice four or more are needed for improved accuracy and to cancel receiver clock errors. The world of electronic wizardry (involving technical stuff like pseudo-random code, carrier-phase, ephemeris adjustments, and time hacks) allows accurate timing to one billionth of a second (.0000000001) and can produce extremely accurate distance measurements in three-dimensional space. Generally speaking, averaged stationary measurements (termed *static mode*) tend to be more accurate than a single reading or sets of readings made while on-the-go (*kinematics mode*).

GPS: Intermediate Stuff

As with everything in the real-world, actual GPS performance depends on several "muddling" factors. First and foremost is GPS's history as a US Department of Defense program. They financed the billions needed to setup the system for military purposes and feel a bit uncomfortable if just anyone (such as terrorists or enemy troops) can simply tap into their system. They purposely degrade the signal using an operational mode called selective availability (S/A) to provide an accuracy of only about 100 meters. With the military muddling turned off, accuracy of about 10 meters is commonplace.

The signal degrading can be overruled by a method termed *differential correction*. A differential GPS unit uses real-time corrections from a local "reference receiver" whose exact location is known. When the reference receiver gets a satellite signal, it calculates its implied position, then quickly "reverse calculates" a direction and distance correction needed to place it where it should be. The correction is broadcast to the field units, or stored for post-processing of field readings back at the office.

There are several companies offering commercial differential correction signals similar to renting a telephone pager. Areas around navigable waterways can receive free correction signals from "beacons" established and maintained by the US Coast Guard. However, these "real-time" signals aren't required since you can download fairly accurate corrections for most areas from the Internet and "post-process" your GPS files the next day.

In general, there are two main hurdles in processing GPS signals—jitters and jumps. As with any instrument, inherent error for a set of readings at a fixed location yields a jittery cluster of possible positions, termed the *sphere of uncertainty*. The cluster is statistically summarized to report the general accuracy of a GPS unit. A frequently used measure, the *circle error probable* (CEP), identifies the radius of a circle capturing 50 percent of the readings around test locations. Another measure reports a radius of a circle having one standard deviation around the actual location. Both measures assume the cluster of points is evenly distributed around the actual point. The worst kind of jitters has a directional bias.

Also, satellites come and go over the horizon with time; as one is dropped and another picked up, the calculated position can take a temporary jump. Although four satellites are technically sufficient, multi-channel receivers lock in on several more satellites and instantaneously switch without a sharp jump. Processing software uses running and dampened averages of several readings to cope with the jitters and jumps. Keep in mind that the silicon in all GPS hardware is about the same—it's creative software that separates the best receivers.

A well-tuned differential GPS system in static mode for use on the farm can easily place you within a meter horizontally and five meters vertically. A simple, inexpensive, autonomous system can place you somewhere within a "football field." That is if atmospheric, ground-cover and terrain factors permit—things quickly deteriorate under a dense vegetation canopy and at the bottom of steep canyons. Also, the satellites are not always available in a nicely dispersed pattern in the sky. That means you need to plan to be in the field at the times the satellites' celestial charts dictate—try explaining that one to your field crew.

A GPS's ability to rapidly and accurately locate positions on the earth's surface is a powerful addition to GIS technology. However, it is important to keep in mind that GPS is not intended to fully replace conventional surveys. It augments *cadastral* records with real-time and real-world positioning. When attached to a vehicle, GPS tracks it better than a hound dog.

The contribution of GPS to generating and updating GIS maps is obvious. Yet, GPS is more than a data collection device—it's a practical tool to navigate GIS results. As GIS matures, more of its applications will involve GIS modeling, such as "variable-width buffers" around streams considering terrain steepness, ground cover and soil erodibility.

Although such relationships follow common sense, their spatial expression is extremely complex. The contractions and expansions of a variable-width buffer on a paper map are impossible to see in the field. However, if you download the coordinates of the buffer into your GPS you can navigate the complicated spatial result, in effect, delineating the spatial expression "on-the-go."

GPS: The Advanced Stuff

You have a grasp of the basics of GPS—"gal-darn positioning system"—that fires-up every time you go near that new Star Trek tractor. It uses a constellation of satellites and mathematical triangulation to tell you where you are—rather precisely, if all goes well. However, there are several complex considerations involved in accurate positioning.

First, the distance from a satellite to you is determined by the time it takes a radio signal to traverse the space. Your typical stopwatch can't cut it: 1/100th of a second equals 1,860 miles for a radio signal traveling at the speed of light. A timing error of just .01 second could put you on another continent. Clock accuracy has to be in the nanosecond range (.000000001 second) that translates to less than one foot and keeps the precision in precision farming.

While the timing signal is the heart of the system, GPS receivers also monitor other information from the satellite. The *almanac* for a satellite reports the timetable of the satellite, path of travel, orbital fine-tuning and horizon setting. This information is augmented by *ephemeris* data identifying the fine-tuning of the satellite's position resulting from small predictable changes in orbit induced by gravitational effects of the sun, moon or other bodies, as well as other effects like the solar wind tugging on the satellite. The almanac information is updated monthly; the ephemeris data get updated every time the satellite passes over a ground control station.

In a sense, the almanac is like a bus schedule's expected position at any time, while the ephemeris data update the estimate tempered by actual conditions. The result is the precise positioning of the satellite every billionth of a second. Knowing the exact location of each satellite and the distance from it to you provides the input to the trigonometry equations that solve for your position—less than a meter if all goes well. All this goes on at speeds, decimal places and minute spacing only electrical engineers and mathematicians truly appreciate.

An even more precise (and concurrently more complex) positioning method isn't based on the measurement of time directly. It counts the number of radio waves between the individual satellites and the receiver. By carefully timing, counting and measuring the arrival of the waves, *carrier phase* receivers provide deca-meter (one-tenth of a meter) accuracy. However, you won't see these GPSs on tractors for awhile since they have a hefty price tag and are notoriously unstable when moving.

The military can muddle civilian use of the system or selectively turn it off during times of crisis. While *differential correction* usually restores the muddled measurements to within a meter, it requires access to base station corrections. Post-processing of a GPS data file can be made via the Internet; however, the turn-around time is several hours. For the real-time positioning needed in precision farming, you need additional hardware to monitor broadcast corrections and a fancier GPS receiver to make the adjustments "on-the-fly."

Even with differential correction, atomic clocks in the satellites and an extra satellite measurement for calculations, there are other sources of error, more subtle and difficult to deal with. Earth's *ionosphere* can subtly affect the speed of the radio waves. As the density of charged particles in the ionosphere increases or the path gets closer to the horizon, the waves slow down and the time increases. While equations adjust for speed variation on an average day, under average ionospheric conditions, more advanced GPS receivers can tweak the calculations by measuring subtle changes in the frequency of the signals from two satellites. *Weather conditions* can disrupt calculations since water vapor affects radio signals in a similar manner. Although this error is almost impossible to correct, fortunately it is usually small.

A localized source of error that can have a significant effect is termed *multi-pathing*. It occurs when signals bounce off nearby objects before getting to the receiver. The effect is similar to "ghosting" on a TV where the signal takes a circuitous route instead of going directly to the receiver as the equations assume. Advanced signal processing techniques and antenna engineering minimize these problems.

A final consideration is the result of the *relative positioning* of the satellites. There are currently 21 satellites (plus three spares) circling the earth every 12 hours. They are dispersed so there are about 11 available to each half of the globe at any time. Of these, three are necessary for two-dimensional positioning (x,y) and four are needed for three-dimensional (x,y,z) positioning. However, the relative positioning of the satellites used for the calculations has a significant effect on accuracy.

If the satellites are clustered in one part of the sky the readings are less reliable. Although the orbits are designed to disperse the pattern, local features, such as a ridge or wind row of trees, often blots-out some of the paths forcing the selection of an alternate set satellites that are more bunched. If the satellites are too close together, the geometry of the overlap of the circles forms a box instead of a point of intersection. If the satellites are too far apart, the increased atmosphere and interference from local terrain can muddle things.

Ideally, the three (or four) satellites should be balanced along a circle 45 degrees above the horizon; with minimal ionospheric, weather and multi-pathing effects at play. If that's the case you are operating at peak precision; if not, things start to degrade. Most of the time, however, you can expect to be navigating within a few feet, which is a heck of a lot better than my golf swing's mark from only 100 yards out.

Remote Sensing: Basic Stuff

The GIS/GPS technologies position both *spatial data* and *spatial relationships* on the landscape. But how to effectively identify, measure and monitor farm conditions is a continuing challenge. A GIS and its closely related field of *remote sensing* form an alliance that greatly enhances the technical toolkit for mapping. Remote sensing is actually GIS's older brother, having its modern roots in World War II. Camouflage detection film was used to distinguish between healthy vegetation and cut branches piled on top of military equipment. To the human eye and normal film the healthy and cut branches were both green (at least for a few days), but on the new film they showed up as two different colors.

Remote sensing uses relative variations in *electromagnetic radiation* (EMR) to identify landscape characteristics and conditions. In fact, so do your eyes. Sunlight (the "visible" form of EMR) starts off with fairly equal parts of blue, green and red light. When sunlight interacts with an object, the object's composition causes it to absorb varying amounts of the different wavelengths of EMR "light." What light isn't absorbed is reflected to your eyes. Your brain interprets the subtle differences in the amount of blue, green and red in the reflected light to recognize the thousands of colors we relate to our surroundings.

Vegetation is particularly "photogenic" because of its structure, pigments and water content. Since sunlight is a plant's source of energy, it goes out of its way to present its leaves in the best possible light. When thousands of plants in a field align themselves, their structure forms an excellent receptor and reflector of sunlight.

The physiology of a leaf determines the relative absorption and reflection of light. The equal portions of blue, green and red light from the sun are basically unaffected by the surface of the leaf, but when it encounters the *chloroplasts* containing *chlorophyll A* and *B* it is radically altered. These pigments absorb most of the blue and red light for the energy needed in photosynthesis used in plant growth and maintenance. Other pigments in the leaf (i.e., *carotines*) absorb lesser amounts of the other wavelengths of light.

As the pigment-altered light continues deeper into the leaf, it interacts with the spongy mesophyll. This bubble-like structure acts like a mirror and reflects the light back toward the sky. Since the blue and red wavelengths have been diminished, we see a predominance of green in the reflected light—a healthy "green" leaf (because blue and red are usurped by the plant).

An unhealthy leaf, however, looks a lot different, particularly in remote sensing imagery. When water pressure changes (i.e., cutting a branch from its stem), the spongy messophyll in the leaves collapse within hours and this area's efficiency of reflecting light is greatly reduced. The chloroplasts, on the other hand, keep on working away at photosynthesis for several days. The result is that we "see" a slight change in reflectance (predominantly green) at first, then a slow progression to brown as the chloroplasts eventually quit preferentially absorbing blue and red light.

However, what makes remote sensing's view different is its ability to look at reflected light beyond visible blue, green and red light. "Invisible" *near-infrared light* (*NIR*) is at wavelengths just beyond the red light your eyes can detect. These wavelengths are unaffected by the plant's pigments and are highly reflected by the spongy mesophyll. When the "bubbles" in this portion of a leaf collapse, there is an immediate and dramatic change in the reflectance of near-infrared light. That's the principle behind camouflage detection film—we see a branch as green for days; remote sensing imagery detects a dramatic difference in near-infrared light in a matter of hours.

What makes remote sensing data so useful is that it encapsulates biological and physical characteristics into an image. The encrypted variations in reflected light emanating from a field provides information about changing conditions and crop status—important stuff you should keep your eye on.

Remote Sensing: Intermediate Stuff

You now have a basic understanding of what happens to light in a plant canopy, let's take a loftier view and see how it is translated into a computer image. An aerial camera operates like your eye, except photographic paper replaces the optical nerves in the retina. The image is focused through the lens, with variation in light recorded by a photochemical process on the film.

The scanner in a satellite operates a bit differently—more like your laser printer that "sees" the world through thousands of dots. Its sensor focuses for an instant at a spot on the ground (a few meters in diameter). Like your eyes, it records the relative amounts of the different types of light it "sees"—a lot of green for a dense healthy crop; much less green and more blue and red for bare ground. In addition to normal light (termed the *visible spectrum*), it can record other types that we can't see, such as *near infrared*, *thermal* and *radar* energy. The sensor sweeps from side to side and the satellite moves forward, recording the relative amounts of light reflected from millions of spots on the ground.

When these spots (termed *pixels* for "picture elements") are displayed on a computer, they form an image similar to an aerial photograph. In fact, a photograph can be "scanned" to generate a digital image—like pulling the satellite out of the sky and passing it over the photo instead of the actual terrain. The important point is that behind any digital image there are millions of numbers recording the various types of light reflected from each instantaneous spot.

Three factors govern the quality and appropriateness of remote sensing data: 1) spatial, 2) spectral and 3) temporal resolutions.

Spatial resolution identifies the smallest thing (*spatial unit*) contained in the data. In a photograph, it is a glob of developed crystals embedded in the emulsion; in a digital image it's the size of the pixel. Up to a point, smaller is better. If there is too much spatial detail, you "can't see the forest for the trees," nor store the burgeoning file.

Spectral resolution refers to the number and width of the *wavelength bands* (colors) contained in the data. Again, more is better, up to a point. The human eye and normal film "see" just three broad bands of light—blue, green and red. Optical scanners can record many more narrowly defined bands that can be "tuned" for specific wavelengths to enhance features of interest. The ability to extend the bands beyond our vision (particularly to near infrared energy) and analyze just the important ones allows us to gain a lot more information from the data than simply viewing an image.

The rub is that there is a tradeoff between spatial and spectral resolutions—pushing both of them to the maximum results in too little energy for a sensor's detector. Early satellite systems required a lot of energy to activate their detectors; therefore, they only had four broad bands and a footprint of about an acre. Modern detectors can record many more narrow bands and commonly have a footprint of only a few meters. At these resolutions (spectral and spatial), even satellite data becomes appropriate for some aspects of site-specific management.

Temporal resolution refers to the time step between images. A series of images collected each week over a field provides far more information than a single image taken at just one point in the crop's growth. You guessed it; more is better, up to a point. But this time the point isn't driven by optical physics but your wallet. By its very nature, site-specific management implies small areas, while most remote sensing systems (particularly satellites) are designed to service large areas. Pricing and distribution channels for digital data in smaller bites (and bytes) and turn-around times needed by farmers are just now coming on line.

While the full potential of remote sensing might be just around the corner, an *aerial photo backdrop* is an essential element of any precision farming system. There's a growing number of ways you can acquire such an image. If you're lucky you can download a "rectified" image from the Internet or pick up one from a governmental agency in your locale. Some farmers have struck a deal with the local flight instructor to snap a few frames over their fields a couple of times a month. The 35 mm slides are scanned for a few dollars at growing numbers of photo shops. The digital images can be aligned in most desktop mapping systems using the GPS coordinates of a set of control points visible in the image.

Once the photo backdrop is in place, it immediately adds reality to the abstract lines and colors of a map. Important features and patterns can be encoded by tracing them directly on the screen (termed *heads-up digitizing*). Linking a differentially-corrected GPS unit to a portable computer allows you to "walk or drive" on a backdrop-photo (really cool!), encoding a map as you go (termed *feet-down digitizing*). Before you start drawing on maps, you need to realize that the patterns you see are the result of complex biological and physical interactions; you might "see" something, but be sure you know what it is before you canonize it as a map.

The organized mountain of numbers forming a digital image can be used to identify both land cover characteristics and their condition. First, the computer is "trained" by locating representative areas of the cover types to be *classified*—sort of rubbing the computer's nose in what it should know. Then, this information is used to determine the classification of other areas with similar responses. The GIS gets a copy of the results of the remote sensing classification—a map indicating crop type and condition. By comparing the maps from two different times, the computer can quickly detect and quantify any changes in a field. Keep in mind that remote sensing and GIS provide "educated guesses" of actual characteristics, conditions and relationships. They are not perfect, but they do provide powerful and compelling insights.

REFERENCES

Gehue, Hazen L. (Hazen Leslie) *GPS integrated systems for precision farming*. Geomatics Engineering, University of Calgary. 1994

Handbook of Precision Agriculture: A comprehensive look at PA, its tools and applications around the world.Haworth Press.

Landers A.J and Steel D. (1994) Precision agriculture a viable option for future farming in Europe? Paper No. 94/1604 St Joseph, MI: American Society of Agricltural Engineers.

National Research Council. *Precision Agriculture in the 21st Century: Geospatial and Information Technologies in Crop Management* Report by the Committee on Assessing Crop Yield: Site-Specific Farming, Information Systems and Research Opportunities. :National Academy Press.

4

Plant Disease Epidemiology

Plant disease epidemiology is the study of disease in plant populations. Much like diseases of humans and animals, plant diseases occur due to pathogens such as bacteria, viruses, fungi, oomycetes, nematodes, phytoplasmas, protozoa, and parasitic plants.Plant disease epidemiologists strive for an understanding of the cause and effects of disease and develop strategies to intervene in situations where crop losses may occur. Typically successful intervention will lead to a low enough level of disease to be acceptable, depending upon the value of the crop.

Plant disease epidemiology is often looked at from a multi-disciplinary approach, requiring biological, statistical, agronomic and ecological perspectives. Biology is necessary for understanding the pathogen and its life cycle. It is also necessary for understanding the physiology of the crop and how the pathogen is adversely affecting it. Agronomic practices often influence disease incidence for better or for worse. Ecological influences are numerous. Native species of plants may serve as reservoirs for pathogens that cause disease in crops. Statistical models are often applied in order to summarize and describe the complexity of plant disease epidemiology, so that disease processes can be more readily understood.For example, comparisons between patterns of disease progress for different diseases, cultivars, management strategies, or environmental settings can help in determining how plant diseases may best be managed. Policy can be influential in the occurrence of diseases, through actions such as restrictions on imports from sources where a disease occurs.

In 1963 J. E. van der Plank published "Plant Diseases: Epidemics and Control", a seminal work that created a theoretical framework for the study of the epidemiology of plant diseases. This book provides a theoretical framework based on experiments in many different hosi pathogen systems and moved the study of plant disease epidemiology forward rapidly, especially for fungal foliar pathogens. Using this framework we can now model and determine thresholds for epidemics that take place in a homogeneous environment such as a mono-cultural crop field.

Disease epidemics in plants can cause huge losses in yield of crops as well threatening to wipe out an entire species such as was the case with Dutch Elm Disease and could occur with Sudden Oak Death. An epidemic of potato late blight, caused by Phytophthora infestans, led to the Great Irish Famine and the loss of many lives.

Commonly the elements of an epidemic are referred to as the "disease triangle": a susceptible host, pathogen, and conducive environment. For disease to occur all three of these must be present. Below is an illustration of this point. Where all three items meet there is disease. The fourth element missing from this illustration for an epidemic to occur, is time. As long as all three of these elements are present disease can initiate, an epidemic will only ensue if all three continue to be present. Any one of the three might be removed from the equation though. The host might out-grow susceptibility as with high-temperature adult-plant resistance, the environment changes and is not conducive for the pathogen to cause disease, or the pathogen is controlled through a fungicide application for instance.

Sometimes a fourth factor of time is added as the time at which a particular infection occurs, and the length of time conditions remain viable for that infection, can also play an important role in epidemics. The age of the plant species can also play a role, as certain species change in their levels of disease resistance as they mature; a process known as ontogenic resistance. If all of the criteria are not met, such as a susceptible host and pathogen are present but the environment is not conducive to the pathogen infecting and causing disease, disease cannot occur. For example, corn is planted into a field with corn residue that has the fungus Cercospora zea-maydis, the causal agent of Grey leaf spot of corn, but if the weather is too dry and there is no leaf wetness the spores of the fungus in the residue cannot germinate and initiate infection. Likewise, it stands to reason if the host is susceptible and the environment favours the development of disease but the pathogen is not present there is no disease. Taking the example above, the corn is planted into a ploughed field where there is no corn residue with the fungus Cercospora zea-maydis, the causal agent of Grey leaf spot of corn, present but the weather means long periods of leaf wetness, there is no infection initiated. When a pathogen requires a vector to be spread then for an epidemic to occur the vector must be plentiful and active.

Monocyclic epidemics are caused by pathogens with a low birth rate and death rate meaning they only have one infection cycle per season. They are typical of soil born diseases such as Fusarium wilt of flax. Polycyclic epidemics are caused by pathogens capable of several infection cycles a season. These are most often caused by airborne diseases such as powdery mildew. Bimodal polycyclic epidemics can also occur. For example in brown rot of stone fruits the blossoms and the fruits are infected at different times. For some diseases it is important to consider the disease occurrence over several growing seasons, especially if growing the crops in monoculture year after year or growing perennial plants. Such conditions can mean that the inoculum produced in one season can be carried over to the next leading to a build of an inoculum over the years. In the tropics there are no clear cut breaks between growing seasons as there are in temperate regions and this can lead to accumulation of innoculum.

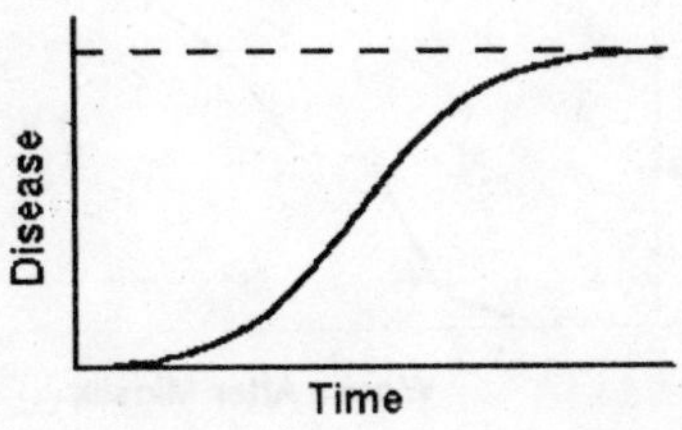

Figure 1. Dynamics of Plant Disease

Epidemics that occur under these conditions are referred to as polyetic epidemics and can be caused by both monocylcic and polycyclic pathogens. Apple powdery mildew is an example of a polyetic epidemic caused by a polycyclic pathogen and Dutch Elm disease a polyetic epidemic caused by a monocyclic pathogen.

DISEASE PROGRESS

Disease on plants usually starts out at a low level, a small number of plants affected and a small amount of plant tissue affected, and it becomes of concern to us only when its incidence and severity increases with time. When we look at some examples of plant disease epidemics from the published literature, we not only notice that the incidence or severity starts near zero and then increases dramatically, but we also can discern some distinct patterns of development with time. For example, in Phytophthora blight of pepper seedlings (*Phytophthora capsici*) and Fusarium kernel rot (*Fusarium moniliforme, currently F. verticilloides*) of maize, disease progress is roughly linear (allowing for some minor deviations that we can consider random error).

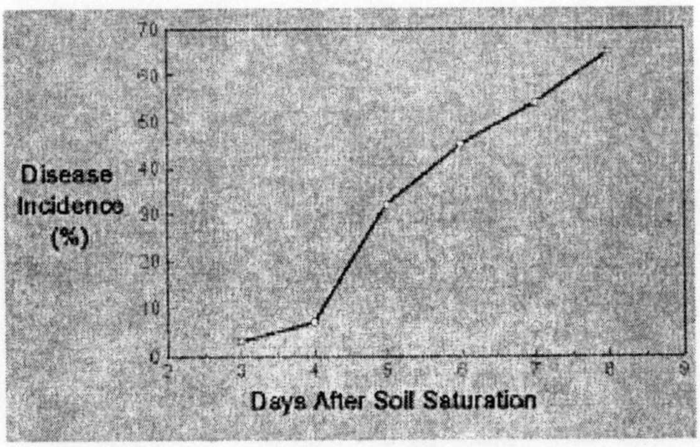

Figure 2. Phytophthora blight of pepper seedlings

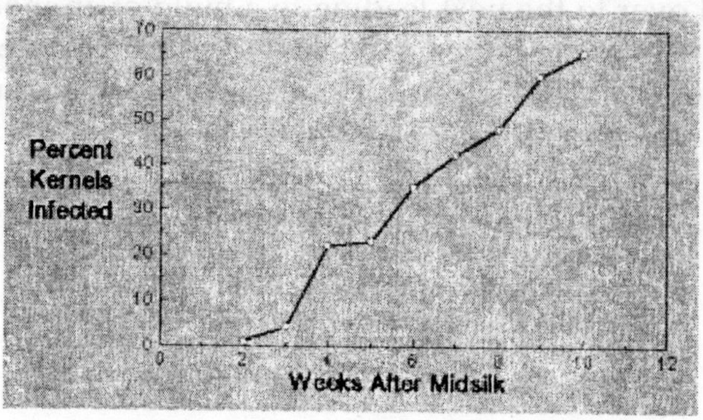

Figure 3. Fusarium kernel rot of maize

On the other hand, in bean rust (*Uromyces phaseoli*) and grey leaf spot of corn (*Cercospora zeae-maydis*), there is a definite upward curve; that is, disease increases at an increasing rate, a curve we could call exponential.

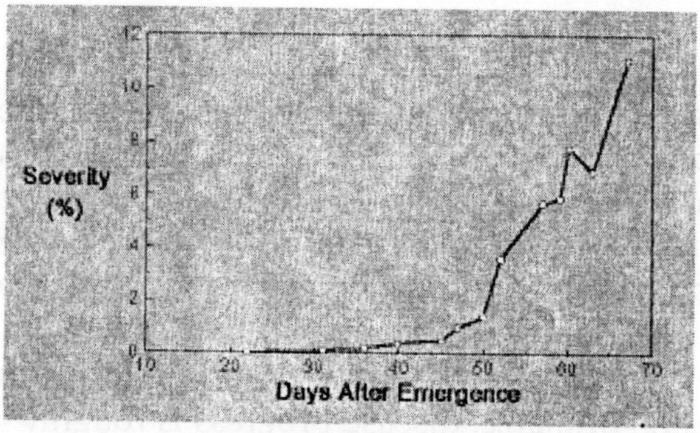

Figure 4. Bean rust

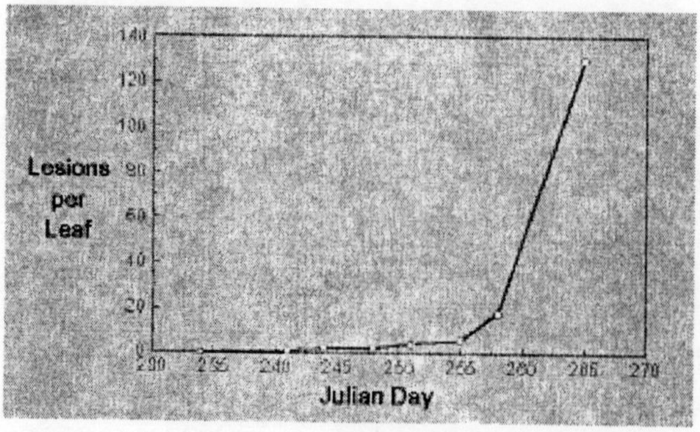

Figure 5. Grey leafspot of maize

Obviously plant disease cannot continue to increase forever, and as the level of disease approaches 100%, the disease progress curve gradually flattens out. For example, in epidemics such as the infection of beans by *Sclerotium rolfsii* or the infection of tobacco by *Phytophthora nicotianae*, disease progress starts out looking linear but slows down as it approaches a maximum.

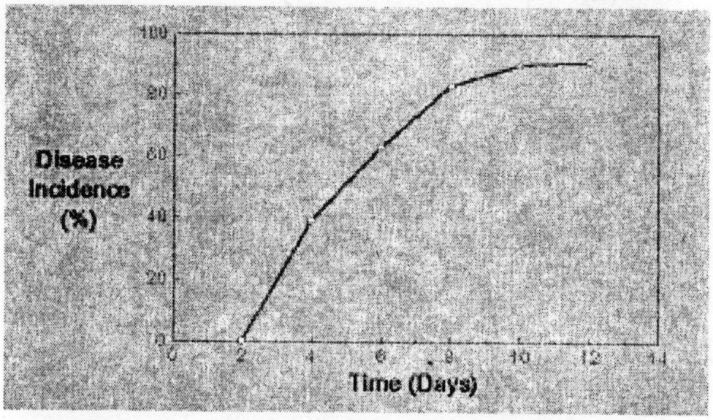

Figure 6. Sclerotium rolfsii on beans

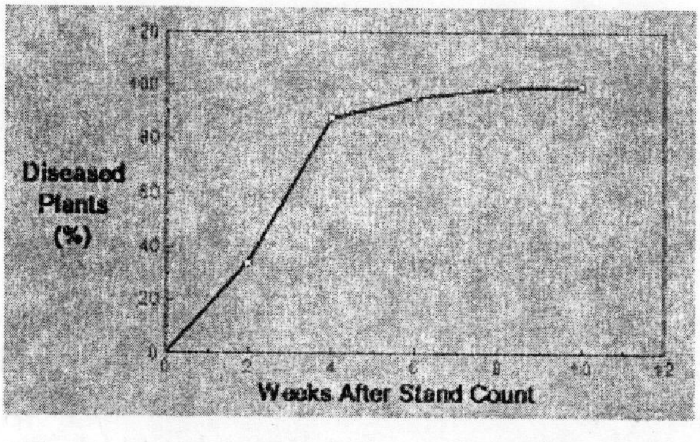

Figure 7. Black shank on tobacco

Likewise, the disease progress curves of *Puccinia graminis* subsp. *graminicola* on ryegrass and *Pyrenophora teres* f. sp. *teres* on barley appear exponential at first, but as time goes on and the incidence and severity of disease approach 100%, the rate of disease progress gradually slows to zero, giving both curves a somewhat sigmoid shape ("S" shape).

To be sure, not all examples of disease progress can be as neatly categorized as these, but in general plant disease epidemics tend to be either roughly linear or exponential in the early stages, and they tend to level off as they approach some limit.

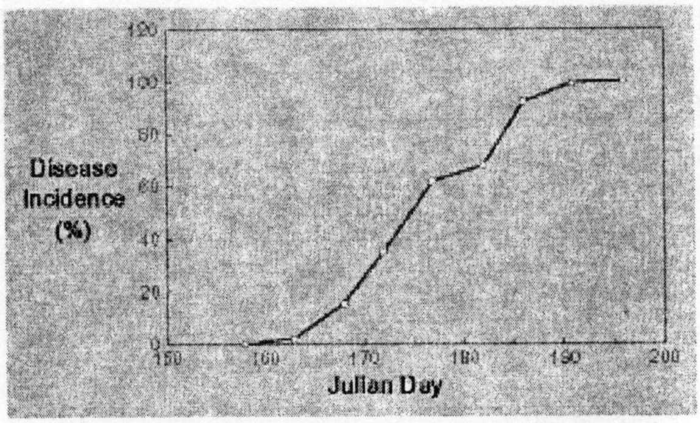

Figure 8. Black stem rust on ryegrass

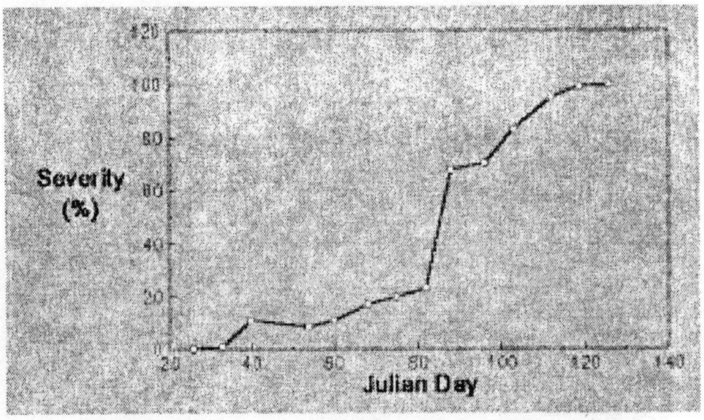

Figure 9. Net blotch on barley

The impact of plant disease and the losses that it causes are a function of disease progress. To reduce this impact, we need not eliminate the disease, we merely need to keep disease development below an acceptable level. That means that the progress of disease and the factors that influence disease progress must be understood in quantitative terms. We have to know what kinds of diseases lead to linear disease progress and what factors affect the slope of the line (the rate of disease progress). Likewise, we have to know what kinds of diseases tend to produce exponential disease progress curves and how we can reduce both the starting level of disease and the rate of epidemic development. Finally, we have to know why epidemics sometimes level off and what imposes limits to their development.

THE CYCLICAL NATURE OF PLANT DISEASE

Plant disease epidemics are cyclical phenomena, that is, they consist of repeated cycles of pathogen development in relation to the host and the environment. The inoculum, which might consist of fungal spores, bacterial cells, nematodes, viruses within an aphid vector, or some other propagules of a pathogen, gains entry into and establishment within the host tissues through the process of infection. The pathogen develops within the host and eventually begins to produce new inoculum, which, in turn, can be dispersed to new susceptible sites to initiate new infections. Pathogens that produce only one cycle of development (one infection cycle) per crop cycle are called monocyclic, while pathogens that produce more than one infection cycle per crop cycle are called polycyclic.

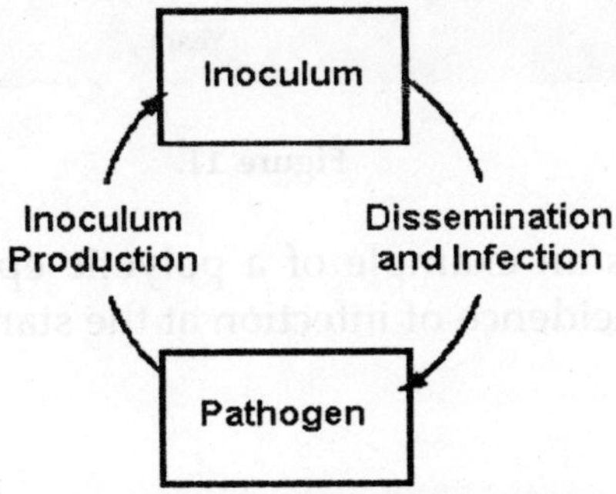

Figure 10. Cyclical Nature of Plant Disease

Generally in temperate climates there is only one crop cycle per year, so the terms "monocyclic" and "polycyclic" are based on the number of cycles per year. In tropical or subtropical climates, however, there can be more than one crop cycle per year, and it is important to remember that "monocyclic" and "polycyclic" are based on a single crop cycle. These same terms are used to describe the epidemics as well as the pathogens, so we often speak of a "monocyclic epidemic" or a "polycyclic epidemic".

For some diseases it is important to consider an epidemic over a period of many growing seasons. This is particularly true for perennial plants (forages, pastures, lawns, orchards, forests, etc.) or for annual crops that are grown in monoculture year after year. In these situations the inoculum produced in one growing season carries over to the next, and there could actually be a buildup of inoculum over a period of years. In the tropics there may not be clear-cut breaks between growing seasons such as we find in the temperate zones, and epidemics can be virtually continuous over periods of many years on such crops as bananas, coffee, and rubber trees. We refer to these epidemics as polyetic epidemics, regardless of whether the pathogen is monocyclic or polycyclic within each season.

The Dutch elm disease is an example of a monocyclic pathogen giving rise to a polyetic epidemic. Note that while there is only one cycle of infection each year and disease progress within each year is roughly linear, the incidence of infected trees increases at an increasing rate from year to year.

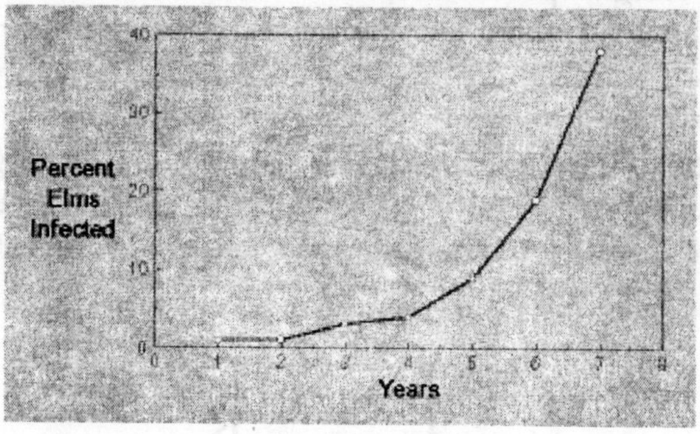

Figure 11.

Apple powdery mildew is an example of a polyetic epidemic caused by a polycyclic pathogen. Note how the incidence of infection at the start of each year tends to increase exponentially.

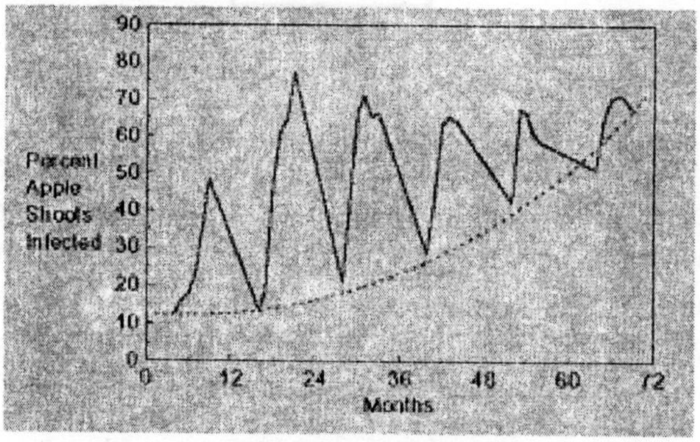

Figure 12.

Sigatoka leaf spot on bananas decreases somewhat during the dry season but otherwise produces more or less continuously repeated cycles of infection. The host, in this case, consists of a population of plants of different ages developing continuously over a long period of time.

Relating Disease Progress to Disease Cycles

If we analyze the disease cycles of the epidemics for which we have plotted disease progress, we see that the epidemics that progress roughly linearly or are linear at the start tend to be monocyclic epidemics. On the other hand, the diseases that increase at an increasing rate during the early part of the epidemic tend to be polycyclic epidemics.

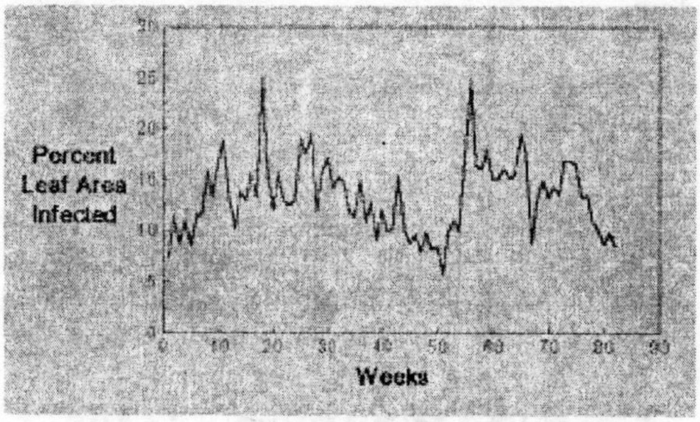

Figure 13.

MONOCYCLIC EPIDEMICS

In general, there are three types of plant diseases that tend to produce only one infection cycle per host cycle (1) postharvest diseases, (2) diseases caused by soil-borne plant pathogens, and (3) rusts without a urediniospore stage.

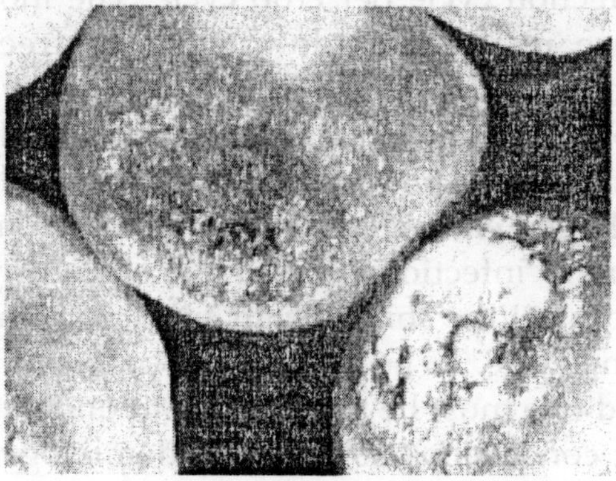

Figure 14. Brown rot of peach

Postharvest Diseases

Not all postharvest diseases produce monocyclic epidemics, but in many cases the infections that result in storage rots have either already occurred before harvest or occur during the harvest and postharvest handling before the product goes into storage. The rot progresses during storage, and new inoculum may be continuously produced, but unless the stored product is handled again to disperse the inoculum and create new infection sites by making small wounds, there are no new infection cycles. Note that the pathogen is not inherently monocyclic but is constrained by the environment to produce only a single cycle of infection. In other environments, these same pathogens might produce polycyclic epidemics.

Figure 15. Fusarium root rot of bean

Diseases Caused by Soilborne Plant Pathogens

Many of the root rots, vascular wilts, and other diseases caused by soilborne pathogens also produce only one infection cycle per crop cycle. The inoculum generally is some kind of survival structure resistant to desiccation or freezing, such as sclerotia, chlamydospores, or oospores in the soil or mycelium in crop residues. This inoculum is dispersed in the soil by plowing and fitting the land and turning under the crop residues. As the roots of the newly planted crop grow through the soil, they encounter the propagules of the pathogen embedded in the soil matrix and become infected. The epidemic progresses as new infections occur, but since any new inoculum that is produced is not dispersed until the soil is again cultivated, there is only one complete infection cycle per cropping cycle. To be sure, not all soilborne plant pathogens produce monocyclic epidemics, and one must be very careful to understand the life cycle of each pathogen before drawing conclusions about its epidemiology.

Demicyclic Rusts

Some rusts produce no urediniospore stage (repeating stage) on a single host, and the inoculum produced on one host species generally must infect a different host species. This alternation of hosts appears to have evolved in adaptation to the annual growth cycles of the hosts, and we see one infection cycle on each host per year. An example is the cedar-apple rust, where all the inoculum that infects apples comes from galls on red cedars (junipers), and all the inoculum that infects red cedars comes from apple leaves and fruit. The epidemic on apples occurs during a four- to six-week period of basidiospore production in the spring. A second monocyclic epidemic on red cedars occurs during a brief period of aeciospore production in late summer.

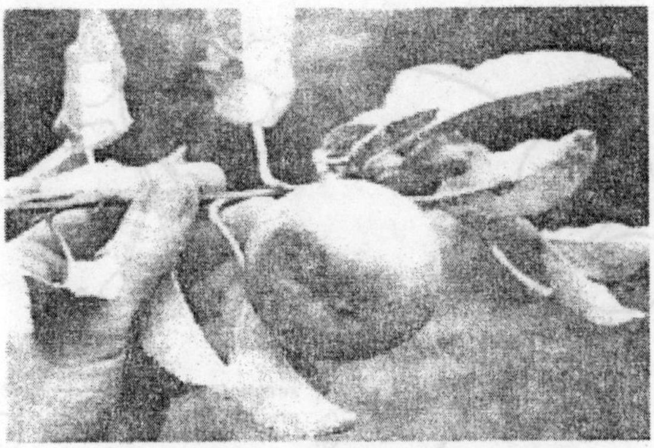

Figure 16. Cedar-apple rust

POLYCYCLIC EPIDEMICS

In order for an epidemic to be considered polycyclic, there must be repeated complete infection cycles, that is, infection followed by pathogen development, new inoculum production, dispersal to new susceptible sites, and new infections, all within a single crop cycle. A good example is potato late blight, where a single cycle of infection, lesion development, sporulation, sporangium dispersal, and new infection can occur in as little as five days, and many overlapping cycles occur simultaneously during periods of favorable weather.

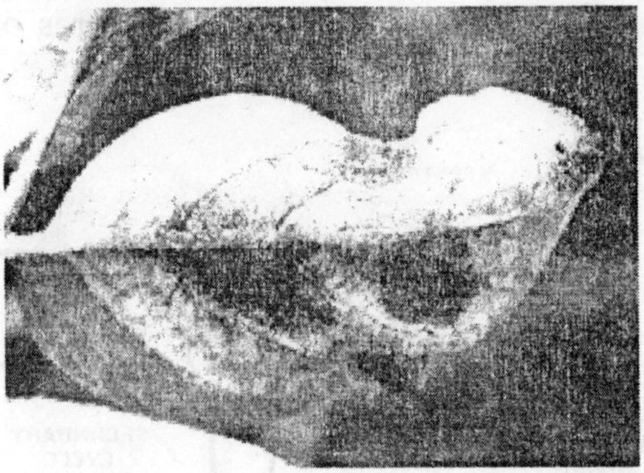

Figure 17. Late blight of potato

Each cycle can produce more than a ten-fold increase in numbers of sporangia landing on susceptible sites, and an explosive epidemic results. Cereal rusts behave similarly; a single urediniospore can infect to produce a pustule from which hundreds of new urediniospores can be released to infect and produce hundreds of new pustules repeatedly throughout the season. Most plant diseases caused by bacteria are polycyclic, and many plant viruses, with the aid of their insect vectors, also can produce repeated cycles of infection in one season.

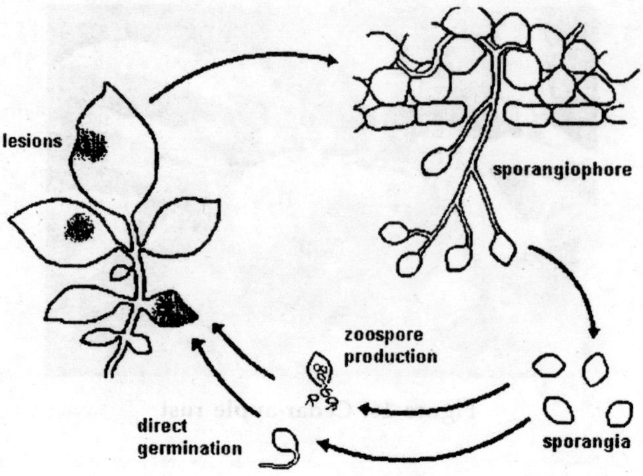

Figure 18. Late blight disease cycle

COMBINATIONS OF MONOCYCLIC AND POLYCYCLIC EPIDEMICS

Not every plant disease epidemic is clearly either monocyclic or polycyclic. Epidemics produced by fungi with two spore stages can have elements of both, sometimes in distinct phases, and sometimes occurring simultaneously. For example, the fungus Venturia inaequalis, causal agent of apple scab, produces ascospores on the dead, infected.

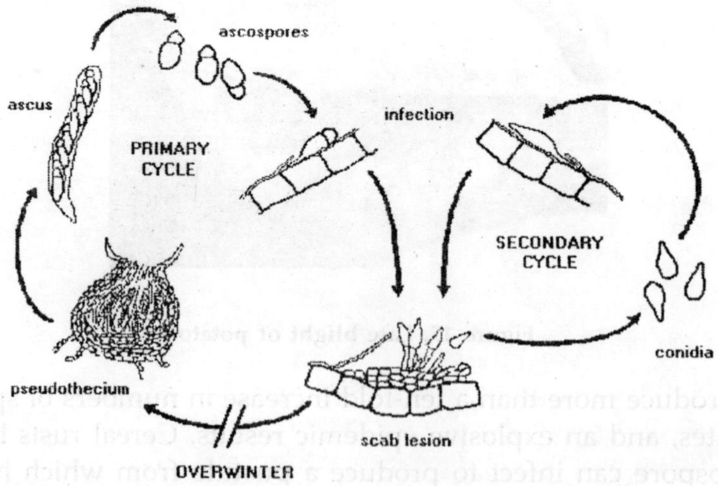

Figure 19. Apple scab disease cycle

An epidemic of tan spot on wheat, caused by Pyrenophora tritici-repentis, also comprises both polycyclic and monocyclic elements. The initial inoculum occurs on infected seed, and the epidemic starts as these seeds germinate to yield seedlings bearing leaf lesions. Conidia are disseminated by wind and splashing rain in repeated cycles of leaf infection throughout the period of crop development (the polycyclic phase).

By the time flowering begins, the inoculum for the second phase of the epidemic has built up on the upper leaves of the plant, principally the flag leaf. The flowers and developing grains are susceptible for a relatively brief period to infection by the inoculum extant at the time of flowering, and the lesions on the glumes of the developing grains do not produce inoculum capable of infecting more grains. Seed infection, therefore, is a monocyclic phenomenon. It is these infected seeds that provide a major means of carryover of this pathogen from one crop to the next and the initial inoculum for the new crop.

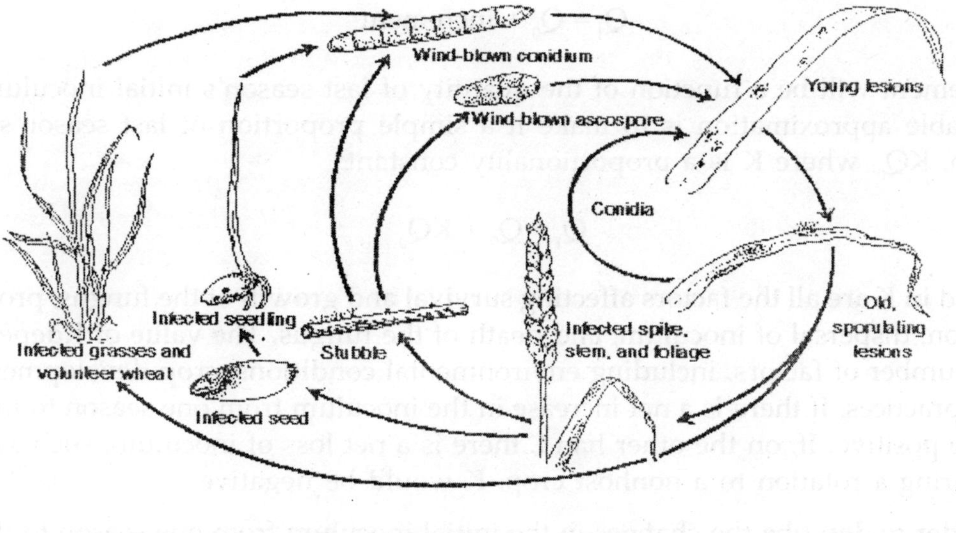

Figure 20. Foliar disease of Wheat

MATHEMATICAL MODELS

If, as we have stated, our goal in plant disease management is to keep disease development below an acceptable level and therefore to understand disease progress in quantitative terms, it will be necessary to find some kind of mathematical model to describe the epidemic.

Our model should show how some variable, preferably something that we can count or measure in the real world, changes with time. From the management perspective, keeping track of new infections perhaps would be the most useful, but actually observing these microscopic events in most cases would be too difficult to be practical. Monitoring inoculum is also very useful for disease management, and for many diseases practical techniques have been developed for quantifying inoculum. We will begin our discussion, therefore, with changes through time in the amount of inoculum available to cause disease.

MONOCYCLIC INOCULUM PRODUCTION

In monocyclic epidemics we are interested primarily in the inoculum present at the beginning of each season (initial inoculum). If we let Q_1 represent the quantity of initial inoculum at the beginning of the current season, it will be equal to the quantity of initial inoculum at the beginning of the previous season, Q_0 plus the increment that has resulted from the pathogen's growth and development during the season:

$$Q_1 = Q_0 + \text{increment}$$

The increment will be a function of the quantity of last season's initial inoculum, and a reasonable approximation is to make it a simple proportion of last season's initial inoculum, KQ_0, where K is a proportionality constant:

$$Q_1 = Q_0 + KQ_0$$

Embodied in K are all the factors affecting survival and growth of the fungus, propagule production, dispersal of inoculum, and death of the fungus. The value of Kdepends on a large number of factors, including environmental conditions, crop development, and cultural practices. If there is a net increase in the inoculum from one season to the next, K will be positive. If, on the other hand, there is a net loss of inoculum, such as might occur during a rotation to a nonhost crop, K would be negative.

In order to describe the chañges in the initial inoculum from one season to the next in a polyetic epidemic, we will generalize the subscript that indicates the season.

$$Q_{T+1} = Q_T + KQ_T$$

We solve this equation repetitively, changing the subscript T, indicating time, with each successive season and making the current value of Q_{T+1} the value of Q_T in the subsequent season. In order to simplify the equation, we assume a constant K (an average over many seasons). The above equation gives the following graph:

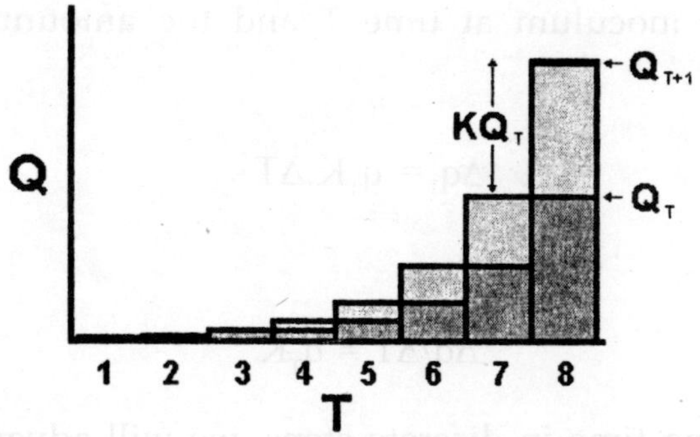

Note that if K is positive, the increment (the light grey area on each bar) increases with the initial inoculum in each successive season, and the graph appears to curve upward.

Polycyclic Inoculum Production

For a polycyclic epidemic we can use essentially the same model as for a monocyclic pathogen viewed over several seasons, where instead of repeating the cycle season after season, we have many repeated cycles within the same season. The time step becomes days or weeks instead of years, and since the time steps are no longer necessarily one time unit (years), the time increment is given as ΔT.

$$q_{T+\Delta T} = q_T + q_T K.\Delta T$$

As a matter of notation, we will use a lower case q to represent the quantity of inoculum during the epidemic and a lower case K to represent the proportion by which inoculum increases in each time step. The units of K correspond to the units of T. For example, if the time is measured in days, the units of K would be proportion/day. Inoculum production actually tends to occur irregularly in discontinuous, discrete infection periods of differing lengths, depending on the weather, and the value of K would likely be different for each infection period.

However, in keeping with our objective of developing the simplest model possible to be useful as a management tool, we will simplify the above model by using uniform time steps and assuming a constant K. (Instead of having a K that varies according to the environmental conditions, we will use a value of K "averaged" over the whole epidemic.) We will first rearrange the above equation to get:

$$q_{T+\Delta T} = -q_T + q_T K.\Delta T$$

The change in the amount of inoculum in one time step, Δq, is simply the difference between the amount of inoculum at time T and the amount of inoculum at time T+ΔT:

$$\Delta q = q_T K.\Delta T$$

Rearranging we get:

$$\Delta q/\Delta T = q_T K$$

Now instead of advancing time in discrete steps, we will advance time continuously, making ΔT infinitesimally small:

$$dq/dt = qK$$

In this differential equation, dq is an infinitesimally small change in the quantity of inoculum, and dt is an infinitesimally small change in time. It tells us that the rate of change of the quantity of inoculum is proportional to the quantity of inoculum at any point in time. Using the calculus, this equation can be integrated to:

$$q = q_0 e^{Kt}$$

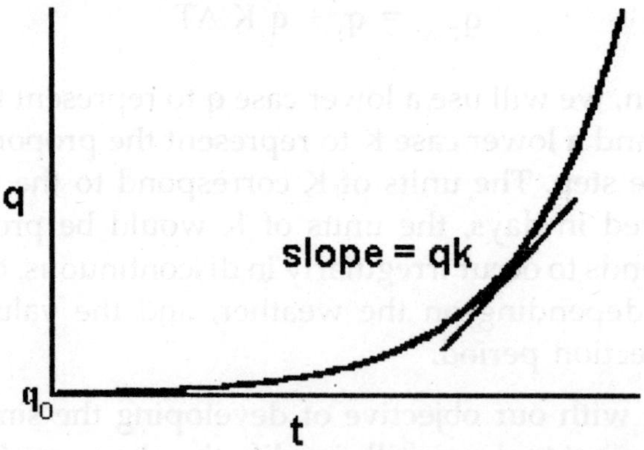

This, we see, is the familiar exponential function, where q_0 is the initial inoculum, and e is the base of the natural (Napierian) logarithm. The instantaneous rate of change in q is dq/dt the slope of the tangent to the curve at any point.

MODELLING DISEASE PROGRESS

While measurements of the quantity of inoculum provide a good assessment of the progress of the epidemic, direct observation of inoculum is not always feasible or even possible. With most plant diseases, however, we can see some kind of symptoms that allow us to count the numbers of plants infected or estimate the amount of tissue infected. To be sure, there is a time lag between infection and the appearance of visible symptoms, and the manifestation of symptoms can be affected by the environmental conditions, but in general the progress of symptoms parallels the progress of inoculum production. Therefore, the same models that we have used to describe the buildup of inoculum also can be used to describe disease progress. We will follow the notation convention of Vanderplank and use x to represent the proportion of infected plants or the proportion of infected plant tissue. Using a dimensionless proportion between zero and one to represent disease makes the mathematics far simpler than using specific units, such as the total population of infected plants or the total area of infected tissue, for example.

Monocyclic Disease Progress

If disease progress in monocyclic epidemics is linear, the slope of the disease progress curve is constant. Furthermore, if disease progress in a monocyclic epidemic is proportional to the amount of initial inoculum (which is itself constant during the epidemic), we can make the slope of the disease progress curve the product of initial inoculum and a proportionality constant. Therefore, we can describe a monocyclic epidemic with linear disease progress using the differential equation:

$$dx/dt = QR$$

where dx is an infinitesimally small increment in the proportion of disease, dt is an infinitesimally small time step, Q is the amount of initial inoculum, and R is a proportionality constant that represents the rate of disease progress per unit of inoculum. Since Q and R are both constant during the course of an epidemic, the slope, dx/dt, is constant, and disease progress is linear. Just as with the K constant in the model of monocyclic inoculum production, R has a value that represents the "average" for the whole epidemic, a value that depends upon many factors, such as aggressiveness of pathogen, susceptibility of the host, the environmental conditions, etc. The units of R are proportion per unit initial inoculum per unit time.

If we integrate the above differential equation, we get:

$$x = QRt$$

Graphically we see a straight line with an intercept of zero and a slope of QR

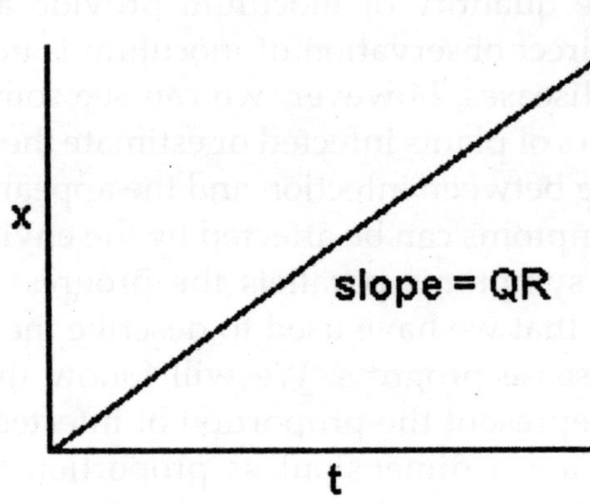

Polycyclic Disease Progress

In a model analogous to that of polycyclic inoculum production, the rate of change in disease is proportional to amount of disease at any point in time. Therefore, in differential form, the equation to describe polycyclic epidemics is:

$$dx/dt = xr$$

As with the monocyclic model, x is a dimensionless proportion between zero and one, and r is a constant that depends on the aggressiveness of pathogen, the susceptibility of the host, the environmental conditions, etc., averaged over the course of the epidemic. In this case, the slope, dx/dt, is proportional to x, and therefore disease progress increases with time at an increasing rate.

In the integrated form the model is:

$$x = x_0 e^{rt}$$

where x_0 is the proportion of disease at the start of the epidemic and e is the base of the natural logarithm. Vanderplank called r the "apparent infection rate" because it is based on the appearance of disease symptoms, which lag behind the actual infections. It is defined as the rate of disease increase per unit of disease and has the units of proportion per unit of time.

The parameter x_0 is sometimes carelessly called initial inoculum, to which it is quantitatively related, but strictly speaking it is the initial disease (a proportion). Graphically the model has the familiar form of the exponential model:

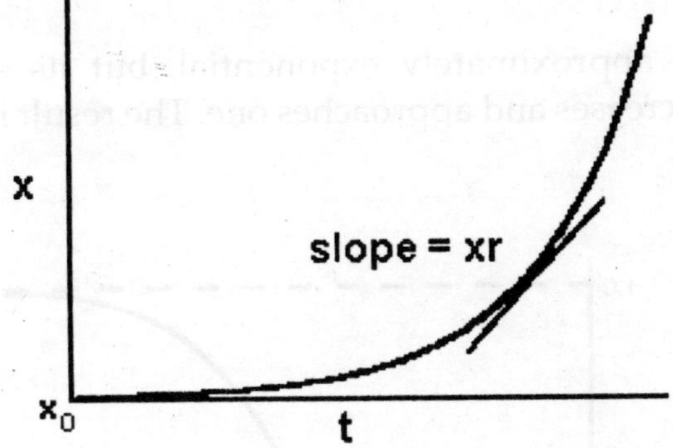

Upper Limit to Disease

The above models assume unlimited growth of disease, which, of course, is impossible; the proportion of diseased plants or of diseased tissue cannot exceed one. We can adjust our models to address this issue by using a correction factor (1-x) to represent the proportion of healthy tissue remaining. A decrease in the remaining healthy tissue reduces the chance for new infections and therefore reduces the rate of disease progress. As x approaches one, there is no healthy tissue left, and the rate of the epidemic slows to zero.

The monocyclic model of disease progress, adjusted for the limit to disease is:

$$dx/dt = QR(1-x)$$

Graphically we see an epidemic that starts out looking linear, but as x approaches 1, the slope decreases to zero.

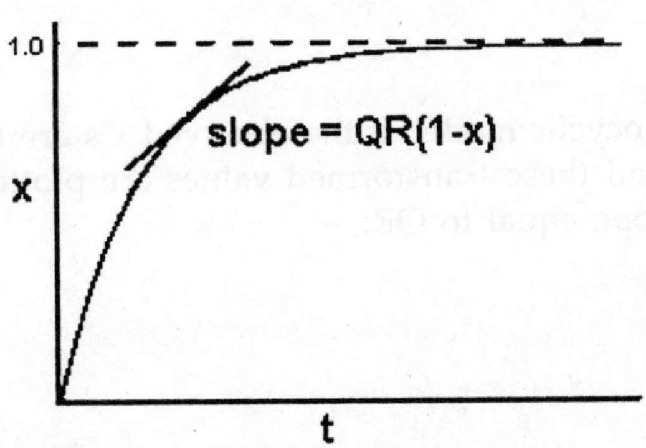

In the polycyclic model we make a similar adjustment:

$$dx/dt = QR(1-x) = xr\ (1-x)$$

This model starts out approximately exponential, but its slope also decreases and approaches zero as x increases and approaches one. The result is a sigmoid-shaped curve:

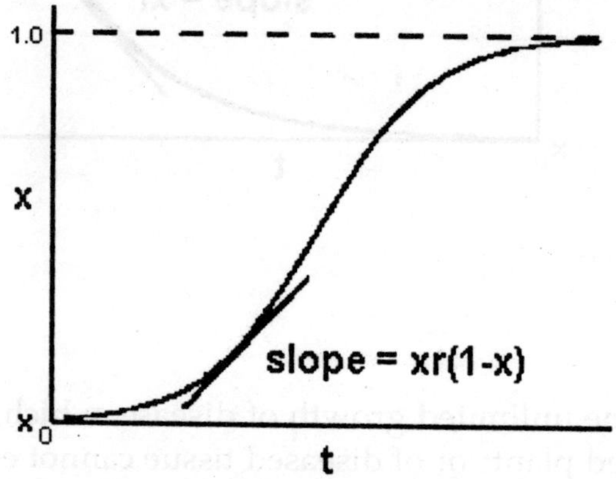

In reality, it is rare that disease incidence or severity exceeds 50%, and when it does, the disease progress curve is usually not quite sigmoid.

Estimating Model Parameters Transformations

To apply our models to real epidemics, we would like to come up with numerical values for their parameters (R in the monocyclic model and $X_0$and r in the polycyclic model). To estimate the parameters, we observe the epidemic, measure x at several times during the epidemic, and then plot Xversus t. The difficulty comes in trying to fit nonlinear models to the data sets. It is far easier to transform the x's to get a linear model that can be fit with simple linear regression.

The monocyclic model

In the case of the monocyclic model, if the observed x's are transformed to the natural logarithm of 1/(1-x), and these transformed values are plotted against t, we will get a straight line with a slope equal to QR.

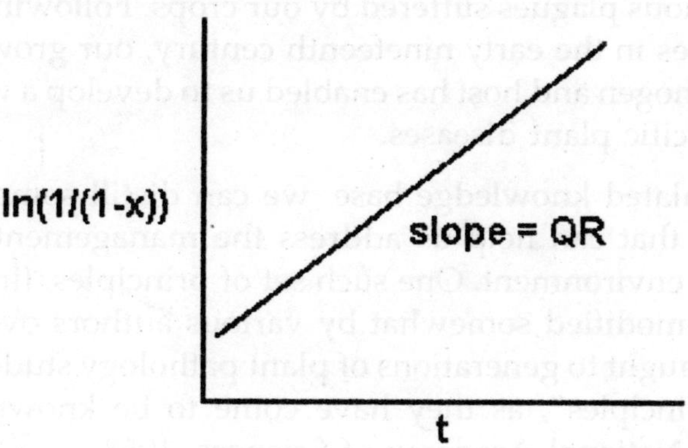

Then with an independent estimate of the initial inoculum, Q, we can calculate R.

The polycyclic model

If the observed x's in a polycyclic epidemic are transformed to the natural logarithm of x/(1-x), and the transformed values are plotted against t, the result will be a straight line with a slope equal to r and an intercept equal to the natural log of x0/(1-x0).

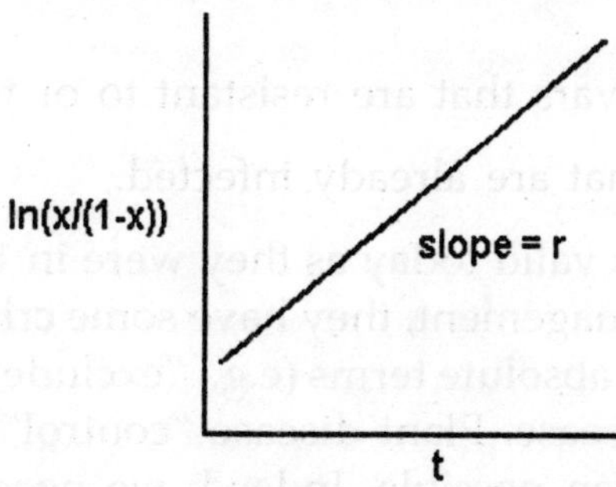

In fitting models to observed data, it is important to select the model based on the known biology of the pathogen rather than simply on the shape of the curve. Likewise, one must not attempt to make inferences about the biology of the pathogen based on the shape of the curve and which model gives the "best fit" of the data. There is random variability in each observation, and in the transformed models the data points on the ends of the line have undue weight in determining the fit to the model. It is quite possible to have a data set that fits both the monocyclic and polycyclic models equally well or to have a data set from a known monocyclic epidemic that gives a better fit to a polycyclic model and vice versa.

PLANT DISEASE MANAGEMENT STRATEGIES

Since the beginning of agriculture, generations of farmers have been evolving practices for combating the various plagues suffered by our crops. Following our discovery of the causes of plant diseases in the early nineteenth century, our growing understanding of the interactions of pathogen and host has enabled us to develop a wide array of measures for the control of specific plant diseases.

From this accumulated knowledge base, we can distill some general principles of plant disease control that can help us address the management of new problems on whatever crop in any environment. One such set of principles, first articulated by H. H. Whetzel in 1929 and modified somewhat by various authors over the years, has been widely adopted and taught to generations of plant pathology students around the world. These "traditional principles", as they have come to be known, were outlined by a committee of the US National Academy of Sciences, 1968.

Traditional Principles of Plant Disease Control

1. Avoidance—prevent disease by selecting a time of the year or a site where there is no inoculum or where the environment is not favorable for infection.
2. Exclusion—prevent the introduction of inoculum.
3. Eradication—eliminate, destroy, or inactivate the inoculum.
4. Protection—prevent infection by means of a toxicant or some other barrier to infection.
5. Resistance—utilize cultivars that are resistant to or tolerant of infection.
6. Therapy—cure plants that are already infected.

While these principles are as valid today as they were in 1929, in the context of modern concepts of plant disease management, they have some critical shortcomings. First of all, these principles are stated in absolute terms (e.g., "exclude", "prevent", and "eliminate") that imply a goal of zero disease. Plant disease "control" in this sense is not practical, and in most cases is not even possible. Indeed, we need not eliminate a disease; we merely need to reduce its progress and keep disease development below an acceptable level. Instead of plant disease control, we need to think in terms of plant disease management.

A second shortcoming is that the traditional principles of plant disease control do not take into consideration the dynamics of plant disease, that is, the changes in the incidence and severity of disease in time and space. Furthermore, considering that different diseases differ in their dynamics, they do not indicate the relative effectiveness of the various tactics for the control of a particular disease. They also fail to show how the different disease control measures interact in their effects on disease dynamics. We need some means of assessing quantitatively the effects of various control measures, singly and in combination, on the progress of disease.

Finally, the traditional principles of plant disease control tend to emphasize tactics without fitting them into an adequate overall strategy.

Does this mean that we should abandon the traditional principles? Of course not! We merely have to fit them into an appropriate overall strategy based on epidemiological principles.

Strategies versus Tactics

Ask a handful of pest management experts to name the major plant disease control strategies, and you are sure to find disagreement. The problem is generally one of semantics rather than of fundamental disagreement over the important means of disease control. The dictionary definitions for the two terms are similar, but generally speaking, an overall plan for reaching a particular objective is called a strategy, while the specific means for implementing a given strategy are called tactics. Like the goals and objectives that they are intended to achieve, strategies and tactics tend to occur in hierarchies. What is a "strategy" at one level of focus could be called a "tactic" at another level of focus.

The important point to remember is that countless human undertakings, be they military operations, political campaigns, football games, or any other kind of organized effort, have failed, despite flawless tactics, for lack of a sound strategy. Any endeavor that requires a series of connected tasks for its completion also requires some kind of overall plan. Each individual task, no matter how skillfully executed or how successful its outcome, will not advance progress toward the final objective unless it has a coherent relationship with all of the other necessary tasks.

EPIDEMIOLOGICAL BASIS OF DISEASE MANAGEMENT

Plant disease epidemics can be classified into two basic types, monocyclic and polycyclic, depending on the number of infection cycles per crop cycle. The early stages of a monocyclic epidemic can be described quite well by a linear model, while the early stages of a polycyclic epidemic can be described with an exponential model. Since we are concerned with keeping disease levels well below 100%, there is no need to adjust the models for approaching the upper limit, and we can use the simple linear and exponential models to plan strategies:

Monocyclic Model

$$x = \overset{\substack{1\\\downarrow}}{Q}\,\overset{\substack{2\\\downarrow}}{R}\,\overset{\substack{3\\\downarrow}}{t}$$

Polycyclic Model

$$x = \overset{1}{\overset{\downarrow}{x_0}}\, e^{\overset{2}{\overset{\downarrow}{r}}\overset{3}{\overset{\downarrow}{t}}}$$

Examining these models, we can see that in both there are three ways in which we can reduce x at any point in the epidemic:

1. Reduce the initial inoculum (Q in the monocyclic model and x_0 in the polycyclic model). (Actually x_0 is the initial incidence of disease, which is proportional to the initial inoculum.)
2. Reduce the rate of infection (R in the monocyclic model and r in the polycyclic model)
3. Reduce the duration of the epidemic (the time, t, at the end of the epidemic)

These, then, can be used as three major strategies for managing plant disease epidemics, and we can organize our plant disease control tactics under one or more of these overall strategies. Furthermore, by means of the model we can assess the quantitative impact of each strategy, not only by itself, but in its interaction with others.

The monocyclic model

It is clear from the above model of a monocyclic epidemic that Q, R, and t have equal weight in their effect on x. A reduction in the initial inoculum or the rate of infection will result in a reduction in the level of disease by the same proportion at any time, t, throughout the epidemic. If t can be reduced (for example, by shortening the season), disease will be reduced proportionately.

The polycyclic model

- If r is very high, the apparent effect of reducing x_0 is to delay the epidemic.
- If r is very high, x_0 must be reduced to very low levels to have a significant effect on the epidemic.
- Reducing r has a relatively greater effect on the epidemic than reducing x0.
- Reducing x_0 makes good strategic sense only if r is low or if r is also being reduced.

It is easier to understand (and remember!) these concepts if we actually select different values for x_0 and r, plug them into the model, and graph the outcome. This can be done easily with a calculator that has an exponential function, or with the accompanying simulation.

Clearly developing a sound disease management strategy requires enough knowledge of the biology of the pathogen and host to select the appropriate epidemiological model. It also requires at least "ball-park" estimates of the model parameters and the magnitude of the impact of each specific tactic on the initial inoculum or the apparent infection rate. Failure to adopt such a quantitative approach can lead to some embarrassing or even very costly errors.

Traditional Principles Revisited

To make the conceptual leap from disease control to disease management, the traditional principles can be modified by fitting them as tactics within each of the three major disease management strategies and by slightly changing the wording to reflect the quantitative impact of the action rather than an absolute effect:

Tactics for the Reduction of Initial Inoculum

- Avoidance—reduce the level of disease by selecting a season or a site where the amount of inoculum is low or where the environment is unfavorable for infection
- Exclusion—reduce the amount of initial inoculum introduced from outside sources
- Eradication—reduce the production of initial inoculum by destroying or inactivating the sources of initial inoculum (sanitation, removal of reservoirs of inoculum, removal of alternate hosts, etc.)
- Protection—reduce the level of initial infection by means of a toxicant or other barrier to infection
- Resistance—use cultivars that are resistant to infection, particularly the initial infection
- Therapy—use thermotherapy, chemotherapy and/or meristem culture to produce certified seed or vegetative planting stock

Tactics for the Reduction of the Infection Rate

- Avoidance—reduce the rate of production of inoculum, the rate of infection, or the rate of development of the pathogen by selecting a season or a site where the environment is not favorable
- Exclusion—reduce the introduction of inoculum from external sources during the course of the epidemic
- Eradication—reduce the rate of inoculum production during the course of the epidemic by destroying or inactivating the sources of inoculum (roguing)

- Protection—reduce the rate of infection by means of a toxicant or some other barrier to infection
- Resistance—plant cultivars that can reduce the rate of inoculum production, the rate of infection, or the rate of pathogen development
- Therapy—cure the plants that are already infected or reduce their production of inoculum

Tactics for the Reduction of the Duration of the Epidemic

- Avoidance—plant early maturing cultivars or plant at a time that favors rapid maturation of the crop
- Exclusion—delay the introduction of inoculum from external sources by means of plant quarantine

REFERENCES

Campbell, C. L. and L. V. Madden. 1990. *Introduction to Plant Disease Epidemiology*. Wiley, New York.

Francl, L. F. and D. A. Neher (eds.) 1997. *Exercises in Plant Disease Epidemiology*. American Phytopathological Society, APS Press.

Jones, D. G. (ed.) 1998. *The Epidemiology of Plant Diseases*. Kluwer Academic Publishers. Dordrecht, Boston.

Anonymous. 1968. Plant Disease Development and Control. National Academy of Sciences, Washington, D. C.

Gilligan, C.A. 2002. An epidemiological framework for disease management. Advances in Botanical Research 38:1-64

Neher, D.A., and Campbell, C.L. 1992. Underestimation of disease progress rates with the logistic monomolecular and Gompertz models when maximum disease intensity is less than 100 percent. Phytopathology 82:811-814

Maloy, O. C. 1993. Plant Disease Control: Principles and Practice. John Wiley and Sons, Inc., New York.

Vanderplank, J. E. 1963. *Plant Diseases: Epidemics and Control*. Academic Press, New York.

Whetzel, H. H. 1929. The terminology of plant pathology. Proc. Int. Cong. Plant Science, Ithaca, NY, 1926:1204-1215.

Zadoks, J. C. and Schein, R. D. 1979. *Epidemiology and Plant Disease Management*. Oxford University Press, London and New York.

5

Pests and Diseases in Plantations

Pests and diseases can damage trees in all stages of development and affect the ability of both natural forests and plantations to meet their management objectives. Some fungi and insects are pests of tree seedlings in nurseries while others attack older trees in a wide variety of ways. While invasive plants compete with seedlings and animals damaging forests are also sometimes problems, the focus here is on insects and diseases. Many examples exist of where one or more species of either indigenous or exotic pests or diseases have caused devastating losses to forests, requiring changes in management regimes or forcing forest managers to switch to alternative tree species.

PEST AND DISEASE OCCURRENCE

Plantations vs Natural Forests

Damaging pests and diseases are found in plantations and natural forests in both the temperate and tropical zones. A fundamental concept of ecosystem dynamics is that as diversity increases, so does stability. The greater the number of plants and animals that occupy an ecosystem, the greater are the checks and balances that prevent any one species from increasing to the point where other ecosystem components are threatened. With pests and diseases, the greatest single deterrent to population increase is the amount of available host material. For example, in a complex tropical rain forest, where as many as a thousand plant species might occupy a single hectare, a population of a host specific, phytophagus caterpillar will find only a limited amount of suitable host material. Therefore its numbers will remain fairly stable. On the other end of the diversity scale is the agro-ecosystem. Dominated by a single plant species, agro-ecosystems often cover large areas and provide a virtually unlimited amount of suitable host biomass. This can result in population explosions of organisms that use the plant as host material. Furthermore, the lack of ecosystem diversity will, at best, provide only a marginal habitat for the natural enemies of the organisms that are using the crop plant as host material.

In terms of diversity and stability, forest plantations tend to be more like agro-ecosystems than natural forests. This would lead one to conclude that plantation forests are more susceptible to damage by pests and disease than are natural forests. While this may be generally true, it does not necessarily mean that natural forests are immune to damage from these agents. The relative lack of pest outbreaks in mixed tropical forests is often cited as evidence for the importance of diversity in stabilising plant communities. Many natural forests in the northern hemisphere boreal and temperate zones however, tend to be relatively simple ecosystems when compared to tropical forests and are susceptible to outbreaks of insects. Examples include:

- the spruce budworm, *Choristoneura fumiferana*, a defoliator of *Abies balsamea* and *Picea* spp in the boreal conifer forests of eastern Canada and adjoining states in the USA,
- western spruce budworm, *Choristoneura occidentalis*, and Douglas-fir tussock moth, *Orgyia pseudotsugata*, in natural conifer forests of western North America,
- the processionary caterpillar, *Thaumetopoea pityocampa*, in Mediterranean Europe and North Africa
- and several species of conifer attacking bark beetles, *Dendroctonus* spp and *Ips* spp (Coleoptera: Scolytidae) in the forests of Canada, the United States, Mexico, Central America and the Himalayas.

Natural forests are also subject to damage by a variety of diseases caused by fungi, bacteria, virus or parasitic mistletoes and they are vulnerable to introductions of exotic pests and disease. The introduction of white pine blister rust, *Cronartium ribicola*, into North America, the European gypsy moth, *Lymantria dispar*, into the mixed broadleaf forests of the eastern United States, of two *Juniperus* infesting scale insects, *Carulaspis visci* and *Lepidosaphis newsteadi*, into Bermuda and the apparent introduction of the soil fungus *Phytophthora cinnamomi* into *Eucalyptus marginata* forests in Western Australia are just a few examples. Moreover, the recent introduction of pitch canker disease, caused by the fungus *Fusarium subglutinans* f. sp. *pini*, into forests of *Pinus radiata*, along the California, USA coast, now is a source of concern. A more recent example is the discovery in 1994-5 of the pink or hibiscus mealybug, *Maconellicoccus hirsutus*, an insect capable of feeding on between 125 and 150 different plant species, in the Caribbean islands of Grenada, Trinidad and Tobago and St. Kitts-Nevis. This insect poses a severe threat to natural mixed tropical forests as well as vegetable crops and ornamental plants in the region.

Exotic vs Indigenous Tree Species

Exotic plantations

A typical scenario in the use of exotics in forest plantation programmes, both in tropical and temperate zones, is that they enjoy a pest free "honeymoon" for a period of time. During this period, management practices often evolve that make the plantations especially susceptible to pests and disease once they begin to appear. These include:

1. Failure to give proper attention to species/site matching.
2. Use of planting stock from a narrow genetic base.
3. Failure to maintain optimum stocking levels and tree vigor through intermediate cuttings.
4. Dependence on one or two species in a plantation programme, resulting in an unlimited supply of host material for potentially damaging organisms.

When pests and diseases do begin to appear, they may come from two sources. A complex of indigenous agents may adapt to the new host or exotic pests may be accidentally introduced.

A number of examples of indigenous insects adapting to exotic forest plantations are known. In Colombia, indigenous insects have been detected in that country's exotic *Cupressus, Eucalyptus* and *Pinus* plantations, with the most frequently encountered groups being defoliating caterpillars of the family Geometridae, leaf cutting ants and stem infesting insects. Rodas P. reports 30 species of defoliating insects feeding on exotic forest plantations in the Andean region of Colombia. A similar situation exists in Chile, where several species of indigenous pests are now associated with exotic forest plantations including *Ormiscodes cinnamomea,* a large Saturniid that occasionally defoliates *Pinus radiata*. Ohmart and Edwards report 96 indigenous insects feeding on various species of *Eucalyptus* in China, 94 in India, 223 in Brazil, 31 in New Zealand, 105 in Papua New Guinea and 62 on the island of Sumatra, Indonesia. Termites and leaf cutting ants are the major damaging agents.

Indigenous disease causing organisms are also capable of adapting to new hosts and causing severe damage. In Kenya, plantations of *Cupressus macrocarpa* have been severely damaged by *Monochaetia unicornis,* a stem canker causing fungus. During the early 1990s, several species of foliar and twig pathogens, including Cylindocladium leaf blight, Coniella leaf spot, Kirramyces leaf spot and pink disease, *Corticum salmonicolor,* damaged *Eucalyptus* plantings in Vietnam.

Exotic pests and diseases

The greatest hazard of pest and disease damage to exotic forest plantations is from the accidental introduction of exotic pests and diseases. Exotic pests and diseases agents, in the absence of natural enemies, and provided with a large supply of suitable host material, that may have little resistance to the new pest or disease, can build up rapidly and cause devastating losses.

One of the best-documented examples of exotic pests and diseases causing damage to exotic plantations is to pine plantations in the Southern Hemisphere, especially *Pinus radiata*. *P. radiata* were first planted in the late nineteenth century in Australia, Chile, New Zealand and South Africa. The excellent growth and form of the tree in these locations encouraged additional plantings. These provided the basis for thriving lumber and paper industries, especially in Chile and New Zealand. One of the first exotic insects to appear in New Zealand's radiata pine plantations was the European wood wasp, *Sirex noctilio*, during the early 1900's. This insect was probably introduced on unprocessed pine logs imported from Europe. Native to southern Europe and North Africa, where it causes little or no damage, *S. noctilio* normally breeds in recently killed pines. Plantations established in New Zealand during 1920-30 stagnated because there was no market for small logs from thinning operations. This made the plantations susceptible to *S. noctilio* and its associated fungus, *Amylostereum areolatum*. By 1947, high levels of tree mortality were occurring, primarily in the unthinned plantations. However, today *S. noctilio* is not a problem because of biological control methods and better stand management.

S. noctilio subsequently appeared in *P. radiata* plantations in Australia in 1952. In 1980 infestations were detected in South America where this insect is damaging plantations of *P. elliotii* and *P. taeda* in Argentina, southern Brazil and Uruguay. In 1994, infestations of *S. noctilio* were discovered in *P. radiata* plantations in the Cape Peninsula of South Africa.

Other damaging insects that have been introduced into *P. radiata* plantations include the bark beetle, *Ips grandicollis*, a North American species, into Australia and the European pine shoot moth, *Rhyacionia bouliana* into Argentina and Chile. The latter is now regarded as Chile's number one forest insect pest. The European bark beetles, *Hyalstes ater* and *Hylurgus ligniperda* were introduced into Chilean radiata pine plantations during the late 1980s and are capable of killing young pine seedlings.

Several introduced fungi have also caused severe damage to *P. radiata* plantations. In 1957, a needle fungus, *Dothistroma septospora* (*D. pini*), became a major concern in *P. radiata* plantations in Chile, parts of East Africa and New Zealand, although there are records of its occurrence in these areas as early as 1940. The stem disease, *Sphaeropsis sapinea* (*Diplodia pini*) has also been introduced in a number of countries. This fungus has been especially damaging in Kenya, where it first appeared in 1973 and proliferated on *P. radiata*, a species adapted to a winter rainfall regime and is especially susceptible to fungal attacks during periods of rain and warm temperatures.

Another example of damage to exotic forest plantations by an introduced pest is that of the leucaena psyllid, *Heteropsylla cubana* (Homoptera: Psyllidae) on *Leucaena leucocephala*, a fast growing legume native to Mexico and Central America. This tree was introduced into the Philippines as early as 1565 and was quickly established throughout the tropics where local people found it an excellent fuelwood and later as shade under which crops such as coffee, cocoa, pepper and vanilla could be grown. Genetically improved varieties were eventually introduced into the tropics where it was regarded as having the widest range of uses of all of the tropical legumes. In 1983, leucaena psyllid, a somewhat obscure insect of *L. leucocephala* in its natural range, was detected in southern Florida, USA and during the following year it was found in the Hawaiian Islands. The insect spread rapidly across the Asia and Pacific Region between 1985 and 1988. In 1991, infestations were found on the southern Indian Ocean islands of Reunion and Mauritius and one year later in several eastern African countries, including Burundi, Kenya, Tanzania and Uganda. This insect feeds on the shoots and young foliage of Leucaena causing wilting and growth loss.

Ironically, attempts to introduce new genetic material into a country have occasionally resulted in introductions of damaging pests. In 1968, a pine woolly aphid, *Pineus borneri (P. pini)*, was introduced into Kenya on *Pinus caribaea* grafts imported from Australia. Twenty years later, a scale insect, *Oracella acuta*, was introduced into China on *P. elliotii* scion material introduced from the United States. By 1995, over 212,500 ha of *P. elliotii* plantations in Guangdong Province in southern China were infested by this insect.

Indigenous species plantations

Throughout much of Asia, Europe and North America, forest plantations are typically established with indigenous species (e.g. *Pinus massoniana* in China, *Picea abies, Pinus sylvestris* and *Populus* spp in Europe, *Pinus taeda* and *Pinus elliotii* in the southeastern USA and *Pseudotsuga menziesii* in western Canada and the USA). These plantations are also susceptible to damage by a complex of both indigenous and exotic pests and disease.

Indigenous pests and diseases

In China and Vietnam, the defoliating caterpillar, *Dendrolimus punctatus*, is a major pest of indigenous pine plantations. This insect has two generations per year in east-central China where it defoliates *Pinus massoniana*. In Vietnam, *D. punctatus* has four generations per year and outbreaks have occurred in plantations of *P. merkusii*, often established on steep deforested slopes where they are an important source of lumber, pulpwood, fuel wood and resin.

Nun moth, *Lymantria monacha*, is a defoliating caterpillar of both *Picea abies* and *Pinus sylvestris* in central Europe. This insect was originally a major pest of natural *Picea abies* forests but readily adapted to the extensive pine plantations. In Poland, plantations of *P. sylvestris* are established on low nutrient, sandy soils or abandoned marginal agricultural lands. Between 1978 and 1983, a major outbreak of this caterpillar necessitated aerial spraying of 2.5 million ha, over 25% of Poland's forests. In 1991, this insect again reached outbreak proportions and large areas of forest were treated between 1992 and 1994. Other defoliating insects of *P. sylvestris* plantations in Europe include *Bupalis piniarius, Panolis flammea, Dendrolimus pini* and sawflies, *Diprion* spp.

Fusiform rust, caused by the fungus *Cronartium quercuum* f.sp. *fusiform*, has become a serious pest of *P. elliottii* and *P. taeda* plantations in the southeastern USA. Prior to 1900, the disease was rare, but because of intensive cultural practices associated with plantation forestry, including planting of infected seedlings, fire management, genetic selection of fast growing trees and expansion of the range of susceptible hosts, this disease has become a major problem. Recent estimates indicate that $US130 million in damage is occurring annually and the disease is increasing at the rate of 2-3% per year.

In China, species of *Paulownia*, are widely planted either as ornamentals in villages or in intercropping systems. Pawlonias are commonly infected by a mycoplasma-like organism (MLO), which produces large witches brooms in tree crowns. This disease causes significant reductions in increment, leading to a gradual decline and death of infected trees. The wood of infected trees is of reduced quality.

The pine wood nematode, *Bursaphelenchus xylophilus* is the cause of pine wilt disease. This disease has been a major cause of tree mortality in pine plantations in Japan since the early 1900's. The nematode is probably native to North America and has been introduced into Japan and more recently China, where it is causing extensive tree mortality in plantations of *Pinus massoniana*. The nematode is spread from tree to tree by longhorn wood boring beetles, *Monochamus* spp, when the adult beetles feed on pine shoots.

Mixed vs Single Species Plantations

The concept of diversity leading to ecosystem stability would suggest that mixed species plantations would be more pest and disease resistant than single species plantation. Indeed, work by Klimetzek in Germany indicates that the inclusion of broadleaf trees in pine forests, either as a mixture or in the understory, tends to reduce the incidence of outbreaks of *Lymantria monacha*. On the other hand, planting *Picea abies* under a plantation of *P. sylvestris*, could intensify or prolong outbreaks of this insect since both trees are hosts.

Wormald indicates that data on relative susceptibility of mixed and pure species plantations in subtropical and tropical regions is, at best, inclusive and confusing. While examples of where pests and diseases have become problems in pure species plantations are cited, many result from accidental introductions or of indigenous pest species adapting to new conditions. Furthermore he cites examples where establishment of mixed species plantations has had little or no effect on reducing pest caused losses.

Some temperate zone pests and diseases have adapted to mixed-species ecosystems by having evolved life stages on alternate hosts. Many rust diseases fall into this category, including fusiform rust of *P. elliotii* and *P. taeda*, *Cronartium quercuum* f.sp. *fusiform*, which has alternate spore stages on oak, *Quercus* spp. Therefore, removal of oaks from the vicinity of pine nurseries in the southeastern USA is a common although not altogether effective practice. Cooley spruce gall aphid, *Adelges cooleyi*, an insect indigenous to western North America, has alternating life stages on *Picea* and *Pseudotsuga menziesii*, trees that are often found growing in close proximity to one another in natural forests.

Perhaps a greater concern with regard to pests and diseases is not mixed vs. single species plantations, but the number of species used in a country's plantation programme. Reliance on one or two closely related species may meet a country's needs for wood products and foreign exchange. However, should a pest or complex of pests appear in these plantations, the results could be devastating. Yet, many countries have fallen into this dilemma. Examples include Kenya, where 46% of its industrial plantation area is composed of *Cupressus lusitanica*; Colombia with 66% of its plantations *Pinus patula*; Poland with 70% of its plantation forests *P. sylvestris*; and Chile and New Zealand with 90% of their plantations composed of *P. radiata*. New Zealand experience has been that deliberately planting a wide range of species to reduce risk from pests and diseases did not work. Rather, good site matching and silviculture, coupled with emphasis on disease prevention and control, backed by research, has been used as an alternative strategy.

MANAGING PESTS AND DISEASES

The objective of managing pests and diseases in forests and forest plantations should be to keep them in a healthy, productive condition. What is a healthy forest? The concept of a healthy forest became a popular concept during the mid- to late 1980s. In economic terms, a "healthy" forest can be defined as "a forest in which pests and diseases remain at low levels and do not interfere with management objectives." In ecological terms, a healthy forest is "a fully functional ecosystem; one in which all of its parts can interact in a mutually beneficial way."

The healthy forest concept directs forest managers to focus on the forest rather than its pests and diseases and takes into account the natural role of insects, fungi, fire and other so called "damaging agents" and their interactions in forest dynamics. Under an overall policy of forest health protection, pests and diseases are looked upon as a symptom of an unhealthy forest rather than as the problem. This directs forest managers and forest protection specialists to address the underlying causes of the pest or disease - factors such as overstocking, over-maturity, poor site/species matching, excessive fuels and single species forests with little diversity. Striving for healthy forests also involves anticipating pests and diseases based on historical records of their occurrence and the knowledge of forest and climatic conditions that favour their abundance. This allows time to implement management practices that will make these forests inhospitable for build-up of damaging pests and disease.

The healthy forest concept may sound like an idyllic approach to managing forest pests and diseases, especially in cases where an introduced pest is causing widespread damage in a simple plantation ecosystem in the absence of its normal complex of natural enemies. Yet even the most damaging of exotic pests and diseases respond to stand and site conditions and there is evidence that they focus their activities on "unhealthy" forests. A classic example is that of *Sirex noctilio* in exotic pine plantations in the Southern Hemisphere. In New Zealand, it was noted that stressed trees in unthinned plantations were the first to be attacked by this insect. Therefore, by its action, the plantations were effectively thinned. Thinning is now an integral part of managing plantations threatened by *S. noctilio* in places where this insect has been introduced. In another example, when a cypress aphid, originally identified as *Cinara cupressi*, appeared in eastern and southern Africa, most severe damage occurred in plantations of *Cupressus lustianica* that were either overmature or had been established on low nutrient soils.

Integrated Pest Management

The healthy forest concept provides the umbrella or objective for addressing damage caused by pests and diseases. Integrated pest management (IPM), on the other hand, provides the tools to accomplish this task.

The concept or philosophy of IPM as a "rational" approach to pest control was formalised during the 1960s. Crop protection specialists had become aware of the adverse side effects of dependence on chemical pesticides, including pesticide resistance, occurrence of secondary pests, environmental damage and human health hazards. This led to the realisation that alternative approaches, including cultural, biological and genetic tactics, used either alone or in combination, were also needed to provide long-term, effective protection against damaging pests.

Numerous definitions of IPM appear in the literature. Smith *et al.* refer to IPM as "*a process based on ecological principles and integrates multi-disciplinary methodologies in developing agro-ecosystem management strategies that are practical and effective and protect both public health and the environment.*" Pimentel describes IPM as a "*pest control method that includes judicious use of pesticide and non-chemical technologies - all of which are based on sound ecological principles.*"

IPM can be looked upon as consisting of two basic elements; a decision process and an action process. The decision process establishes the basis for any subsequent actions to be undertaken, including no action. The action program may consist of one or more ecologically, economically and socially acceptable tactics designed to reduce pest populations to non-damaging levels.

The decision process

The decision process is often the most time consuming and complex aspect of IPM. It requires careful consideration of the pest, its host, resource management objectives and the ecological, economic and social consequences of the various available tactics.

Population levels of pests are estimated and anticipated resource losses are projected, as are the costs of treatment and its anticipated benefits. If treatment costs exceed losses, a rational decision may be to not treat and accept the losses. Other questions to address include: will natural controls take over within a short enough time so that artificial controls will be unnecessary; or will the effects of proposed treatments be so adverse that they would outweigh the benefits of treatment?

Monitoring of forest pests and diseases and their resultant damage is a critical input to the IPM decision process. Pest monitoring is becoming a sophisticated process that makes use of many technologies. Pheromones and other chemical regulators are often used to monitor insect population levels. Remote sensing technologies such as aerial sketch-mapping, aerial photography and airborne video are used to map and assess forest damage. Geographic information systems (GIS) can be used to relate the location of affected areas to key resource values, terrain features, land ownerships and environmentally sensitive areas. Mathematical models can predict resultant damage caused by certain levels of pest numbers and their consequences. In some cases, pest, growth and yield and economic models are linked to make projections of pest and disease impacts. Data visualisation techniques can display the expected results of alternative action tactics.

The action process

IPM action programmes consist of two overall strategies: prevention or direct control (suppression). Specific pest management tactics exist under each of these strategies.

Prevention

Prevention consists of tactics designed to either reduce the probability of the occurrence of a pest or disease or to create environmental conditions inhospitable for its buildup into damaging numbers. Regulatory, cultural or genetic tactics are examples of prevention strategies.

Regulatory tactics are designed to prevent introductions of exotic pests and diseases and to prevent their spread once established. Examples include inspection of wood products and wooden containers at ports of entry to intercept pest species, conduct of pest risk analyses when new trade agreements are made and establishment of quarantine zones when a pest species is first discovered in a new location. Cultural tactics are designed to create conditions inhospitable for the development of damaging numbers of pests and diseases. These include matching tree species selected for planting to suitable growing sites, controlling stocking through intermediate harvests to maintain tree vigor and timely harvesting of plantations when they reach maturity. A drastic but sometimes necessary cultural approach is to simply eliminate a tree species from a plantation programme because of its high susceptibility to certain pests and diseases. An example is Kenya's policy to discontinue planting of *Cupressus macrocarpa* due to the stem canker fungus, *Monochaetia unicornis*, and *P. radiata* because of its susceptibility to *Sphaeropsis sapinea*.

Genetic tactics make use of varieties of host plants that are either more tolerant to damage or less palatable to the pest. Identification and testing of varieties of *Leucaena leucocephala* and hybrids with other species of *Leucaena* for resistance or tolerance to the psyllid, *Heteropsylla cubana*, was a major line of investigation following this insect's introduction into the Asia-Pacific region. In the southeastern United States, screening and development of pine seedlings resistant to fusiform rust is an integral part of the management of this disease. Tree breeding programmes in Brazil address resistance to the canker *Cryphonectria cubensis* on eucalypts and breeding of *P. radiata* in New Zealand includes a rating for resistance to the needle fungus *Dothistroma pini*.

Suppression

Tactics directed against the pest or disease are referred to as direct control or suppression tactics. Examples include various types of biological, mechanical or chemical methods.

Biological control involves the use of natural enemies of a pest or disease to help keep its numbers in check. Classic biological control, a technique widely used in agriculture, involves the importation of natural enemies to help control an exotic pest. In Colombia, the geometrid *Oxydia trychiata*, an indigenous species attacking pines and cypress, was successfully controlled by the introduction and release of an egg parasitoid *Telenomus alsophilae*. In another successful example of classic biological control, releases of a parasitoid, *Pauesia bicolor*, resulted in the collapse of an aphid, *Cinara cronartii*, a North American species accidentally introduced into pine plantations in South Africa. In Chile, state-of-the-art mass rearing facilities have been established by forest industry for production of the European pine shoot moth parasitoid, *Orgilis obscurator*. One such laboratory, *Controladora de Plagas Forestales*, was established in 1992. In 1996, this facility produced 1.2 million parasitised larvae and 22,600 adult female parasitoids for field release.

A key concern about biological control is the possibility that the introduced natural enemy might also attack innocuous or beneficial insects in the ecosystem. Therefore it is necessary to thoroughly evaluate candidate species prior to release to ensure their relative host specificity. Another concern is the hazard of accidentally introducing hyperparasites, natural enemies of the biological control agents, which might eventually affect the agent's efficacy. For example, colonies of the leaucana psyllid predator *Olla v-nigrum*, released on the Indian Ocean island of Reunion, were subsequently discovered infested by three species of hyperparsites. These could severely affect the ability of this predator to function as an effective biocontrol agent.

Augmentative biological control is a process designed to increase the efficiency of natural enemies already in place. Mass release of *Trichogramma dendrolimi*, an indigenous egg parasitoid of the defoliator *Dendrolimus punctatus* in pine plantations in China is an example of this tactic. Another biological control tactic is the use of biological insecticides such as the bacterial agent, *Bacillus thuringiensis* (Bt), nuclear polyhedrosis viruses or fungal preparations to control insects. Bt is widely used for control of lepidopterous defoliators in both natural and plantation forests. In China and Vietnam, a fungus, *Beauvaria bassiana*, is used for control of the pine caterpillar, *Dendolimus punctatus*.

Mechanical tactics include removal and destruction or rapid removal of infested or infected trees with the objective of destroying the pest. Examples include cutting and burning of trees infested by bark beetles or rapid salvage of infested trees and destruction of infested bark at the sawmill. When the pine woolly aphid, *Pineus borneri*, was first discovered in Kenya, the initial response was to destroy the infested pines. This insect is easily carried on air currents, however, and it was soon realised that the infestation had spread far beyond the designated area where trees were being destroyed.

Use of chemical pesticides applied either from the ground or by low flying aircraft is still considered to be an integral part of IPM. However these materials are used more sparingly, are applied at reduced intervals and with greater precision. In IPM, chemical pesticides are often considered to be the tactic of last resort.

Integration of new technologies

An important part of IPM is that no matter how advanced, sophisticated or effective an IPM system for a specific pest or pest complex may be, there is always room for the introduction of new technologies. These may include more accurate pest monitoring, prediction of new pest management tactics, and more efficacious treatments with fewer undesirable side effects.

Integrated pest management systems

IPM systems consist of a combination of decision-making and pest management tools directed against a pest or pest complex and are in various stages of development. An example of an evolving IPM system is the approach being taken to manage the European wood wasp, *Sirex noctilio*, in pine plantations in southern Brazil. Early detection of this insect is accomplished by baiting suppressed trees in plantations with an herbicide, a procedure that attracts attacking wasps. Infestations are treated either through inoculation of a parasitic nematode, *Daladenus siricidola*, which renders the female wood wasps incapable of producing eggs, or thinning plantations to reduce stocking and maintain tree vigor. Damage assessment technologies such as aerial sketch-mapping and aerial photos are currently being evaluated. Landsat satellite imagery is being tested to determine its capacity to map the location of small, non-industrial private pine plantations that might serve as reservoirs for populations of this damaging insect. Introduction of additional parasitic insects is another potential tactic for managing this insect.

REFERENCES

Doane, C.C. & McManus, M.L. 1981. *The gypsy moth: Research toward integrated pest management*. USDA Forest Service, Science and Education Agency and Animal and Plant Health Inspection Service, Technical Bulletin 1584, 757 pp.

Drooz, A.T. 1985. *Insects of eastern forests*. USDA Forest Service, Miscellaneous Publication 1426, 608 pp.

Furniss, R.L. & Carolin, V.M. 1977. *Western forest insects*. USDA Forest Service, Miscellaneous Publication 1339, 654 pp.

Gilbert, J.M. & Miller, L.W. 1952. An outbreak of *Sirex noctilio* (F.) in Tasmania. *Australian Forestry*, 16: 63-69.

Manion, P.D. 1991. *Tree disease concepts*. Prentice Hall, Engelwood Cliffs, NJ (USA), 402 pp.

Ohmart, C.P. & Edwards, P.B. 1991. Insect herbivory on *Eucalyptus*. *Annual Review of Entomology* 36: 637-657.

Speight, M.R. & Wainhouse, D. 1989. *Ecology and management of forest insects*. Clarendon Press, Oxford, UK, 374 pp.

Wormald, T.J., 1992. *Mixed and pure forest plantations in the tropics and subtropics*. FAO, Rome, Italy, Forestry Paper 103, 152 pp.

6

Biological Control of Plant Pathogens

Plant diseases need to be controlled to maintain the quality and abundance of food, feed, and fiber produced by growers around the world. Different approaches may be used to prevent, mitigate or control plant diseases. Beyond good agronomic and horticultural practices, growers often rely heavily on chemical fertilizers and pesticides. Such inputs to agriculture have contributed significantly to the spectacular improvements in crop productivity and quality over the past 100 years. However, the environmental pollution caused by excessive use and misuse of agrochemicals, as well as fear-mongering by some opponents of pesticides, has led to considerable changes in people's attitudes towards the use of pesticides in agriculture. Today, there are strict regulations on chemical pesticide use, and there is political pressure to remove the most hazardous chemicals from the market. Additionally, the spread of plant diseases in natural ecosystems may preclude successful application of chemicals, because of the scale to which such applications might have to be applied. Consequently, some pest management researchers have focused their efforts on developing alternative inputs to synthetic chemicals for controlling pests and diseases. Among these alternatives are those referred to as biological controls.

The terms "biological control" and its abbreviated synonym "biocontrol" have been used in different fields of biology, most notably entomology and plant pathology. In entomology, it has been used to describe the use of live predatory insects, entomopathogenic nematodes, or microbial pathogens to suppress populations of different pest insects. In plant pathology, the term applies to the use of microbial antagonists to suppress diseases as well as the use of host-specific pathogens to control weed populations. In both fields, the organism that suppresses the pest or pathogen is referred to as the biological control agent (BCA). More broadly, the term biological control also has been applied to the use of the natural products extracted or fermented from various sources. These formulations may be very simple mixtures of natural ingredients with specific activities or complex mixtures with multiple effects on the host as well as the target pest or pathogen. And, while such inputs may mimic the activities of living organisms, non-living inputs should more properly be referred to as biopesticides or biofertilizers, depending on the primary benefit provided to the host plant. The various definitions offered in the scientific literature have sometimes caused confusion and controversy.

For example, members of the U.S. National Research Council took into account modern biotechnological developments and referred to biological control as "the use of natural or modified organisms, genes, or gene products, to reduce the effects of undesirable organisms and to favor desirable organisms such as crops, beneficial insects, and microorganisms", but this definition spurred much subsequent debate and it was frequently considered too broad by many scientists who worked in the field. Because the term biological control can refer to a spectrum of ideas, it is important to stipulate the breadth of the term when it is applied to the review of any particular work.

Published definitions of biocontrol differ depending on the target of suppression; number, type and source of biological agents; and the degree and timing of human intervention. Most broadly, biological control is the suppression of damaging activities of one organism by one or more other organisms, often referred to as natural enemies. With regards to plant diseases, suppression can be accomplished in many ways. If growers' activities are considered relevant, cultural practices such as the use of rotations and planting of disease resistant cultivars (whether naturally selected or genetically engineered) would be included in the definition.

Because the plant host responds to numerous biological factors, both pathogenic and non-pathogenic, induced host resistance might be considered a form of biological control. More narrowly, biological control refers to the purposeful utilization of introduced or resident living organisms, other than disease resistant host plants, to suppress the activities and populations of one or more plant pathogens. This may involve the use of microbial inoculants to suppress a single type or class of plant diseases. Or, this may involve managing soils to promote the combined activities of native soil- and plant-associated organisms that contribute to general suppression. Most narrowly, biological control refers to the suppression of a single pathogen (or pest), by a single antagonist, in a single cropping system. Most specialists in the field would concur with one of the narrower definitions presented above.

INTERACTIONS CONTRIBUTING TO BIOLOGICAL CONTROL

Throughout their lifecycle, plants and pathogens interact with a wide variety of organisms. These interactions can significantly affect plant health in various ways. In order to understand the mechanisms of biological control, it is helpful to appreciate the different ways that organisms interact. Note, too, that in order to interact, organisms must have some form of direct or indirect contact. Odum proposed that the interactions of two populations be defined by the outcomes for each. The types of interactions were referred to as mutualism, protocooperation, commensalism, neutralism, competition, amensalism, parasitism, and predation. While the terminology was developed for macroecology, examples of all of these types of interactions can be found in the natural world at both the macroscopic and microscopic level. And, because the development of plant diseases involves both plants and microbes, the interactions that lead to biological control take place at multiple levels of scale.

From the plant's perspective, biological control can be considered a net positive result arising from a variety of specific and non-specific interactions. Using the spectrum of Odum's concepts, we can begin to classify and functionally delineate the diverse components of ecosystems that contribute to biocontrol. Mutualism is an association between two or more species where both species derive benefit. Sometimes, it is an obligatory lifelong interaction involving close physical and biochemical contact, such as those between plants and mycorrhizal fungi. However, they are generally facultative and opportunistic. For example, bacteria in the genus Rhizobium can reproduce either in the soil or, to a much greater degree, through their mutualistic association with legume plants. These types of mutualism can contribute to biological control, by fortifying the plant with improved nutrition and/or by stimulating host defenses. Protocooperation is a form of mutualism, but the organisms involved do not depend exclusively on each other for survival.

Many of the microbes isolated and classified as BCAs can be considered facultative mutualists involved in protocooperation, because survival rarely depends on any specific host and disease suppression will vary depending on the prevailing environmental conditions. Further down the spectrum, commensalism is a symbiotic interaction between two living organisms, where one organism benefits and the other is neither harmed nor benefited. Most plant-associated microbes are assumed to be commensals with regards to the host plant, because their presence, individually or in total, rarely results in overtly positive or negative consequences to the plant. And, while their presence may present a variety of challenges to an infecting pathogen, an absence of measurable decrease in pathogen infection or disease severity is indicative of commensal interactions. Neutralism describes the biological interactions when the population density of one species has absolutely no effect whatsoever on the other.

Related to biological control, an inability to associate the population dynamics of pathogen with that of another organism would indicate neutralism. In contrast, antagonism between organisms results in a negative outcome for one or both. Competition within and between species results in decreased growth, activity and/or fecundity of the interacting organisms. Biocontrol can occur when non-pathogens compete with pathogens for nutrients in and around the host plant. Direct interactions that benefit one population at the expense of another also affect our understanding of biological control. Parasitism is a symbiosis in which two phylogenetically unrelated organisms coexist over a prolonged period of time. In this type of association, one organism, usually the physically smaller of the two (called the parasite) benefits and the other (called the host) is harmed to some measurable extent. The activities of various hyperparasites, i.e., those agents that parasitize plant pathogens, can result in biocontrol. And, interestingly, host infection and parasitism by relatively avirulent pathogens may lead to biocontrol of more virulent pathogens through the stimulation of host defense systems. Lastly, predation refers to the hunting and killing of one organism by another for consumption and sustenance.

While the term predator typically refer to animals that feed at higher trophic levels in the macroscopic world, it has also been applied to the actions of microbes, e.g. protists, and mesofauna, e.g. fungal feeding nematodes and microarthropods, that consume pathogen biomass for sustenance. Biological control can result in varying degrees from all of these types of interactions, depending on the environmental context within which they occur. Significant biological control, as defined above, most generally arises from manipulating mutualisms between microbes and their plant hosts or from manipulating antagonisms between microbes and pathogens.

MECHANISMS OF BIOLOGICAL CONTROL

Because biological control can result from many different types of interactions between organisms, researchers have focused on characterizing the mechanisms operating in different experimental situations. In all cases, pathogens are antagonized by the presence and activities of other organisms that they encounter. Here, we assert that the different mechanisms of antagonism occur across a spectrum of directionality related to the amount of interspecies contact and specificity of the interactions (Table 1). Direct antagonism results from physical contact and/or a high-degree of selectivity for the pathogen by the mechanism(s) expressed by the BCA(s). In such a scheme, hyperparasitism by obligate parasites of a plant pathogen would be considered the most direct type of antagonism because the activities of no other organism would be required to exert a suppressive effect. In contrast, indirect antagonisms result from activities that do not involve sensing or targeting a pathogen by the BCA(s).

Stimulation of plant host defense pathways by non-pathogenic BCAs is the most indirect form of antagonism. However, in the context of the natural environment, most described mechanisms of pathogen suppression will be modulated by the relative occurrence of other organisms in addition to the pathogen. While many investigations have attempted to establish the importance of specific mechanisms of biocontrol to particular pathosystems, all of the mechanisms described below are likely to be operating to some extent in all natural and managed ecosystems. And, the most effective BCAs studied to date appear to antagonize pathogens using multiple mechanisms. For instance, pseudomonads known to produce the antibiotic 2,4-diacetylphloroglucinol (DAPG) may also induce host defenses). Additionally, DAPG-producers can aggressively colonize roots, a trait that might further contribute to their ability to suppress pathogen activity in the rhizosphere of wheat through competition for organic nutrients.

Table 1. Types of interspecies antagonisms leading to biological control of plant pathogens.

Type	Mechanism
Direct antagonism	Hyperparasitism/predation
Mixed-path antagonism	Antibiotics
	Lytic enzymes
	Unregulated waste products
	Physical/chemical interference
Indirect antagonism	Competition
	Induction of host resistance

Hyperparasites and Predation

In hyperparasitism, the pathogen is directly attacked by a specific BCA that kills it or its propagules. In general, there are four major classes of hyperparasites: obligate bacterial pathogens, hypoviruses, facultative parasites, and predators. Pasteuria penetrans is an obligate bacterial pathogen of root-knot nematodes that has been used as a BCA. Hypoviruses are hyperparasites. A classical example is the virus that infects *Cryphonectria parasitica*, a fungus causing chestnut blight, which causes hypovirulence, a reduction in disease-producing capacity of the pathogen.

The phenomenon has controlled the chestnut blight in many places. However, the interaction of virus, fungus, tree, and environment determines the success or failure of hypovirulence. There are several fungal parasites of plant pathogens, including those that attack sclerotia (e.g. *Coniothyrium minitans*) while others attack living hyphae (e.g. *Pythium oligandrum*). And, a single fungal pathogen can be attacked by multiple hyperparasites. For example, *Acremonium alternatum, Acrodontium crateriforme, Ampelomyces quisqualis, Cladosporium oxysporum*, and *Gliocladium virens* are just a few of the fungi that have the capacity to parasitize powdery mildew pathogens. Other hyperparasites attack plant-pathogenic nematodes during different stages of their life cycles (e.g. *Paecilomyces lilacinus* and *Dactylella oviparasitica*).

In contrast to hyperparasitism, microbial predation is more general and pathogen non-specific and generally provides less predictable levels of disease control. Some BCAs exhibit predatory behavior under nutrient-limited conditions. However,*Trichoderma* produce a range of enzymes that are directed against cell walls of fungi. However, when fresh bark is used in composts, *Trichoderma* spp. do not directly attack the plant pathogen, *Rhizoctonia solani*. But in decomposing bark, the concentration of readily available cellulose decreases and this activates the chitinase genes of *Trichoderma* spp., which in turn produce chitinase to parasitize *R. solani*.

Antibiotic-mediated suppression

Antibiotics are microbial toxins that can, at low concentrations, poison or kill other microorganisms. Most microbes produce and secrete one or more compounds with antibiotic activity. In some instances, antibiotics produced by microorganisms have been shown to be particularly effective at suppressing plant pathogens and the diseases they cause. To be effective, antibiotics must be produced in sufficient quantities near the pathogen to result in a biocontrol effect.

In situ production of antibiotics by several different biocontrol agents has been measured; however, the effective quantities are difficult to estimate because of the small quantities produced relative to the other, less toxic, organic compounds present in the phytosphere. And while methods have been developed to ascertain when and where biocontrol agents may produce antibiotics, detecting expression in the infection court is difficult because of the heterogenous distribution of plant-associated microbes and the potential sites of infection. In a few cases, the relative importance of antibiotic production by biocontrol bacteria has been demonstrated, where one or more genes responsible for biosynthesis of the antibiotics have been manipulated.

For example, mutant strains incapable of producing phenazines or phloroglucinols have been shown to be equally capable of colonizing the rhizosphere but much less capable of suppressing soilborne root diseases than the corresponding wild-type and complemented mutant strains. Several biocontrol strains are known to produce multiple antibiotics which can suppress one or more pathogens. For example, Bacillus cereus strain UW85 is known to produce both zwittermycin and kanosamine. The ability to produce multiple antibiotics probably helps to suppress diverse microbial competitors, some of which are likely to be plant pathogens. The ability to produce multiple classes of antibiotics, that differentially inhibit different pathogens, is likely to enhance biological control. More recently, *Pseudomonas putida* WCS358r strains genetically engineered to produce phenazine and DAPG displayed improved capacities to suppress plant diseases in field-grown wheat.

Lytic Enzymes and other Byproducts of Microbial Life

Diverse microorganisms secrete and excrete other metabolites that can interfere with pathogen growth and/or activities. Many microorganisms produce and release lytic enzymes that can hydrolyze a wide variety of polymeric compounds, including chitin, proteins, cellulose, hemicellulose, and DNA. Expression and secretion of these enzymes by different microbes can sometimes result in the suppression of plant pathogen activities directly. For example, control of *Sclerotium rolfsii* by *Serratia marcescens* appeared to be mediated by chitinase expression. And, a b-1,3-glucanase contributes significantly to biocontrol activities of *Lysobacter enzymogenes* strain C3. While they may stress and/or lyse cell walls of living organisms, these enzymes generally act to decompose plant residues and nonliving organic matter. Currently, it is unclear how much of the lytic enzyme activity that can be detected in the natural environment represents specific responses to microbe-microbe interactions. It seems more likely that such activities are largely indicative of the need to degrade complex polymers in order to obtain carbon nutrition. Nonetheless, microbes that show a preference for colonizing and lysing plant pathogens might be classified as biocontrol agents. *Lysobacter* and *Myxobacteria* are known to produce copious amounts of lytic enzymes, and some isolates have been shown to be effective at suppressing fungal plant pathogens. So, the lines between competition, hyperparasitism, and antibiosis are generally blurred.

Furthermore, some products of lytic enzyme activity may contribute to indirect disease suppression. For example, oligosaccharides derived from fungal cell walls are known to be potent inducers of plant host defenses. Interestingly, *Lysobacter enzymogenes* strain C3 has been shown to induce plant host resistance to disease, though the precise activities leading to this induction are not entirely clear. The quantitative contribution of any and all of the above compounds to disease suppression is likely to be dependent on the composition and carbon to nitrogen ratio of the soil organic matter that serves as a food source for microbial populations in the soil and rhizosphere. However, such activities can be manipulated so as to result in greater disease suppression.

For example, in post-harvest disease control, addition of chitosan can stimulate microbial degradation of pathogens similar to that of an applied hyperparasite. Chitosan is a non-toxic and biodegradable polymer of beta-1,4-glucosamine produced from chitin by alkaline deacylation. Amendment of plant growth substratum with chitosan suppressed the root rot caused by *Fusarium oxysporum* f. sp. *radicis-lycopersici* in tomato. Although the exact mechanism of action of chitosan is not fully understood, it has been observed that treatment with chitosan increased resistance to pathogens.

Other microbial byproducts also may contribute to pathogen suppression. Hydrogen cyanide (HCN) effectively blocks the cytochrome oxidase pathway and is highly toxic to all aerobic microorganisms at picomolar concentrations. The production of HCN by certain fluorescent pseudomonads is believed to be involved in the suppression of root pathogens. *P. fluorescens* CHA0 produces antibiotics, siderophores and HCN, but suppression of black rot of tobacco caused by Thielaviopsis basicola appeared to be due primarily to HCN production.

Howell et al. reported that volatile compounds such as ammonia produced by *Enterobacter cloacae* were involved in the suppression of *Pythium ultimum*-induced damping-off of cotton. While it is clear that biocontrol microbes can release many different compounds into their surrounding environment, the types and amounts produced in natural systems in the presence and absence of plant disease have not been well documented and this remains a frontier for discovery.

Competition

From a microbial perspective, soils and living plant surfaces are frequently nutrient limited environments. To successfully colonize the phytosphere, a microbe must effectively compete for the available nutrients. On plant surfaces, host-supplied nutrients include exudates, leachates, or senesced tissue. Additionally, nutrients can be obtained from waste products of other organisms such as insects (e.g. aphid honeydew on leaf surface) and the soil. While difficult to prove directly, much indirect evidence suggests that competition between pathogens and non-pathogens for nutrient resources is important for limiting disease incidence and severity. In general, soilborne pathogens, such as species of Fusarium and Pythium, that infect through mycelial contact are more susceptible to competition from other soil- and plant-associated microbes than those pathogens that germinate directly on plant surfaces and infect through appressoria and infection pegs.

Genetic work of Anderson et al. revealed that production of a particular plant glycoprotein called agglutinin was correlated with potential of *P. putida* to colonize the root system. *P. putida* mutants deficient in this ability exhibited reduced capacity to colonize the rhizosphere and a corresponding reduction in Fusarium wilt suppression in cucumber. The most abundant nonpathogenic plant-associated microbes are generally thought to protect the plant by rapid colonization and thereby exhausting the limited available substrates so that none are available for pathogens to grow. For example, effective catabolism of nutrients in the spermosphere has been identified as a mechanism contributing to the suppression of *Pythium ultimum* by *Enterobacter cloacae*. At the same time, these microbes produce metabolites that suppress pathogens. These microbes colonize the sites where water and carbon-containing nutrients are most readily available, such as exit points of secondary roots, damaged epidermal cells, and nectaries and utilize the root mucilage.

Biocontrol based on competition for rare but essential micronutrients, such as iron, has also been examined. Iron is extremely limited in the rhizosphere, depending on soil pH. In highly oxidized and aerated soil, iron is present in ferric form, which is insoluble in water (pH 7.4) and the concentration may be as low as 10-18 M. This concentration is too low to support the growth of microorganisms, which generally need concentrations approaching 10-6 M. To survive in such an environment, organisms were found to secrete iron-binding ligands called siderophores having high affinity to sequester iron from the micro-environment.

Almost all microorganisms produce siderophores, of either the catechol type or hydroxamate type. Kloepper et al. were the first to demonstrate the importance of siderophore production as a mechanism of biological control of *Erwinia carotovora* by several plant-growth-promoting *Pseudomonas fluorescens* strains A1, BK1, TL3B1 and B10. And, a direct correlation was established in vitro between siderophore synthesis in fluorescent pseudomonads and their capacity to inhibit germination of chlamydospores of *F. oxysporum*. As with the antibiotics, mutants incapable of producing some siderophores, such as pyoverdine, were reduced in their capacity to suppress different plant pathogens. The increased efficiency in iron uptake of the commensal microorganisms is thought to be a contributing factor to their ability to aggressively colonize plant roots and an aid to the displacement of the deleterious organisms from potential sites of infection.

Induction of Host Resistance

Plants actively respond to a variety of environmental stimuli, including gravity, light, temperature, physical stress, water and nutrient availability. Plants also respond to a variety of chemical stimuli produced by soil- and plant-associated microbes. Such stimuli can either induce or condition plant host defenses through biochemical changes that enhance resistance against subsequent infection by a variety of pathogens. Induction of host defenses can be local and/or systemic in nature, depending on the type, source, and amount of stimuli.

Recently, phytopathologists have begun to characterize the determinants and pathways of induced resistance stimulated by biological control agents and other non-pathogenic microbes. The first of these pathways, termed systemic acquired resistance (SAR), is mediated by salicylic acid (SA), a compound which is frequently produced following pathogen infection and typically leads to the expression of pathogenesis-related (PR) proteins. These PR proteins include a variety of enzymes some of which may act directly to lyse invading cells, reinforce cell wall boundaries to resist infections, or induce localized cell death. A second phenotype, first referred to as induced systemic resistance (ISR), is mediated by jasmonic acid (JA) and/or ethylene, which are produced following applications of some nonpathogenic rhizobacteria.

Interestingly, the SA- and JA- dependent defense pathways can be mutually antagonistic, and some bacterial pathogens take advantage of this to overcome the SAR. For example, pathogenic strains of *Pseudomonassyringae* produce coronatine, which is similar to JA, to overcome the SA-mediated pathway. Because the various host-resistance pathways can be activated to varying degrees by different microbes and insect feeding, it is plausible that multiple stimuli are constantly being received and processed by the plant. Thus, the magnitude and duration of host defense induction will likely vary over time. Only if induction can be controlled, i.e. by overwhelming or synergistically interacting with endogenous signals, will host resistance be increased.

A number of strains of root-colonizing microbes have been identified as potential elicitors of plant host defenses. Some biocontrol strains of Pseudomonas sp. and Trichoderma sp. are known to strongly induce plant host defenses. In several instances, inoculations with plant-growth-promoting rhizobacteria (PGPR) were effective in controlling multiple diseases caused by different pathogens, including anthracnose (Colletotrichum lagenarium), angular leaf spot (Pseudomonas syringae pv. lachrymans and bacterial wilt (Erwinia tracheiphila). A number of chemical elicitors of SAR and ISR may be produced by the PGPR strains upon inoculation, including salicylic acid, siderophore, lipopolysaccharides, and 2,3-butanediol, and other volatile substances. Again, there may be multiple functions to such molecules blurring the lines between direct and indirect antagonisms.

Table 2. Bacterial determinants and types of host resistance induced by biocontrol agents

Bacterial strain	Plant species	Bacterial determinant
Bacillus mycoides strain Bac J	Sugar beet	Peroxidase, chitinase and â-1,3-glucanase
Bacillus subtilis GB03 and IN937a	*Arabidopsis*	2,3-butanediol
Pseudomonas fluorescens strains		
CHA0	Tobacco	Siderophore
	Arabidopsis	Antibiotics (DAPG)
WCS374	Radish	Lipopolysaccharide
		Siderophore
		Iron regulated factor
WCS417	Carnation	Lipopolysaccharide
	Radish	Lipopolysaccharide
		Iron regulated factor
	Arabidopsis	Lipopolysaccharide
	Tomato	Lipopolysaccharide
Pseudomonas putida strains	*Arabidopsis*	Lipopolysaccharide
WCS 358	*Arabidopsis*	Lipopolysaccharide
		Siderophore
BTP1	Bean	Z,3-hexenal
Serratia marcescens 90-166	Cucumber	Siderophore

More generally, a substantial number of microbial products have been identified as elicitors of host defenses, indicating that host defenses are likely stimulated continually over the course of a plant's lifecycle. Excluding the components directly related to pathogenesis, these inducers include lipopolysaccharides and flagellin from Gram-negative bacteria; cold shock proteins of diverse bacteria; transglutaminase, elicitins, and ?-glucans in Oomycetes; invertase in yeast; chitin and ergosterol in all fungi; and xylanase in Trichoderma. These data suggest that plants would detect the composition of their plant-associated microbial communities and respond to changes in the abundance, types, and localization of many different signals. The importance of such interactions is indicated by the fact that further induction of host resistance pathways, by chemical and microbiological inducers, is not always effective at improving plant health or productivity in the field.

Microbial Diversity and Disease Suppression

Plants are surrounded by diverse types of mesofauna and microbial organisms, some of which can contribute to biological control of plant diseases. Microbes that contribute most to disease control are most likely those that could be classified competitive saprophytes, facultative plant symbionts and facultative hyperparasites. These can generally survive on dead plant material, but they are able to colonize and express biocontrol activities while growing on plant tissues.

A few, like avirulent Fusarium oxysporum and binucleate Rhizoctonia-like fungi, are phylogenetically very similar to plant pathogens but lack active virulence determinants for many of the plant hosts from which they can be recovered. Others, like Pythium oligandrum are currently classified as distinct species. However, most are phylogenetically distinct from pathogens and, most often, they are subspecies variants of the same microbial groups. Due to the ease with which they can be cultured, most biocontrol research has focused on a limited number of bacterial (Bacillus, Burkholderia, Lysobacter, Pantoea, Pseudomonas, and Streptomyces) and fungal (Ampelomyces, Coniothyrium, Dactylella, Gliocladium, Paecilomyces, and Trichoderma) genera. Still, other microbes that are more recalcitrant to in vitro culturing have been intensively studied. These include mycorrhizal fungi, e.g. Pisolithus and Glomus spp. that can limit subsequent infections, and some hyperparasites of plant pathogens, e.g. Pasteuria penetrans which attack root-knot nematodes. Because multiple infections can and do take place in field-grown plants, weakly virulent pathogens can contribute to the suppression of more virulent pathogens, via the induction of host defenses. Lastly, there are the many general micro- and meso-fauna predators, such as protists, collembola, mites, nematodes, annelids, and insect larvae whose activities can reduce pathogen biomass, but may also facilitate infection and/or stimulate plant host defenses by virtue of their own herbivorous activities.

While various epiphytes and endophytes may contribute to biological control, the ubiquity of mycorrhizae deserves special consideration. Mycorrhizae are formed as the result of mutualist symbioses between fungi and plants and occur on most plant species. Because they are formed early in the development of the plants, they represent nearly ubiquitous root colonists that assist plants with the uptake of nutrients (especially phosphorus and micronutrients). The vesicular arbuscular mycorrhizal fungi (VAM, also known as arbuscular mycorrhizal or endomycorrhizal fungi) are all members of the zygomycota and the current classification contains one order, the Glomales, encompassing six genera into which 149 species have been classified. Arbuscular mycorrhizae involve aseptate fungi and are named for characteristic structures like arbuscles and vesicles found in the root cortex. Arbuscules start to form by repeated dichotomous branching of fungal hyphae approximately two days after root penetration inside the root cortical cell. Arbuscules are believed to be the site of communication between the host and the fungus. Vesicles are basically hyphal swellings in the root cortex that contain lipids and cytoplasm and act as storage organ of VAM. These structures may present intra- and inter- cellular and can often develop thick walls in older roots. These thick walled structures may function as propagules. During colonization, VAM fungi can prevent root infections by reducing the access sites and stimulating host defense. VAM fungi have been found to reduce the incidence of root-knot nematode. Various mechanisms also allow VAM fungi to increase a plant's stress tolerance. This includes the intricate network of fungal hyphae around the roots which block pathogen infections. Inoculation of apple-tree seedlings with the VAM fungi Glomus fasciculatum and G. macrocarpum suppressed apple replant disease caused by phytotoxic myxomycetes.

VAM fungi protect the host plant against root-infecting pathogenic bacteria. The damage due to Pseudomonas syringae on tomato may be significantly reduced when the plants are well colonized by mycorrhizae. The mechanisms involved in these interactions include physical protection, chemical interactions and indirect effects. The other mechanisms employed by VAM fungi to indirectly suppress plant pathogens include enhanced nutrition to plants; morphological changes in the root by increased lignification; changes in the chemical composition of the plant tissues like antifungal chitinase, isoflavonoids, etc.; alleviation of abiotic stress and changes in the microbial composition in the mycorrhizosphere.

In contrast to VAM fungi, ectomycorrhizae proliferate outside the root surface and form a sheath around the root by the combination of mass of root and hyphae called a mantle. Disease protection by ectomycorrhizal fungi may involve multiple mechanisms including antibiosis, synthesis of fungistatic compounds by plant roots in response to mycorrhizal infection and a physical barrier of the fungal mantle around the plant root. Ectomycorrhizal fungi like Paxillus involutus effectively controlled root rot caused by Fusarium oxysporum and Fusarium moniliforme in red pine. Inoculation of sand pine with Pisolithus tinctorius, another ectomycorrhizal fungus, controlled disease caused by Phytophthora cinnamomi.

Because plant diseases may be suppressed by the activities of one or more plant-associated microbes, researchers have attempted to characterize the organisms involved in biological control. Historically, this has been done primarily through isolation, characterization, and application of individual organisms. By design, this approach focuses on specific forms of disease suppression. Specific suppression results from the activities of one or just a few microbial antagonists. This type of suppression is thought to be occurring when inoculation of a biocontrol agent results in substantial levels of disease suppressiveness. Its occurrence in natural systems may also occur from time to time. For example, the introduction of Pseudomonas fluorescens that produce the antibiotic 2,4-diacetylphloroglucinol can result in the suppression of various soilborne pathogens. However, specific agents must compete with other soil- and root-associated microbes to survive, propagate, and express their antagonistic potential during those times when the targeted pathogens pose an active threat to plant health. In contrast, general suppression is more frequently invoked to explain the reduced incidence or severity of plant diseases because the activities of multiple organisms can contribute to a reduction in disease pressure. High soil organic matter supports a large and diverse mass of microbes resulting in the availability of fewer ecological niches for which a pathogen competes. The extent of general suppression will vary substantially depending on the quantity and quality of organic matter present in a soil. Functional redundancy within different microbial communities allows for rapid depletion of the available soil nutrient pool under a large variety of conditions, before the pathogens can utilize them to proliferate and cause disease. For example, diverse seed-colonizing bacteria can consume nutrients that are released into the soil during germination thereby suppressing pathogen germination and growth. Manipulation of agricultural systems, through additions of composts, green manures and cover crops is aimed at improving endogenous levels of general suppression.

BIOCONTROL RESEARCH, DEVELOPMENT, AND ADOPTION

Biological control really developed as an academic discipline during the 1970s and is now a mature science supported in both the public and private sector. Research related to biological control is published in many different scientific journals, particularly those related to plant pathology and entomology. Additionally, three academic journals are specifically devoted to the discipline. In the United States, research funds for the discipline are provided primarily by several USDA programs. These include the Section 406 programs, regional IPM grants, Integrated Organic Program, IR-4, and several programs funded as part of the National Research Initiative. Monies also exist to stimulate the development of commercial ventures through the small business innovation research (SBIR) programs. Such ventures are intended to be conduits for academic research that can be used to develop new companies.

Much has been learned from the biological control research conducted over the past forty years. But, in addition to learning the lessons of the past, biocontrol researchers need to look forward to define new and different questions, the answers to which will help facilitate new biocontrol technologies and applications. Currently, fundamental advances in computing, molecular biology, analytical chemistry, and statistics have led to new research aimed at characterizing the structure and functions of biocontrol agents, pathogens, and host plants at the molecular, cellular, organismal, and ecological levels.

Over the past fifty years, academic research has led to the development of a small but vital commercial sector that produces a number of biocontrol products. The current status of commercialization of biological control products has been reviewed recently. As in most industries, funding in the private sector research and development goes through cycles, but seems likely to increase in the years ahead as regulatory and price pressures for agrochemical inputs increase. Most of the commercial production of biological control agents is handled by relatively small companies, such as Agraquest, BioWorks, Novozymes, Prophyta, Kemira Agro. Occasionally, such companies are absorbed by or act as subsidiaries of multi-billion dollar agrochemical companies, such as Bayer, Monsanto, Syngenta, and Sumitomo. Total revenues of products used for biocontrol of plant diseases represented just a small fraction of the total pesticide market during the first few years of the 21st century with total sales on the order of \$10 to 20 million dollars annually. However, significant expansion is expected over the next 10 years due to increasing petroleum prices, the expanded demand for organic food, and increased demand for "safer" pesticides in agriculture, forestry, and urban landscapes.

Growers are interested in reducing dependence on chemical inputs, so biological controls (defined in the narrow sense) can be expected to play an important role in Integrated Pest Management (IPM) systems. A model describing the several steps required for a successful IPM has been developed. In this model, good cultural practices, including appropriate site selection, crop rotations, tillage, fertility and water management, provide the foundation for successful pest management by providing a fertile growing environment for the crop. The use of pest- and disease-resistant cultivars, developed through conventional breeding or genetic engineering, provides the next line of defense. However, such measures are not always sufficient to be productive or economically sustainable. In such cases, the next step would be to deploy biorational controls of insect pests and diseases These include BCAs, introduced as inoculants or amendments, as well as active ingredients directly derived from natural origins and having a low impact on the environment and non-target organisms. If these foundational options are not sufficient to ensure plant health and/or economically sustainable production, then less specific and more harmful synthetic chemical toxins can be used to ensure productivity and profitability. With the growing interest in reducing chemical inputs, companies involved in the manufacturing and marketing of BCAs should experience continued growth. However, stringent quality control measures must be adopted so that farmers get quality products. New, more effective and stable formulations also will need to be developed.

Most pathogens will be susceptible to one or more biocontrol strategies, but practical implementation on a commercial scale has been constrained by a number of factors. Cost, convenience, efficacy, and reliability of biological controls are important considerations, but only in relation to the alternative disease control strategies. Cultural practices (e.g. good sanitation, soil preparation, and water management) and host resistance can go a long way towards controlling many diseases, so biocontrol should be applied only when such agronomic practices are insufficient for effective disease control. As long as petroleum is cheap and abundant, the cost and convenience of chemical pesticides will be difficult to surpass. However, if the infection court or target pathogen can be effectively colonized using inoculation, the ability of the living organism to reproduce could greatly reduce application costs. In general, though, regulatory and cultural concerns about the health and safety of specific classes of pesticides are the primary economic drivers promoting the adoption of biological control strategies in urban and rural landscapes. Self-perpetuating biological controls (e.g. hypovirulence of the chestnut blight pathogen) are also needed for control of diseases in forested and rangeland ecosystems where high application rates over larger land areas are not economically-feasible. In terms of efficacy and reliability, the greatest successes in biological control have been achieved in situations where environmental conditions are most controlled or predictable and where biocontrol agents can preemptively colonize the infection court. Monocyclic, soilborne and post-harvest diseases have been controlled effectively by biological control agents that act as bioprotectants (i.e. preventing infections). Specific applications for high value crops targeting specific diseases (e.g. fireblight, downy mildew, and several nematode diseases) have also been adopted. As research unravels the various conditions needed for successful biocontrol of different diseases, the adoption of BCAs in IPM systems is bound to increase in the years ahead.

BIOLOGICAL CONTROL OF PLANT PATHOGENS: RESEARCH, COMMERCIALIZATION, AND APPLICATION

Biological control of plant pests and pathogens continues to inspire research and development in many fields. Plant pathogens are just one class of targets of biological control, which also is designed to limit other pests such as insects, parasitic nematodes, and weeds. In the narrowest sense, biocontrol involves suppressing pest organisms with other organisms. However, the interrelationships of many environmental variables can result in multiple interactions among organisms and their environment, several of which might contribute to effective biological control. Furthermore, natural products and chemical compounds discovered as a result of basic research into the molecular mechanisms of pathogenesis and biological control have led to the development of "biorational" pesticides. Here, we use the term biological control in the broader sense as we describe the current status of research, commercial development, and application of biocontrol strategies targeted at plant pathogens. We conclude by describing future prospects for using biological control to limit the damage of plant pathogens in both conventional and organic agriculture.

Research

Over the past one hundred years, research has repeatedly demonstrated that phylogenetically diverse microorganisms can act as natural antagonists of various plant pathogens. The interactions between microorganisms and plant hosts can be complex. Interactions that lead to biocontrol can include antibiosis, competition, induction of host resistance, and predation. When testing bacterial and fungal isolates from the environment for biocontrol activities, between 1 and 10% show at least some capacity to inhibit the growth of pathogens *in vitro*. However, fewer isolates can suppress plant diseases under diverse growing conditions and fewer still have broad-spectrum activity against multiple pathogenic taxa. Nonetheless, intensive screens have yielded numerous candidate organisms for commercial development. Some of the microbial taxa that have been successfully commercialized and are currently marketed as EPA-registered biopesticides in the United States include bacteria belonging to the genera *Agrobacterium, Bacillus, Pseudomonas,* and *Streptomyces* and fungi belonging to the genera *Ampelomyces, Candida, Coniothyrium,* and *Trichoderma*.

Screening is a critical step in the development of biocontrol agents. The success of all subsequent stages depends on the ability of a screening procedure to identify an appropriate candidate. Many useful bacterial biocontrol agents have been found by observing zones of inhibition in Petri plates. However, this method does not identify biocontrol agents with other modes of action such as parasitism, induced plant resistance, or some forms of competition. Screening methods for parasitism include burying and retrieving propagules of the pathogen to isolate parasites. For competition, methods include looking for microbes that quickly colonize sterilized soil and have the ability to exclude other organisms that attempt to invade the space, and looking for microbes that colonize the infection court. Primary screens for new biocontrol microbes are still undertaken, and it seems likely that continued prospecting will be required to diversify the potential applications of biocontrol as well as replace more widely used biocontrol products should resistance develop.

In addition, other areas of research in the field of plant pathology have opened up new opportunities for disease control. Research on the mechanisms of biocontrol employed by effective bacterial strains has revealed a variety of natural products that can be exploited for the development of chemical control measures. One well-known example is pyrrolnitrin, a natural product produced by some *Pseudomonas* spp. That compound provided the chemical model for development of fludioxonil, a broad-spectrum fungicide used as seed treatment, foliar spray, or soil drench. Research into the mechanism by which plants resist bacterial pathogens led to the discovery of harpin, a protein that is now being used to activate crop defenses prior to pathogen attack. Indeed, a variety of pathogenic and non-pathogenic microorganisms can induce plant defenses and may be useful as biocontrol agents. Other research, on the effects of organic amendments, suggests that both chemical and biological components of compost-amended soils can contribute to disease suppression.Such discoveries point to the substantial potential of diverse programs in basic research to lead to improvements in various biological control strategies.

A variety of research questions remain to be fully answered about the nature of biological control and the means to most effectively manage it under production conditions. Advanced molecular techniques are now being used to characterize the diversity, abundance, and activities of microbes that live in and around plants, including those that significantly impact plant health. Still, much remains to be learned about the microbial ecology of both plant pathogens and their microbial antagonists in different agricultural systems. Fundamental work remains to be done on characterizing the different mechanisms by which organic amendments reduce plant disease. More studies on the practical aspects of mass-production and formulation need to be undertaken to make new biocontrol products stable, effective, safer and more cost-effective.

Commercial Development

Interest in biological control research continues reflecting the desire of multiple constituencies to develop sustainable methods for controlling plant disease. Growing concerns about environmental health and safety have led to substantial regulatory changes in the past several years, including the Food Quality Protection Act of 1996. The FQPA reconciles previous discrepancies between pesticide regulation under the Federal Insecticide, Fungicide, and Rodenticide Act (FIFRA) and the Federal Food Drug, and Cosmetic Act (FFDCA). Such concerns have led to increased restrictions on a variety of chemical pesticides, including some of those used to suppress plant diseases. The IR4 program also encourages the development of biorational pesticides. The CREES Biological Control Working Group, which brings together representatives of state and federal agencies and the private sector, have published a white paper defining a new paradigm for biocontrol and outlining action steps to foster further development of the industry. In addition, the majority of growers continue to express interest in biologically-based pest management strategies of all types as central components of an IPM approach. Such market realities promote the development of biocontrol products. Still, the path to developing and applying effective biocontrol methods is a long one, fraught with a variety of difficulties. Scientific, regulatory, business management, and marketing issues must all be handled effectively for a biocontrol product to be successful in the marketplace.

Currently in the US, a number of biologically-based products are being sold for the control of plant diseases. A growing number of companies are also developing new products that are in the process of being registered. Many of these companies are small, privately owned firms with a limited product-line (e.g., Galltrol produced by AgBiochem). Others are publicly traded and have substantial capitalization values of between 10 and 100 million dollars (e.g., Eden Bioscience). In addition, larger, billion-dollar companies with more diverse product lines that include a variety of agricultural chemicals and biotechnologies have played a significant role in the development and marketing of products for the control of plant pathogens. Because the cost of registering a product with the EPA as a biopesticide can be between $200,000 and $500,000, a minimum capitalization close to 1 million dollars is required for establishing a start-up company. The time to complete registration currently averages under 18 months, but the data generation requirements for different types of products and the level of experience a company has with past registrations can significantly impact the time to registration. To assist companies with the registration process, the EPA sponsors workshops each year. Biocontrol products are either marketed as stand-alone products or formulated as mixtures with other microbials or synthetic fungicides. Some products with biocontrol properties may not be registered, but are sold instead as plant strengtheners or growth promoters without any specific claims regarding disease control. To help improve the global market perception of biopesticides as effective products, the Biopesticide Industry Alliance is establishing a certification process to ensure industry standards for efficacy, quality, and consistency.

More complete information on commercially available biocontrol products can be obtained from a number of sources. The U.S. Environmental Protection Agency's Office of Pesticide Programs manages a website containing substantial amount of information on biopesticides. In addition, listings of products and sources can be freely obtained from sites managed by ATTRA and the APS Biological Control Committee. Several companies also have had the Organic Materials Review Institute (OMRI) certify their products as suitable for organic production. Product labels and material safety data sheets for biopesticides as well as many other pesticides can be freely obtained from CDMS, a private information service. Additional contact information can be obtained from a variety of business information sources for a fee. Efficacy data are generally available from the producers of biopesticides, most of whom who have invested substantial sums of money into the development and testing of their products. Independent assessments of efficacy can be found scattered in a variety of sources, including extension reports and bulletins as well-as scientific journals published by the American Phytopathological Society including *Plant Disease* and *Biological and Cultural Tests* which are available in print and on-line formats.

Application

Successful application of biological controls requires more knowledge-intensive management. To this end, extension personnel and growers must also become more fully aware of the costs and benefits that biopesticides can provide. Understanding when and where biological control of plant pathogens can be profitable requires an appreciation of its place within integrated pest management systems. For this purpose, the model of a pyramid can be used to highlight the way growers might productively construct their pest management programs.In this model, the foundation of a sound pest and disease management program in an annual cropping system begins with sound cultural practices that alter the farm landscape to promote crop health. These include crop rotations that limit the availability of host material used by plant pathogens. Judicious use of tillage can disrupt pest and pathogen life cycles, bury weeds, and prepare seed beds of optimal moisture and bulk density. Careful management of soil fertility and moisture can also limit plant disease by minimizing plant stress. In nurseries and greenhouses, environmental control can be more tightly regulated in terms of temperature, light, moisture, and soil composition (i.e., peat-based potting mixes), but the design of such systems cannot wholly eliminate pest problems.

The second layer of defense against pests consists of the quality of crop germplasm. Breeding for pathogen resistance contributes substantially to crop success in most regions. Newer technologies that directly incorporate genes into crop genomes, commonly referred to as genetic modification or genetic engineering, are bringing new traits into crop germplasm. The most-widely distributed of these plant-incorporated protectants (PIPs) are the different insecticidal proteins derived from *Bacillus thuringiensis*. Other technologies, such as seed washing, testing for pathogens, and treatment are also used to keep germplasm pathogen-free. In perennial cropping systems, such as orchards and forests, germplasm quality may be more important than cultural practices, because rotation and tillage cannot be used as regularly. Upon these two layers, growers can further reduce pest and pathogen pressure by considering both biological and chemical inputs.

Biologically-based inputs such as pheromone traps and microbial pesticides can be used to interfere with pest activities. Registered biopesticides are generally labeled with short reentry intervals (REI) and pre-harvest intervals (PHI), giving greater flexibility to growers who need to balance their operational requirements and pest management goals. When living organisms are introduced, they may also augment natural beneficial populations to further reduce the damage caused by targeted pathogens or pests.

Finally, synthetic chemicals may be added to limit crop damage by pests and pathogens. Because of growing concerns about health and environmental safety, the use of toxic, carcinogenic, and/or environmentally damaging chemicals is currently being discouraged. Within the context of IPM programs, the use of such chemicals should be considered only after other management options have been fully implemented.

Clearly, the different levels described by this model are not entirely independent of one another. Indeed, the numerous biotic and abiotic variables that influence crop health are rarely independent of one another. Such complexities point to the need for more integrative, multidisciplinary research on agricultural systems that will provide a deeper understanding of the conditions under which microbial biocontrol agents might be most productively applied.

The goals of an IPM program are simple: minimize the economic and environmental costs of controlling plant pests and achieve sustainable agricultural production. All disease management strategies have associated costs. Growers need to carefully assess the potential costs and benefits of different disease management strategies. Conventional growers may have more options available to them in terms of chemical alternatives, but economic pressures to limit rotations or reduce tillage may promote more serious disease problems. In contrast, certified organic farmers are prevented from using most pesticides, including those commonly used as seed treatments, but longer and more diverse rotations may limit disease problems. In both cases though, thorough disease assessments are needed to determine the amount of disease pressure present and the causal agents. If the severity of a particular disease problem justifies management action beyond the first two tiers of the IPM model described above, then on-site testing of biocontrol products can generally be recommended.

REFERENCES

Anderson, A. J., Tari, P. H., and Tepper, C. S. 1988. Genetic studies on the role of an agglutinin in root colonization by *Pseudomonas putida*. Appl. Environ. Microbiol. 54:375-380.

Maurhofer, M., Hase, C., Meuwly, P., Métraux, J. P., and Defago, G. 1994. Induction of systemic resistance to tobacco necrosis virus by the root-colonizing *Pseudomonas fluorescens* strain CHA0: influence of the *gac*A gene and of pyoverdine production. Phytopathology 84:139-146.

Notz, R., Maurhofer, M., Schnider-Keel, U., Duffy, B., Haas, D., and Defago, G. 2001. Biotic factors affecting expression of the 2,4-diacetylphloroglucinol biosynthesis gene phlA in *Pseudomonas fluorescens* biocontrol strain CHA0 in the rhizosphere. Phytopathology 91:873-881.

Ordentlich, A., Elad, Y., Chet, I. 1988. The role of chitinase of *Serratia marcescens* in the biocontrol of *Sclerotium rolfsii*. Phytopathology 78:84-88.

7

Genetic Technology and Disease-resistant Plants

Most plant diseases are caused by fungi, bacteria, and viruses. Although the term *disease* is usually used only for the destruction of live plants, the action of dry rot and the rotting of harvested crops in storage or transport is similar to the rots of growing plants; both are caused by bacteria and fungi. Any environmental factor that favors the growth of parasites or disease transmitters or that is unfavorable to the growth of the plants will lead to increases in the likelihood of infection and the amount of destruction caused by parasitic disease. Parasitic diseases are spread by dissemination of the agent itself (bacteria and viruses) or of the reproductive structures (the spores of fungi). Wind, rain, insects, humans, and other animals may provide the means for dissemination.

Most names for plant diseases are descriptive of the physical appearance of the affected plant, e.g., blight (a rapid death of foliage, blossom, or the whole plant); leaf spot, fruit spot and scab, and stem canker (localized death of an organ); wilt (loss of turgor); gall (overgrowth of cells); witches'-broom (growth of abnormal shoots); stunting (underdevelopment); and leaf curl, mosaic, and yellows (resulting from chlorosis, or lack of chlorophyll). Many of these abnormalities are caused by different agents on different plants; when parasites are involved, each individual parasite usually invades only certain plant species and specific organs. Dutch elm disease, chestnut blight, rust, smut, certain mildews, and ergot are caused by various fungi. Clubroot diseases are caused by slime molds, and water molds cause downy mildew (a disease of grapes), blue mold of tobacco, and sudden oak death (also known as ramorum leaf blight or ramorum dieback). Sudden oak death, caused by *Phytophthora ramorum*, was identified in 1995 in California, where it caused the deaths of many oaks. The disease, which affects many plant species besides oaks, has since been found in Oregon, and is also found in Europe; there, it was identified in 1993 in Germany, where it affected rhododendrons and viburnums. The water mold *P. infestans* was the cause of the late blight of potatoes that resulted in the Great Potato Famine in Ireland (1845–49). Both slime molds and water molds are now usually considered protists, rather than fungi. Most mosaic diseases and many other types of chlorosis are caused by viruses.

Plant diseases are more often classified by their symptoms than by the agent of disease, because the discovery of microscopic agents such as bacteria dates only from the 19th cent. The Irish potato blight stimulated the development of plant pathology. The identification of tobacco mosaic virus in 1892 was the starting point of all modern knowledge about viruses.

Plants can suffer from bacterial, viral and fungal attack just as we can ourselves. The organisms themselves (pathogens) are different, but at the microbial level the infection is much the same since one cell is as good a host as another. No matter which part of the plant is attacked the effect is usually to weaken or kill it. By infecting the leaves the plant's ability to produce its food is reduced. Some pathogens block the vessels in the stems which supply the leaves and by attacking the roots, the uptake of water and nutrients is reduced or stopped completely.When a plant is attacked by one of these microorganisms the damage caused provides an opportunity for the others to get in and it is the combined onslaught which deals the final blow. Also if it is under stress, such as through drought or poor nutrition it is more susceptible.

Sometimes thc 'infection' is symbiotic where both organisms derive a benifit. A good example of this is the nitrogen fixing bacteria (Rhizobium) which reside in nodules on the roots of leguminous (pea family) plants, the plant provides food and protection, the bacteria takes nitrogen from the air and converts it to a form usable by the host. Also the Mycorrhizae are a whole Order of fungi which have a symbiotic relationship with plant roots.

In other cases of interest to gardeners, the plant is not benefited but the changes caused produce more attractive features. This is what happens with Aucuba japonica where a viral infection produces the mottled leaves in the 'Variegata' variety. The Tulipomania craze in the 17th century was caused when 'Broken' varieties of tulips began to appear with streaked and mottled petals. This occurred at random and increased the desire and fascination for the bizarre effects. Prized examples were valued at more than "a mansion with servants", although it was not discovered until the early 20th century that viruses were to blame for the unusual colours and effects. In Holland, where the craze caused great hardship when it crashed in 1637, they now grow the pure forms with no streaks or frills.

Fungi are essential in breaking down dead organic matter to produce the humus which is needed for good soil structure - saprophytes. They do not have any chlorophyll so cannot use light to capture energy, instead they derive their energy by breaking down plant and animal material - alive or dead. They can also live in a symbiotic relationship, eg. the micorrhiza in the fine roots of conifers which cannot survive without them to take up vital nutrients. The widespread use of chemical control can damage the balance of these beneficial fungi and this forms part of the principals of Organic management. Lichens are an algae and a fungus growing together as a symbiotic conjunction, ie. the fungus provides physical support for the algae and the algae produces food. There are some less welcome fungi which attack living plants and weaken or kill them - these are the ones which are mentioned in more detail here.

Viruses dwell inside the cells and cannot be treated with chemicals so affected plants must be destroyed (special microculture techniques may overcome the infection by taking cells from the growing tip, but this is restricted to the laboratory). There are no antibiotics for plants, so bacterial attacks, eg. fireblight, are untreatable as well. Fungi can be killed with chemicals without damaging the host because their growth habit is different, ie. they tend to grow on the plant and not in it, using root-like structures to extract nurishment.

Since killing the pathogens is difficult or impossible, "prevention is better than cure". By observing good hygiene when propogating and growing your plants, you can prevent a lot of diseases from taking hold.

- Destroy diseased plants, clear up dead leaves and other debris.
- As the spores can have tough outer coatings, do not add diseased material to the domestic compost heap as they do not usually achieve high enough temperatures to destroy them.
- Prune fruit trees and bushes regularly to keep an open structure allowing a good air flow and to remove damaged branches.
- Disinfect secateurs, saws or knives used for cutting out diseased branches with methylated spirits or a flame (a cigarette lighter comes in handy for this). Household bleach or other disinfectants can also be used as a dip for shears and clippers, including electric or petrol machines. This also helps when taking cuttings.
- Only use new or well-washed containers when growing cuttings and sowing seeds.
- Crop rotation in the vegetable plot will prevent a build up of disease.
- Space plants well apart especially crops where similar plants are growing together, to allow good air flow. Fungal diseases in particular, thrive in still, damp air and there is a greater chance for them to be transmitted to surrounding plants if they are in close proximity.
- Catching disease early is important so keep an eye out for it at all times.
- Plants are more susceptible to disease if they are not growing well. This can be due to poor soil, drought or both. So prepare the site well adding plenty of organic matter and give plants an occasional feed. As gardeners we are usually trying to grow plants which most likely are not native to the region or in a place they would not choose for themselves, so there is a greater probability that they could be under stress.

Of course not all plant diseases can be prevented by good hygiene as some are transmitted by insects and others are wind-borne. Aphids and other sap-sucking insects are the main vectors of viruses. Tobacco mosaic virus can be harboured in cigarettes, so smokers are banned from greenhouses where Surfinia Petunias (susceptible to the virus) are raised from cuttings - all stocks had to be destroyed in 1995 when they became infected. The spores of fungal diseases are carried in the air, and in rain drops and splashes. Be careful when bringing plant material into the garden. Diseases can be in the soil as well as the plants themselves. Some gardeners will only raise their plants from seed, especially vegetables as some onion sets may have White Rot or brassicas may have Clubroot and this would infest the plot for years to come.

INSECT PESTS AND DISEASES

Damaging pests and diseases are found in plantations and natural forests in both the temperate and tropical zones. A fundamental concept of ecosystem dynamics is that as diversity increases, so does stability. The greater the number of plants and animals that occupy an ecosystem, the greater are the checks and balances that prevent any one species from increasing to the point where other ecosystem components are threatened. With pests and diseases, the greatest single deterrent to population increase is the amount of available host material. For example, in a complex tropical rain forest, where as many as a thousand plant species might occupy a single hectare, a population of a host specific, phytophagus caterpillar will find only a limited amount of suitable host material. Therefore its numbers will remain fairly stable. On the other end of the diversity scale is the agro-ecosystem. Dominated by a single plant species, agro-ecosystems often cover large areas and provide a virtually unlimited amount of suitable host biomass. This can result in population explosions of organisms that use the plant as host material. Furthermore, the lack of ecosystem diversity will, at best, provide only a marginal habitat for the natural enemies of the organisms that are using the crop plant as host material.

In terms of diversity and stability, forest plantations tend to be more like agro-ecosystems than natural forests. This would lead one to conclude that plantation forests are more susceptible to damage by pests and disease than are natural forests. While this may be generally true, it does not necessarily mean that natural forests are immune to damage from these agents. The relative lack of pest outbreaks in mixed tropical forests is often cited as evidence for the importance of diversity in stabilising plant communities. Many natural forests in the northern hemisphere boreal and temperate sones however, tend to be relatively simple ecosystems when compared to tropical forests and are susceptible to outbreaks of insects. Examples include:

— the spruce budworm, *Choristoneura fumiferana*, a defoliator of *Abies balsamea* and *Picea* spp in the boreal conifer forests of eastern Canada and adjoining states in the USA.

— western spruce budworm, *Choristoneura occidentalis*, and Douglas-fir tussock moth, *Orgyia pseudotsugata*, in natural conifer forests of western North America,

— the processionary caterpillar, *Thaumetopoea pityocampa,* in Mediterranean Europe and North Africa

— and several species of conifer attacking bark beetles, *Dendroctonus* spp and *Ips* spp in the forests of Canada, the United States, Central America and the Himalayas.

Natural forests are also subject to damage by a variety of diseases caused by fungi, bacteria, virus or parasitic mistletoes and they are vulnerable to introductions of exotic pests and disease. The introduction of white pine blister rust, *Cronartium ribicola,* into North America, the European gypsy moth, *Lymantria dispar,* into the mixed broadleaf forests of the eastern United States, of two *Juniperus* infesting scale insects, *Carulaspis visci* and *Lepidosaphis newsteadi,* into Bermuda and the apparent introduction of the soil fungus *Phytophthora cinnamomi* into *Eucalyptus marginata* forests in Western Australia are just a few examples. Moreover, the recent introduction of pitch canker disease, caused by the fungus *Fusarium subglutinans* f. sp. *pini,* into forests of *Pinus radiata,* along the California, USA coast, now is a source of concern. A more recent example is the discovery in 1994-5 of the pink or hibiscus mealybug, *Maconellicoccus hirsutus,* an insect capable of feeding on between 125 and 150 different plant species, in the Caribbean islands of Grenada, Trinidad and Tobago and St. Kitts-Nevis. This insect poses a severe threat to natural mixed tropical forests as well as vegetable crops and ornamental plants in the region.

DISEASES OF EXOTIC PLANTS

A typical scenario in the use of exotics in forest plantation programmes, both in tropical and temperate zones, is that they enjoy a pest free "honeymoon" for a period of time. During this period, management practices often evolve that make the plantations especially susceptible to pests and disease once they begin to appear. These include:

1. Failure to give proper attention to species/site matching.
2. Use of planting stock from a narrow genetic base.
3. Failure to maintain optimum stocking levels and tree vigor through intermediate cuttings.
4. Dependence on one or two species in a plantation programme, resulting in an unlimited supply of host material for potentially damaging organisms.

When pests and diseases do begin to appear, they may come from two sources. A complex of indigenous agents may adapt to the new host or exotic pests may be accidentally introduced. The greatest hazard of pest and disease damage to exotic forest plantations is from the accidental introduction of exotic pests and diseases. Exotic pests and diseases agents, in the absence of natural enemies, and provided with a large supply of suitable host material, that may have little resistance to the new pest or disease, can build up rapidly and cause devastating losses.

One of the best-documented examples of exotic pests and diseases causing damage to exotic plantations is to pine plantations in the Southern Hemisphere, especially *Pinus radiata*. *P. radiata* were first planted in the late nineteenth century in Australia, Chile, New Zealand and South Africa. The excellent growth and form of the tree in these locations encouraged additional plantings. These provided the basis for thriving lumber and paper industries, especially in Chile and New Zealand. One of the first exotic insects to appear in New Zealand's radiata pine plantations was the European wood wasp, *Sirex noctilio*, during the early 1900's. This insect was probably introduced on unprocessed pine logs imported from Europe. Native to southern Europe and North Africa, where it causes little or no damage, *S. noctilio* normally breeds in recently killed pines. Plantations established in New Zealand during 1920-30 stagnated because there was no market for small logs from thinning operations. This made the plantations susceptible to *S. noctilio* and its associated fungus, *Amylostereum areolatum*. By 1947, high levels of tree mortality were occurring, primarily in the unthinned plantations. However, today *S. noctilio* is not a problem because of biological control methods and better stand management.

S. noctilio subsequently appeared in *P. radiata* plantations in Australia in 1952. In 1980 infestations were detected in South America where this insect is damaging plantations of *P. elliotii* and *P. taeda* in Argentina, southern Brazil and Uruguay. In 1994, infestations of *S. noctilio* were discovered in *P. radiata* plantations in the Cape Peninsula of South Africa. Other damaging insects that have been introduced into *P. radiata* plantations include the bark beetle, *Ips grandicollis*, a North American species, into Australia and the European pine shoot moth, *Rhyacionia bouliana* into Argentina and Chile. The latter is now regarded as Chile's number one forest insect pest. The European bark beetles, *Hyalstes ater* and *Hylurgus ligniperda* were introduced into Chilean radiata pine plantations during the late 1980s and are capable of killing young pine seedlings.

Several introduced fungi have also caused severe damage to *P. radiata* plantations. In 1957, a needle fungus, *Dothistroma septospora* (*D. pini*), became a major concern in *P. radiata* plantations in Chile, parts of East Africa and New Zealand, although there are records of its occurrence in these areas as early as 1940. The stem disease, *Sphaeropsis sapinea* (*Diplodia pini*) has also been introduced in a number of countries. This fungus has been especially damaging in Kenya, where it first appeared in 1973 and proliferated on *P. radiata*, a species adapted to a winter rainfall regime and is especially susceptible to fungal attacks during periods of rain and warm temperatures. Another example of damage to exotic forest plantations by an introduced pest is that of the leucaena psyllid, *Heteropsylla cubana* (Homoptera: Psyllidae) on *Leucaena leucocephala*, a fast growing legume native to Mexico and Central America. This tree was introduced into the Philippines as early as 1565 and was quickly established throughout the tropics where local people found it an excellent fuelwood and later as shade under which crops such as coffee, cocoa, pepper and vanilla could be grown. Genetically improved varieties were eventually introduced into the tropics where it was regarded as having the widest range of uses of all of the tropical legumes. In 1983, leucaena psyllid, a somewhat obscure insect of *L. leucocephala* in its natural range, was detected in southern Florida, USA and during the following year it was found in the Hawaiian Islands. The insect spread rapidly across the Asia and Pacific Region between 1985 and 1988. In 1991, infestations were found on the southern Indian Ocean islands of Reunion and Mauritius and one year later in several eastern African countries, including Burundi, Kenya, Tanzania and Uganda. This insect feeds on the shoots and young foliage of Leucaena causing wilting and growth loss. Ironically, attempts to introduce new genetic material into a country have occasionally resulted in introductions of damaging pests. In 1968, a pine woolly aphid, *Pineus borneri*, was introduced into Kenya on *Pinus caribaea* grafts imported from Australia. Twenty years later, a scale insect, *Oracella acuta*, was introduced into China on *P. elliotii* scion material introduced from the United States. By 1995, over 212,500 ha of *P. elliotii* plantations in Guangdong Province in southern China were infested by this insect.

A number of examples of indigenous insects adapting to exotic forest plantations are known. In Colombia, indigenous insects have been detected in that country's exotic *Cupressus*, *Eucalyptus* and *Pinus* plantations, with the most frequently encountered groups being defoliating caterpillars of the family Geometridae, leaf cutting ants and stem infesting insects. Rodas P. reports 30 species of defoliating insects feeding on exotic forest plantations in the Andean region of Colombia. A similar situation exists in Chile, where several species of indigenous pests are now associated with exotic forest plantations including *Ormiscodes cinnamomea*, a large Saturniid that occasionally defoliates *Pinus radiata*. Ohmart and Edwards report 96 indigenous insects feeding on various species of *Eucalyptus* in China, 94 in India, 223 in Brazil, 31 in New Zealand, 105 in Papua New Guinea and 62 on the island of Sumatra, Indonesia. Termites and leaf cutting ants are the major damaging agents. Indigenous disease causing organisms are also capable of adapting to new hosts and causing severe damage. In Kenya, plantations of *Cupressus macrocarpa* have been severely damaged by *Monochaetia unicornis*, a stem canker causing fungus. During the early 1990s, several species of foliar and twig pathogens, including Cylindocladium leaf blight, Coniella leaf spot, Kirramyces leaf spot and pink disease, *Corticum salmonicolor*, damaged *Eucalyptus* plantings in Vietnam.

DISEASES OF INDIGENOUS PLANTS

Throughout much of Asia, Europe and North America, forest plantations are typically established with indigenous species (e.g. *Pinus massoniana* in China, *Picea abies*, *Pinus sylvestris* and *Populus* spp in Europe, *Pinus taeda* and *Pinus elliotii* in the southeastern USA and *Pseudotsuga menziesii* in western Canada and the USA). These plantations are also susceptible to damage by a complex of both indigenous and exotic pests and disease. In China and Vietnam, the defoliating caterpillar, *Dendrolimus punctatus*, is a major pest of indigenous pine plantations. This insect has two generations per year in east-central China where it defoliates *Pinus massoniana*. In Vietnam, *D. punctatus* has four generations per year and outbreaks have occurred in plantations of *P. merkusii*, often established on steep deforested slopes where they are an important source of lumber, pulpwood, fuel wood and resin.

Nun moth, *Lymantria monacha,* is a defoliating caterpillar of both *Picea abies* and *Pinus sylvestris* in central Europe. This insect was originally a major pest of natural *Picea abies* forests but readily adapted to the extensive pine plantations. In Poland, plantations of *P. sylvestris* are established on low nutrient, sandy soils or abandoned marginal agricultural lands. Between 1978 and 1983, a major outbreak of this caterpillar necessitated aerial spraying of 2.5 million ha, over 25% of Poland's forests. In 1991, this insect again reached outbreak proportions and large areas of forest were treated between 1992 and 1994. Other defoliating insects of *P. sylvestris* plantations in Europe include *Bupalis piniarius, Panolis flammea, Dendrolimus pini* and sawflies, *Diprion* spp.

Fusiform rust, caused by the fungus *Cronartium quercuum* f.sp. *fusiform,* has become a serious pest of *P. elliottii* and *P. taeda* plantations in the southeastern USA. Prior to 1900, the disease was rare, but because of intensive cultural practices associated with plantation forestry, including planting of infected seedlings, fire management, genetic selection of fast growing trees and expansion of the range of susceptible hosts, this disease has become a major problem. Recent estimates indicate that $US130 million in damage is occurring annually and the disease is increasing at the rate of 2-3% per year.

In China, species of *Paulownia,* are widely planted either as ornamentals in villages or in intercropping systems. Pawlonias are commonly infected by a mycoplasma-like organism (MLO), which produces large witches brooms in tree crowns. This disease causes significant reductions in increment, leading to a gradual decline and death of infected trees. The wood of infected trees is of reduced quality. The pine wood nematode, *Bursaphelenchus xylophilus* is the cause of pine wilt disease. This disease has been a major cause of tree mortality in pine plantations in Japan since the early 1900's. The nematode is probably native to North America and has been introduced into Japan and more recently China, where it is causing extensive tree mortality in plantations of *Pinus massoniana*. The nematode is spread from tree to tree by longhorn wood boring beetles, *Monochamus* spp, when the adult beetles feed on pine shoots.

DISEASES IN MIXED PLANTATIONS

The concept of diversity leading to ecosystem stability would suggest that mixed species plantations would be more pest and disease resistant than single species plantation. Indeed, work by Klimetzek in Germany indicates that the inclusion of broadleaf trees in pine forests, either as a mixture or in the understory, tends to reduce the incidence of outbreaks of *Lymantria monacha.* On the other hand, planting *Picea abies* under a plantation of *P. sylvestris*, could intensify or prolong outbreaks of this insect since both trees are hosts.

Wormald indicates that data on relative susceptibility of mixed and pure species plantations in subtropical and tropical regions is, at best, inclusive and confusing. While examples of where pests and diseases have become problems in pure species plantations are cited, many result from accidental introductions or of indigenous pest species adapting to new conditions. Furthermore he cites examples where establishment of mixed species plantations has had little or no effect on reducing pest caused losses.

Some temperate zone pests and diseases have adapted to mixed-species ecosystems by having evolved life stages on alternate hosts. Many rust diseases fall into this category, including fusiform rust of *P. elliotii* and *P. taeda, Cronartium quercuum* f.sp. *fusiform,* which has alternate spore stages on oak, *Quercus* spp. Therefore, removal of oaks from the vicinity of pine nurseries in the southeastern USA is a common although not altogether effective practice. Cooley spruce gall aphid, *Adelges cooleyi,* an insect indigenous to western North America, has alternating life stages on *Picea* and *Pseudotsuga menziesii,* trees that are often found growing in close proximity to one another in natural forests.

Perhaps a greater concern with regard to pests and diseases is not mixed vs. single species plantations, but the number of species used in a country's plantation programme. Reliance on one or two closely related species may meet a country's needs for wood products and foreign exchange. However, should a pest or complex of pests appear in these plantations, the results could be devastating. Yet, many countries have fallen into this dilemma. Examples include Kenya, where 46% of its industrial plantation area is composed of *Cupressus lusitanica*; Colombia with 66% of its plantations *Pinus patula*; Poland with 70% of its plantation forests *P. sylvestris*; and Chile and New Zealand with 90% of their plantations composed of *P. radiata*. New Zealand experience has been that deliberately planting a wide range of species to reduce risk from pests and diseases did not work. Rather, good site matching and silviculture, coupled with emphasis on disease prevention and control, backed by research, has been used as an alternative strategy.

DEVELOPING DISEASE-RESISTANT PLANTS THROUGH GENETIC ENGINEERING

Since the earliest days of agriculture, farmers have taken advantage of genetic differences between plants. Those varieties of wheat or corn that were more resistant to disease, pests or temperature variation, or which yielded more or better products were selected and cultivated. Those varieties that were susceptible to disease or demonstrated lower quality yields, were screened out. These chara-cteristics can be traced to gene expression. By cultivating plants with desirable characteristics, farmers caused certain desirable genes to be more abundant than undesirable ones. Until recently, this process of selecting plants with the optimum genetic characteristics required years of breeding and cross-breeding.

Using biotechnology, scientists can now develop crops with desirable characteristics more quickly and less expensively by identifying the desired gene in another plant and integrating this desired gene into the recipient plant's genome, thus creating a transgenic plant. Sometimes, a gene from another plant of the same species is used, but most transgenic plants incorporate genes from other species. In addition, scientists can introduce genes that endow plants with characteristics that could never be achieved through cross-breeding, like the capability of producing vaccines or other drugs. Presented below is a description of some of the methods commonly employed in producing transgenic plants, followed by an overview of some of the many ways these novel plants are used.

The introduction of genetically engineered plants has given rise to several concerns worldwide. Proponents claim the technology improves agricultural productivity, has the potential to reduce the use of herbicides and pesticides, to increase yields, and is needed to breed the next generation of crops to meet a wide range of new threats to our food security. On the other hand opponents of the technology suggest that genetically engineered plants and products may harm the environment and human health, may cause unintended genetic changes in other plants, and permit vendors of GM seeds to gain an unacceptable degree of control of the food chain.

The plants we use in agriculture are nearly always specially adapted to provide an abundant and reliable supply of nutrition. The major cereal crops we use today, which provide the bulk of our calories, are descended from plants selected in prehistory for their useful traits. Maize was domesticated in central America from a wild relation called teosinte approximately 8,000 years ago.

In the Middle East wild grasses with large seeds were cultivated as annual crops, and bread wheat was domesticated from hybrids of these wild grasses perhaps as recently as 10,000 years ago. The chance hybridisation of wheat species led to a larger grain that was not encased in a hard cover, making it easier to harvest and mill. Generations of selection have led to modern wheat varieties. Since then a wide range of crop plants have been domesticated for food, fibre, and oil production and a multitude of other uses. This was achieved through continuous adaptations and refinements of local varieties and the movement of species around the globe to develop new industries for the production of sugar, cotton and rubber.

When the principles of genetics were established by studying pea seed and flower characteristics, they were incorporated into the new science of plant breeding during the early 20th century. Today nearly all our crop plants are descended from many generations lines that have been carefully selected for desirable genetic changes, and breeders continue to make new genetic combinations to improve performance and to meet new requirements. The discovery of DNA as the genetic material has helped breeders to select desired progeny from crosses of parents that are increasingly diverse. In this way a wider range of characteristics from wild relations of crop plants, such as disease resistance and fruit size, can be incorporated into crops.

CROP PLANT BREEDING

Plant breeding involves selecting optimal combinations of genes that provide desired characteristics. To understand plant breeding and its relationship to genetic engineering, the function of DNA and genes needs to be considered. All organisms contain many genes in their genetic material that provide the instructions for making the organism and for passing these instructions to the next generation. The genetic material of plants and most other organisms is a very long molecule called DNA that carries instructions (the DNA code) in packages called genes. A gene is a unit of information that usually makes one specific protein that performs a specific function in cells.

The function of cells and complex organisms composed of many cells requires many proteins, and hence genes, to function. For example bacteria have between 2,000-8,000 genes, while plants, mice, humans and flies have between 15,000-50,000 genes. Genes are kept in a specific order on long stretches of DNA called chromosomes that are visible in cells. DNA is extraordinarily stable, but over very many generations changes in the DNA code occur and accumulate. These changes are called mutations and they can affect the functions of proteins, such that they might not perform their function properly, they may change the way the protein works, or they may interfere with another protein's functions.

Most organisms accumulate many changes in their DNA with no outward effects, but some small changes in the DNA code can lead to large changes in the appearance of organisms. Genetic variation is the building material of plant breeding, and understanding how this variation is inherited, and what effects it has on plant performance, are complex and interesting problems for plant scientists.

In plant breeding, progeny with new combinations of genes from each parent are screened for desired characteristics over several generations by sifting out combinations of genes giving unwanted characters while preserving sets that give the desired character. This process typically takes up to 8-10 years for a new variety to make it to the field. It may take even longer to breed new lines in the future because the useful genetic variation available in elite breeding lines is tending to become exhausted.

Wild species related to modern crop plants are a very important source of new genetic material for breeding as wild relatives have often evolved useful adaptive traits such as disease resistance or tolerance of extreme environmental conditions. For example wild grasses collected in the Middle East are useful sources of drought and salt tolerance. The loss of habitat for wild species is therefore of great concern to plant breeders, and much effort is spent on collecting and cataloging wild relatives of crop plants.

Plant breeders can also make hybrid plants that have the entire set of genetic information from each parent. For example oilseed rape (*Brassica napus*) is a combination of Chinese cabbage (*Brassica rapa*) and broccoli (*Brassica oleracea*). These plants are called polyploids (meaning several sets of genomes) and this process happens naturally at a low rate to generate new species and increase diversity. These polyploids can then be bred as described above to select desired progeny. Breeders have also used a variety of other methods to create mixtures of genes from un-related plants from which useful new crop varieties can be selected in breeding programmes.

As our knowledge of the DNA code increases, breeders are using this knowledge to try and speed up the breeding process and to include new sources of genetic variation into breeding programmes. For example, crosses between a cultivated tomato and a wild tomato have been used to make new lines with larger fruit, improved processing quality and pest resistance. The difficulty with crosses between more distantly related plants is the larger degree of variation among the offspring of the cross. To sort out the few desired characteristics from the many unwanted ones, landmarks on DNA are used to track the inheritance of different regions of DNA in the progeny. By knowing which landmarks (or markers to use the correct term) are associated with desired and undesirable characteristics, breeders are now able to sort through the complex genetic make-up of progeny of distantly related plants directly to select improved lines. This new technology is called marker assisted breeding, and large projects are now going on to map precisely the DNA code of crop plants and related species, in order to facilitate breeding.

Identifying potentially useful characteristics in wild relatives of crop plants is a key feature of this strategy; therefore much effort is spent on understanding the genetic make up of wild relatives of wheat and barley from the Middle East, wild relatives of tomato, potato and pepper from the Andes, and legumes from the Mediterranean basin. Consequently the loss of plant biodiversity is a major threat to plant breeding and food security. Another type of breeding involves the production of hybrid varieties. F1 hybrids have generally increased vigour and increased yield, and several important crops such as maize and tomato are usually grown from F1 ·hybrid seed. F1 hybrids can show improved performance over that of each of the parents, but this advantage is lost in subsequent generations. Very little is understood about the mechanisms giving rise to hybrid vigour, and it is quite a difficult method to establish. For example establishing F1 hybrid wheat has been a major challenge for breeders.

Genetic engineering uses three features of DNA, two of which have not been described so far. Genes contain two distinct sets of information in their DNA. The first, described above, provides the code to make a specific protein; one gene usually makes one type of protein. The second type of information comprises genetic switches that dictate the amount, location and timing of protein production by the gene. For example the eye lens needs lots of protein called crystallin to be made a specific time in the right place, and a genetic switch achieves this complex task. The information in DNA is extremely stable, as described above, but DNA molecules are far from inert. DNA can be moved about by a variety of natural processes; for example bacteria can exchange certain types of DNA very readily by a mating process called conjugation. About 20 years ago it was realised that some bacteria naturally use this mechanism to transfer DNA into plant cells. This "transferred DNA" becomes stably incorporated into the plant DNA and is henceforth stably inherited as part of the plant DNA.

The technology of DNA cloning could then be used to harness this natural DNA transfer method by incorporating a plant genetic switch into the transferred DNA and hooking the switch up to the coding part of a gene. The combination of a plant genetic switch with the coding information of a gene in a piece of DNA that can be transformed into a plant provided the basis for the so- called "genetic engineering" of plants. The revolutionary feature of this new technology was that the coding region of any gene could be used successfully—from bacteria to human, because the genetic switch functions in plants to make the new protein. Essentially what was originally a bacterial or human gene had now become a functioning plant gene, and many new opportunities for changing the genetic makeup of crop plants became available.

Since the development of this technology over 20 years ago it has been substantially refined so that many types of plants can be genetically engineered. Novel methods of introducing DNA into plants have been invented, such as the biolistic gun that shoots tiny DNA- coated gold spheres directly into plants. Genetic engineering is widely used in the laboratory to identify genes and understand their function, and is also used to make crop plant varieties with novel characteristics.

Plant breeding uses genetic differences between related plants to identify and select combinations of genes that give desired traits, such as increased disease resistance or drought tolerance. The range of genetic variation available limits breeding to characteristics already present or latent in crops and their near relations, but it permits very complex characteristics to be selected for without any prior knowledge of the genes involved apart from their approximate location on chromosomes.

Genetic engineering attempts to achieve the same objectives as plant breeding more directly by identifying a gene of interest that confers a desired trait and transferring it to the crop plant to be improved. Genetic engineering can use any gene in any organism, and as long as this can be transferred and switched on when required, it will be a suitable target for genetic engineering methods. At the moment the main limitations to the genetic engineering approach to plant breeding involve identifying appropriate genes to transfer.

Methods of Producing Transgenic Plants

Using Agrobacterium Tumefaciens

Commonly used to develop transgenic broadleaf plants like tomato and soybean, this method employs a plant bacterium called *Agrobacterium tumefaciens*. *Agrobacterium* contains a plasmid, called the Ti plasmid, part of which gets integrated into a plant cell's

DNA upon infection. This bacterium, which is capable of inserting new DNA into the host plant cell, is a natural genetic engineer. Biotechno-logists alter the Ti plasmid, to contain the gene they wish to be incorporated into the plant cell. Next, the plasmid is reintroduced into the bacterial cell. When this bacterium infects a plant cell, the new gene will be incorporated into the plant cell's genome. The altered plant cell can then grow into a complete plant, which will contain the new DNA in all of its cells.

Biolistics

The Biolistic Particle Delivery System is sometimes referred to as a gene gun. The gene gun propels tiny gold or tungsten particles (1mm in diameter) coated with DNA into the plant tissue. The particles penetrate the rigid cell wall of many of the plant cells in the tissue, and deliver DNA containing the genes of interest inside. Those cells which incorporate the new DNA into their genome are selected and allowed to grow into full transgenic plants.

Electroporation

Normally, incubating DNA in a solution with plant cells will not result in the plant cells taking up the DNA. This is because each plant cell is surrounded by a cell membrane and a cell wall, which are barriers to penetration by the DNA. However, when short, high intensity electrical pulses are applied to plant cell protoplasts (plant cells with their cell wall removed but which still have a cell membrane), small pores form in the cell membrane. These pores are large enough to allow the DNA in the solution to penetrate the cells. After the pulses, the pores close, trapping the new trapping the new DNA inside the cell. A small number of the cells in the solution will not only take up the DNA, but will also incorporate the DNA into their genome. Those cells are selected and allowed to grow into transgenic plants expressing the desired gene.

Microinjection

This is a method also employed in producing transgenic animal cells. The DNA is first incubated with liposomes. Liposomes are small, hollow spheres of fatty molecules that are capable of carrying DNA inside of them. When the liposomes are incubated with DNA under specific conditions, the DNA is absorbed. Liposomes containing plasmid DNA are called lipoplexes. The lipoplexes are subsequently injected into a plant cell, thus introducing the new DNA into that cell. If the new DNA is incorporated into the cell's genome, it is allowed to grow into the desired transgenic plant.

Applications of Transgenic Plants

Transgenic plants can be used for a variety of applications. Some of these are presented below under four headings: product qualities, increasing yields, producing plants that are resistant to pests and diseases, and producing drugs. When genetic engineering was first developed for plants there were two main challenges. The first was to find ways of making gene transfer work effectively in the major crop plants. This has been achieved for a wide range of plants, from pine to wheat. The second was to identify genes that might confer a useful characteristic in a crop. So far only a very limited number of potentially useful genes have made it into commercially successful crop varieties. Among these are genes isolated from bacteria that confer tolerance to herbicides and resistance to harmful pests.

Genetically engineered crops have different attributes depending on the functions of the introduced genes. So far three main crops have been engineered for two characteristics- herbicide tolerance and insect pest resistance. Insect resistant cotton and maize plants, when grown with appropriate refuges to permit pests to propagate normally and avoid excessive selection pressure, do not require so many applications of insecticide and have on average higher yields. Herbicide tolerant crops require fewer treatments with herbicide and tilling is reduced, which helps to maintain soil quality. Biotech papaya has been grown in Hawaii since 1998 in order to combat a virus that devastated the crop. Currently about half of the fresh fruit crop is biotech, while the other half is maintained as non-biotech to supply export markets. Interestingly the biotech papaya plans appear to act as a buffer that stops virus spread to the susceptible crop.

The 2002 farm gate value of GM produce in the USA was $20bn. The evidence of continued yearly increases in the acreages of genetically engineered crops in the USA suggests farmers benefit from these crops; otherwise they would not spend a premium on seeds or submit to product testing. These benefits include increased yields and reduced inputs. These reports suggest that small- scale producers also benefit as well as larger scale producers. Seed merchants selling these successful crops have seen yearly increases in the proportion of their lines sold with world-wide seed sales worth $4bn. In contrast sales of insecticides used in cotton production have been predicted to decrease from $300m in 2002 to less that $70m in 2012. Possible future gains for other groups of farmers can be anticipated; for example pests and disease accounts for the annual loss of approximately 27 m tons of rice, therefore rice producers will benefit from the introduction of disease resistant rice varieties. About 30% of the 81 million hectares (200 million acres) of global transgenic crops were grown in developing countries in 2004.

Product Qualities

Many transgenic plants have been developed to improve the quality of their products. For example, genes have been inserted into tomato plants which delay tomato ripening and softening. As a result, these tomatoes have a longer shelf life, and spoilage is reduced. Other plants have been engineered to produce more healthy foods. Canola and soybean plants, for example, have been genetically altered to contain more healthy concentrations and types of fats. In addition, the amino acid content of corn and soybeans have been improved and some potatoes have been designed to have an increased starch content. Also, tree genomes are currently being studied to determine the roles of the thousands of genes that are responsible for varying tree characteristics. It is anticipated that this knowledge will allow biotechnologists to develop genetically modified trees that grow faster or which could have a superior wood quality. Royal Dutch/Shell has genetically engineered eucalyptus trees to contain modified lignin, the molecular glue that holds wood fibres together. Removing lignin is one of the most challenging parts of pulp and paper production process. These trees, now growing in test sites in Uruguay and Chile, will contain lignin that is easier to remove.

Transgenic plants can also be used to produce plastics or other materials traditionally synthesized artificially. For example, in 1997, scientists at Agracetus in Wisconsin reported that they had produced cotton plants that were genetically altered to fill the hollow middle of their cotton fibres with a small amount of plastic polyester called polyhydroxybutyrate (PHB). These novel plants had been developed using Biolistics. So far, the yield of PHB is not very high, but Agracetus claims that cotton fabric woven from these cottonpolyester hybrids retains 8.6 percent more heat. The scientists also plan to insert other, more complex forms of PHB, polymers from other plants, and keratin (the protein that makes up human hair) which could help the novel cotton retain heat, and keep colour.

Increasing Yields

Plants that are capable of withstanding climate changes are likely to produce greater yields during inclement weather conditions. By modifying heat-shock factors (heat-shock factors allow misfolded proteins to resume their proper shape when these proteins are damaged by high temperatures during hot summers), geneticists have recently managed to create heat tolerant thale cress plants. The genetically modified plants were found to be more robust, generally tolerating salty soils, hot and cold temperatures and drought. Because crops have similar heat-shock factors, similar alterations could eventually lead to novel crops capable of producing greater yields during unusually hot, dry or cold seasons.

In Sainte Foy, Québec, scientists at Agriculture and Agri-Food Canada have identified two genes in alfalfa that are directly linked to cold tolerance. As a result, new cold tolerant varieties of alfalfa can be developed more quickly and cheaply. The loss of alfalfa from exposure to cold weather is estimated at $10 million a year in Québec alone, this application of biotechnology may help to prevent it.

Another way of increasing crop yields is to trick them into thinking that autumn has not arrived. In autumn, plants usually pull useful proteins out of their leaves for storage in seeds or in their trunk or stem (this causes the leaves change colour) before allowing the leaves to drop off. This process is initiated when plant genes producing specific enzymes are turned on.

To prevent these enzymes from shutting down the plant's production capability, scientists at the University of Wisconsin-Madison added a gene that gets turned on at the same time as the shut down genes. This new gene produces an enzyme that makes a hormone called cytokinin. Cytokinin encourages leaves to stay green. The presence of cytokinin nullifies the "shut down" message, and keeps the leaves green well into autumn, extending the plant's growing season. It was also found that these modified plants stayed greener and fresher for longer after having been cut. This technique could eventually be applied to crops, where the extra weeks of photosynthesis could boost yields of grain and flowers, and could extend the shelf life of leaf vegetables like lettuce and cabbage.

Disease Resistance

A major challenge for farmers has always been in protecting their crops from pathogens like insects, viruses and fungi. Pests are often kept at bay by spraying chemical pesticides on crops, some of which have been shown to have adverse affects on the environment. Biotechnology provides new, and environmentally friendly ways of dealing with these pests. In some instances, crops can be developed that are *naturally resistant* to these pests. Some tomato plants, corn and others have been protected from insect pests by inserting a gene from a bacterium called Bacillus thuringiensis (BT). The gene causes a protein to be produced that is lethal to some leaf-eating insects like the tomato hornworm. When the insect eats the plant, it ingests the protein, and enzymes in its stomach convert the protein into a lethal toxin that causes paralysis and death. Since the necessary enzymes are not found in any other insects or animals, the protein is harmless to them. The use of BT is environmentally friendly, and avoids the use of chemical pesticides.

A challenge in the use of BT is to avoid generating excessive numbers of insect pests that are BT resistant. While the presence of a few insects that are naturally resistant to BT is always inevitable, the concern is that widespread use of BT will kill off all of those insects that are not naturally resistant, leaving only the resistant ones to multiply. After a few generations, nearly all of the insect pests targeted by BT would be BT resistant. Many other attempts are being made to foster disease or pathogen resistance in crops. For example, a new gene that codes for a small, harmless protein from the Tobacco mosaic virus (TMV) has been incorporated in broadleaf plants, like tomato plants. As a result the plant cells express the protein, and this makes them "immune" to infection by the complete TMV virus, which would cause the leaves to wither and die. The immunity occurs because the TMV virus does not infect a cell that is already TMV-infected, and the presence of the single harmless protein in all of the plant cells tricks the TMV viruses into behaving as if the cells of the plant were already infected. Also, scientists at the

Canadian Forest Service's Laurentian Forestry Centre (Sainte Foy, Québec) are trying to introduce insect genes into trees so that they are resistant to some harmful fungi.

Parasites and pathogens of plants are a significant growing threat to crop production worldwide. The of producing crops with increased and durable resistance to a spectrum of diseases is therefore a major focus plant research. In nature, plants are continually challenged by fungi, bacteria, viruses and nematodes, comparatively few of these are successful in gaining entry into a prospective host. That is, disease is rare in nature because plants carry different 'layers' of defence – from structural barriers and pre-formed antimicrobials, adaptive defense mechanisms that encompass non-host, race-specific and race non-specific resistance. However, with cultivation of huge areas of genetically identical crops the situation can be quite different. Here, protection relies on a small number of in-bred disease resistance genes per crop species and on the wide-spread application of pesticides. Unfortunately, control can transient because pathogens can overcome disease resistance genes and/or become resistant to pesticides.

Genetic engineering has the potential to solve these problems inserting carefully selected and possibly multiple genes transgenes and the search is therefore on for genes confer durable broad-spectrum resistance but that also safe for all other organisms. Despite all efforts, however, the development of crops that are resistant fungal and bacterial diseases by the introduction transgenes has generally been unsuccessful. Often failures were not because of the nature of the transgene itself but rather the way in which it was expressed. In many cases, constitutively overexpressed transgenes adversely affected plant size and/or seed production. The simple answer to this problem is to express the transgenes only when and where they are needed – at infection sites. This will limit the cost of resistance by restricting induced defence responses to the infection site but it requires pathogen-inducible promoters and few have been successfully used. Fortunately, advances in promoter technology look set to increase the possible ways in which the regulated expression of transgenes can be achieved in plants.

Increasing knowledge of plant defence has led to moresophisticated transgenic approaches to enhancing resistance. The number of candidate genes put forward by transcriptomics, proteomics and protein interaction studies gives us a large choice of genes to be used. Potentially these genes can be manipulated by overexpression, induced expression, tissue-specific expression, stable gene knockouts or silencing by RNAi and they can come from the plant itself, from other plants, from a pathogen or they could even be completely synthetic.

Plant Surveillance Systems

In addition to the defence offered by structural barriers and pre-formed antimicrobial compounds that are already in place to ward off attack, a plant constantly monitors for pathogen challenge. Important components of this surveillance system are resistance genes (R *genes*). R genes directly or indirectly recognise the pathogen and this triggers a diverse array of defence mechanisms. The degree to which a plant recognises a pathogen determines its level of resistance although the pathogen can also influence the outcome by avoiding or actively suppressing the host defences.

There are four major classes of R genes, the NB-LRR genes, Ser/Thr kinases such as Pto, receptor-like kinases (RLKs) and receptor-like proteins (RLPs). These R genes recognise pathogen avirulence (Avr) determinants and bring about resistance in the classic gene-for-gene manner. Briefly, when corresponding R and Avr genes are present, the result is disease resistance. If either is inactive or absent the result is disease.

The activation of plant defence leads to immediate responses local to the site of invasion that include the production of reactive oxygen species (ROS), nitric oxide (NO) and in some cases a hypersensitive response (HR) and accumulation of phenolic compounds and cellwall reinforcements. There are also local tissue responses, such as the synthesis of pathogenesis-related (PR) proteins, accumulation of the phytohormones salicylic acid (SA), ethylene (ET) and jasmonic acid (JA) and cell wall strengthening. In addition, there are also systemic responses that prime uninfected parts of the plant against potential pathogen attack (systemic acquired resistance, SAR).

A classic tactic for producing plants with increased disease resistance involves the manipulation of R genes and it is a strategy common to both transgenic approaches and classical breeding programs. The idea is to introduce an R gene and thereby confer on the plant the ability to recognise the pathogen and mount an effective defence. There have been some notable reports of success. The Bs2 gene confers durable resistance to bacterial spot disease in pepper. This disease is economically important in tomato and transformation of tomato plants with the pepper Bs2 gene led to resistance to bacterial spot disease.

There are, however, some potential problems with this approach. First, to engineer durable resistance more than one R gene might need to be introduced because resistance could be lost by a single loss-of-function mutation in the corresponding pathogen Avr gene. Oneway to overcome this problem is pyramiding, in which multiple R genes, each recognising a unique range of isolates of a pathogen, can be incorporated into a single cultivar. A good example is the production of rice resistant to the bacterium Xanthomonas oryzae pv. oryzae by the introductionof four differentRgenes. Pyramiding requires several R genes with demonstrated specificities for the disease-causing pathogen. Herein lies a problem.

There are many cloned R genes or resistance gene candidates (RGCs) but the specificities of these genes must be established and activity against the given pathogensmust be demonstrated. This can be amajor barrier to the rapid engineering of durable resistance. In addition,many resistance genes have shown restricted host range, perhaps because of mechanistic differences in recognition and response elements in heterologous plants. Another problem with the transgenic approach to pyramiding R genes is that there are reports that ectopic expression of R genes can sometimes activate defence pathways in the absence of pathogen, something that is likely to reduce crop yields. The cause of this might be the level of expression of the R gene and a possible solution might be the use of weaker promoters. A final problem is common to both transgenic and classical breeding approaches. Notably, the presence of an gene can come at a large fitness cost as demonstrated by reduced fitness of Arabidopsis plants containing the R gene RPM1 in the absence of pathogens compared with plants lacking the gene.

Plants also have broader perception systems that include receptors that recognise pathogen-associated molecular patterns (PAMPs). A good example is the flagellin receptor FLS2 (an LRR-type receptor kinase RLK) that recognises a conserved 22 amino acid portion of bacterial flagellin. Treatment of plants with flagellin induces the expression of numerous defencerelated genes and triggers resistance to pathogenic bacteria in wild-type plants.

Two strategies using RLKs show promise. First, flagellin treatment upregulates the expression of FLS2 and numerous other RLKs. This upregulation suggests that transgenic approaches that upregulate RLK expression might lead to increased resistance. Second, RLK genes can be introduced into species or ecotypes that are deficient in them to confer the ability to recognise a pathogen and impart resistance. For example, the Arabidopsis ecotype Ws-0 is flagellin insensitive, owing to a stop codon in the kinase domain of the FLS2 gene, and Ws-0 plants exhibit faster and more severe disease symptoms after spraying with Pseudomonas syringae. When Ws-0 plants were transformedwith a functional FLS2 gene, under the control of its native promoter, the plants acquired responsiveness to flagellin and became less susceptible to Pseudomonas syringae. This demonstrates that pathogen perception and increased resistance could be engineered by the introduction of a RLK transgene, although, as with R genes, they might also show restricted host range.

Avr Genes and Elicitor Molecules

A promising approach to engineering disease resistance is to express a pathogen component in the plant that the plant can recognise. Recognition of these elicitor molecules then leads to the activation of a full defence response that is sufficient to inhibit the pathogen. Many pathogen components can be recognised by plants, including Avr genes, structural components from bacteria such as flagellin, toxins, Hrp proteins delivered by the bacterial Type III secretion system, cell-wall components such as chitin and melanin, ensymes that degrade plant polymers such as pectate lyase and cutinase, enzymes that function in overcoming plant defences such as tomatinase and various secreted peptides such as Pep13, Avr-Pita and elicitins.

All of these pathogen components are candidates for use as transgenes in the strategy first described by de Wit. This consists of making transgenic crop plants that carry a gene encoding a highly active protein elicitor under the control of a promoter that is specifically inducible by a virulent pathogen. Because the production of these elicitor molecules will result in an activation of the plants defences and even possibly cell death, it is important that the expression is strictly limited to infection sites and therefore the difficulty with this approach is finding a suitable pathogen-inducible promoter. Nevertheless, there are at least three reports of success using this approach. Keller et al. expressed the elicitor cryptogein as a transgene in tobacco under the control of the pathogen-inducible hsr203J promoter. Under non-induced conditions, the transgene was silent, whereas after infection by the virulent oomycete Phytophthora parasitica var. nicotianae localized necrosis similar to a hypersensitive response was seen. Some plant lines displayed broad-spectrum disease resistance. However, this promising strategy met with limited success when the promoter was linked to the popA elicitor gene from the bacterium Ralstonia solanacearum. Increased resistance was achieved but at the price of adversely affecting plant health.

Interfering with Pathogenicity

There are many pathogen components that could potentially be targeted and exploited in the rational design of disease-resistant transgenic plants. They include structural targets toxins, effectors, secreted proteins andpeptides or even ways of suppressing pathogen suppressors of host defence. Recent findings reveal that bacteria can undermine plant defences by diverse means – for example by compromising HR-based programmed cell death (PCD) or cell wall defences, by interfering with JA signal transduction or by perturbing defence gene expression. Much work has been done with Pseudomonas syringae, in which HR is invoked by the introduction of effector proteins into the plant cell via Type III secretion, as in other bacteria. These effectors are essential for pathogenicity and can also suppress HR-based defence. Because they are essential for pathogenicity, they could be targeted by strategies aimed against the pathogen.

Fungi can also suppress plant defence responses by targeting HR-based PCD and cell-wall fortification or by targeting pre-formed antimicrobial compounds. For example, infection by the biotrophic powdery mildew fungus Blumeria graminis can lead to localised areas of green living tissue surrounded by senescent leaf tissue whereas in the tomato leaf spot fungus Septoria lycopersici defence suppression operates rather differently – here the fungus produces a tomatinase enzyme that degrades the plant's preformed antimicrobial saponins. The saponin degradation products, in turn, suppress HR-based defence. Research in this area is in its infancy but some exciting emerging technologies target components of the pathogen. For example, Peschen et al. used a Fusarium-specific antibody linked to antifungal peptides. Transgenic Arabidopsis plants expressing these fusion proteins exhibited high levels of protection against Fusarium oxysporum f.sp. matthiolae. Antibodies could be used to target a range of pathogenicity factors and thereby inhibit the infection process.

RNAi Defence Systems

RNAi is a useful tool in inhibiting the expression of pathogen genes at both the transcriptional and posttranscriptional levels in plants. Indeed, many virusresistant plants have been produced using these methods. The most notable success has been in the development and commercialisation of transgenic coat protein-protected papaya in Hawaii. The 1994 papaya ringspot virus (PRSV) crisis in Hawaii led to the production and subsequent commercialisation of transgenic papaya that express the PRSV coat protein and thereby eliminate expression of this essential protein upon infection.

Master Switch Genes

Overexpression of a single defence-related protein might not be the best way to increase resistance. More enlightened strategies make use of our increasing knowledge of pathogen-induced signalling pathways in plants. One idea is to manipulate the expression of 'masterswitch' genes, such as kinases and transcription factors, which regulate banks of target genes that could boost signalling through large portions of the pathogeninduced signalling network and thereby lead to an increase in disease resistance. The disadvantage with this approach is that manipulation of some master switch genes could be detrimental to plant development.

Data from transcriptome and quantitative trait loci (QTL) analysis suggest that transcription factors are promising candidate genes for engineering increased disease resistance. They might act as master-switches by controlling the expression of several genes in a single pathway, thus producing large changes in a single trait, such as disease resistance, with few side effects on other traits.

A good example is WRKY transcription factors. Since the first demonstration that WRKYs are involved in plant defence much evidence has emerged to show that they play many crucial roles and encouragingly, WRKY transcription factors have also been shown to be important in quantitative resistance to pathogens such as Phytophthora infestans. One problem with using transcription factors to improve crops is the identification of the best candidate gene(s) for manipulation because many consist of large multigene families.

Attempts to assign a function to each gene are hindered by functional redundancy with knockouts of single genes often having no observable phenotypes. Despite this, good candidate WRKY genes have been identified, including the Arabidopsis genes WRKY18, WRKY29 and WRKY70. When these genes were overexpressed the resultant plants showed enhanced resistance to P. syringae and, in the case of WRKY70, also to *Erwinia carotovora subsp carotovora*.

Several transcription factor families that have roles in plant defence could yield useful master switch genes. Overexpression of genes such as ERF1, Pti4 and MYB30 shows promise for increasing resistance to pathogens and exciting new candidates including the Whirly factor Why1, the CGCG box-binding proteins SR1–6 and the tobacco DNA-binding protein DBP1 can also be tested. Another source of potential master-switch genes are protein kinases. MAP kinase (MAPK) signalling cascades are integral parts of many defence-signalling pathways, such as the response to flagellin, Pep-13 and N-mediated resistance to tobacco mosaic virus (TMV). Among their targets are WRKY, MYB and TGA transcription factors and NPR1.

Overexpression of the tobacco MAPK, SIPK, illustrates their potential as it led to activation of defence responses and HR-like cell death. Additionally, transient overexpression of MKK4a, MKK5a or constitutively active MEKK1 resulted in enhanced resistance to virulent P. syringae and Botrytis cinerea. Other protein kinases could also be employed and some of the best candidates could be calciumdependent protein kinases (CDPKs) because they act as calcium sensor proteins that link changes in cytosolic Ca2C to defence responses.

In addition to kinases and transcription factors, other signalling molecules such as NPR1, NDR1, EDS1, PAD4, SGT1, COI1 and JAR1 that might represent important nodes in the signalling networks are candidates for this approach. For example, NPR1 is an important master switch gene because it constitutes a node that links SAR, ISR, R gene-mediated resistance, SA, JA and ethylene. It can activate defence gene expression through interaction with members of the TGA family of bZIP transcription factors in the nucleus. In the case of SAR, induction of SAR leads to morereducing conditions in the cell and as a result, NPR1 molecules present as an inactive oligomeric complex in the cytoplasm are converted into active monomers that become nuclear localised and trigger gene expression via interaction with TGA factors.

In Arabidopsis, overexpression of NPR1 led to enhanced resistance to diverse pathogens and, crucially, this was achieved without a substantial yield penalty. The reason for this appeared to be that the NPR1-overexpressing plants did not constitutively turn on their defences but rather appeared to be primed to respond to pathogen attack. Recently, however, similar rice plants have shown a lesion mimic or cell death phenotype. This would reduce yields. It seems that care must be taken with the level and location of expression of NPR1.

As an increasing number of important signalling components are discovered, so the list of candidate genes for manipulation grows. One exciting new discovery is SAbinding protein 2 (SABP2) that specifically binds SA and displays lipase activity. SABP2 might be a receptor for SA because lipase activity is stimulated by SA binding and this could generate a lipid-derived signal that is important in defence signalling. Several other new discoveries such as DIR1 and SFD1 implicate lipid-derived signals in SAR. Negative regulators of defence also represent good candidates for manipulation. Mutations or knockouts of these might impart resistance even though the loss of activity is felt in all cells of the plant. For example the mlo mutation of barley has conferred durable resistance to all B. graminis isolates for decades. The list of candidates for manipulation also includes proteins such as EDS1, PAD4, SGT1, NDR1, ETR1, RIN4 and SNI1.

Antimicrobial Compounds

Challenge of plants with pathogens causes the coordinated induction of antifungal proteins, phytoalexins (low molecular weight antimicrobial compounds) and enzymes involved in plant cell reinforcement or in the breakdown of pathogen infection structures. A longstanding strategy for engineering durable resistance has therefore been to express proteins with antimicrobial activity in plants. However, these proteins are often effective against only a few pathogens and might not provide broad spectrum resistance. One possible way to broaden the spectrum of resistance is to use stacked antimicrobial peptides. These small lytic peptides interact directly with microbial membranes and their small size facilitates the stacking of multiple activities on single transgenes, thus improving the chances of achieving durable broad-spectrum resistance.

Other candidates that could potentially be used to engineer for increased disease resistance include pathogenesis related (PR) genes that might increase the level of pre-formed barriers against pathogen invasion and genes that are involved in the biosynthesis of hormones such as SA, JA and ET. Of the eleven classes of PR protein, most have now been assigned probable functions. They target the pathogen cell wall, the membrane, pathogen RNA, undefined pathogen proteins (PR-6) or display peroxidase activity (PR-9). There have been numerous reports of transgenic plants with increased disease resistance as a result of the overexpression of PR genes.

Several mutants that activate or suppress defence were found in genes that have a role in the biosynthesis of hormones with known roles in defence signalling. These include SA (eds5/sid1, eds16/sid2), JA (JMT) and ET (eto1). Manipulation of these genes therefore represents another strategy to increase resistance.

Pre-formed barriers and failure to breach the cell wall are a major part of non-host resistance. These barriers include wax composition, the plant cell wall, antimicrobial enzymes and secondary metabolites, and they can be chemical, enzymatic or structural. Recently, it has been reported that syntaxins such as PEN1 and ROR2 have important roles in non-host resistance by forming a binary SNAP receptor (SNARE) with a SNAP-25 homologue. The requirement for SNARE proteins implies a role for membrane fusion and vesicle trafficking at the plant cell wall in resistance.

One of the first strategies for producing transgenic plants with increased disease resistance involved the artificial generation of an 'HR-like' local cell death by the production of a toxic gene product. The success of this strategy relies on the 'HR' being totally restricted to infection sites otherwise uncontrolled cell death will occur in uninfected tissues of the plant. Unfortunately, most promoters show some background expression in uninfected tissues and this does not augur well for this strategy. Moreover, the use of a 'toxic' gene product might also prove disadvantageous should any transgenic product come to market.

Ethical Issues

In comparison to the obvious importance attached to ethical issues surrounding human genetic research, the ethical issues surrounding plant genetics have not yet been systematically explored. Current concerns focus on the growth of genetically engineered plants as an intrusion on nature that may have unknown and irreversible consequences, and on the amassing of unprecedented power by corporations through ownership and control of the supply of food from seed to plate. These and related issues have generated substantial debate in some countries, including those where the crops are not grown. Crop plant genetic engineering has the potential to introduce genes from animals and even humans into food, therefore important ethical concerns arise. For example, would it be appropriate to eat food containing a human gene? Would vegetarians consider crops containing animal genes to be non- vegetable? How do these concerns relate to the fact that plants and animals have many genes in common? Genetically engineered crops so far only contain genes from bacteria, plant viruses and other plants so these particular issues remain hypothetical for the moment. Increased experience of genetic modification including new products and more comprehensive regulatory procedures may contribute to focusing arguments and resolving ethical and social concerns over our food supply.

Environmental Issues

Much concern has been raised over the potential of genetically engineered crop plants to harm the environment. Particular anxiety surrounds the possibility of exchanging genetically engineered characteristics between crop plants and wild plants. Pollen movement can transfer the genetic material of flowering plants over substantial distances to sexually compatible species. This problem is well known to plant breeders, who need to ensure the proper genetic composition of seeds for crop plants. To do this they produce seeds from plants that are an appropriate distance apart that results in minimally acceptable cross- pollination, usually about 1%. In the field pollen exchange between sexually compatible crop and weed species also has the potential to intermingle traits from crops and weeds. Pollen exchange is generally not an important issue in crops harvested from single plantings of genetically pure seedlings. In farm- kept seeds any traits derived from infrequent crosses with wild relatives generally only have a minor effect in the population of crop seeds, but over time favourable characteristics of the crop may start to be lost. Similarly weedy relatives of crop plants may inherit characteristics from crop plants, and in principle this has the potential to enhance weediness, although this has not been observed. Because genetically engineered genes are stably inherited they can also be spread by pollen from crops to weedy species, and this has the potential to confer traits such as herbicide and virus tolerance on weeds. It is conceivable that these traits may confer a selective advantage on populations of weeds, but this requires the close cultivation of genetically engineered crops with sexually compatible weedy species over time, and also requires a relatively undisturbed cultivation of weeds.

Current agricultural practices generally mitigate these conditions. Large- scale experiments have been conducted to measure the movement of genetically engineered traits between crop plants. These have helped to determine minimum separation distances to reduce cross pollination to levels that are below the limits needed to maintain genetic purity.

Cross-pollination between organically cultivated and non-organically cultivated crops can be controlled by implementing separation distances that are dependent on the crop, the likely means of pollen transfer (insect- borne or wind for example) and the degree of cross pollination in each crop type that has been defined as maintaining the genetic purity of the crop.

Of the major crops that may eventually be introduced as genetically engineered varieties, sugar beet and oilseed rape can interbreed with weedy relations. Relatively simple agronomic measures to limit any potential effects of crosses between weedy species and crops can be implemented, such as thorough weed control. For example if herbicide tolerant weeds arise in unwanted numbers in fields, they can be combated by application of another type of herbicide. Outside the field herbicide tolerance in weeds is not likely to confer a selective advantage as herbicides are not widely used in these environments.

The so-called "Terminator Technology" was developed, but not applied, for stopping the flow of pollen from GM plants. The technology makes crop plants sterile and was originally invented as a "technology protection system" to prevent farmers from keeping seeds and growing these in the next season. This technology has the potential to ensure GM plants producing non-food products are genetically isolated in the field.Cross-pollination between crop species and their wild relatives raises the possibility that this process may erase essential differences between wild relations and their cultivated relatives. The loss of these differences has the potential to reduce genetic variation that may be useful for future breeding programmes.

The erosion of genetic diversity is among the most serious threats to agriculture and the environment because of the increasing scale of irreversible loss of genetic variation. For example maize has been cultivated in regions of Mexico where wild varieties also grow for many thousands of years, and there is evidence of widespread crosspollination in overlapping populations of cultivated maize and wild relatives. Nevertheless the wild populations have maintained their distinctive identities over many generations, suggesting cross- pollination is not a major contributor to the loss of genetic diversity. Instead, habitat destruction is the principal and obvious cause of the irretrievable loss of wild species.

Health Issues

To date no effects on human or animal health of consuming soybean or maize genetically engineered for herbicide tolerance or insect resistance have been attributable to the novel genes present in these crops after many years of consumption worldwide. Public concerns about the risks of ingesting DNA from genetically engineered plants have been addressed by phasing out the use of marker genes conferring antibiotic resistance. DNA is broken down during digestion in humans and is not able to pass through the gut wall and function effectively in human cells or be passed on to the next generation. We have all been eating DNA for generations and there is no evidence of transfer of DNA from food to our DNA.

In anticipation of the wider adoption of current and anticipated new genetically engineered crops and products, the standards used to test the potential impact of genetically engineered crops on human health and on the health of farm animals are under review in many countries to ensure greater standardisation and conformity to best scientific practice. New biosafety laws may involve appropriate tests comparing the composition, nutritional value, toxicity and allergenicity of the genetically engineered crop to its non-engineered counterpart. the ecological impact of the crop and any agronomic practices associated with its use can now be judged using standards and methods developed in the recent farm-scale trials. These procedures, when applied consistently to genetically engineered crops, should provide the evidence necessary to either support or reject each case for introduction. These tests could also be applied to conventionally bred plants grown in different agricultural systems to provide consumers with information to help establish and guide their preferences. This information could be provided by labeling to track and audit each step of the supply chain and recording the provenance and processing of food. This validation and labeling is proposed to help consumers make informed choices.

Currently a variety of genetically engineered crop plants are used in commercial production. There appears to be a trend towards increased acceptance of growing genetically engineered crops subject to appropriate local rules and practices after thorough review and consultation processes. Thailand, Brazil, Paraguay, New Zealand, Australia and England have all recently decided on limited case-by-case introduction of genetically engineered crops after the formulation of appropriate biosafety regulations. Planned introductions in Australia include herbicide- tolerate canola and in England herbicide- tolerant maize can be grown for a limited period. Some of the major influences on these decisions are increasingly based on a pragmatic assessment of the added value to the farmer and processors. These include reduced inputs of chemicals and improved yields and quality due to effective resistance to pests and herbicides. A high emphasis is placed on the environmental impact of growing genetically engineered crops because conventional agricultural practices are thought to have had an unacceptably deleterious effect on farmed ecosystems. The farm scale evaluation of the influence of genetically engineered crops provides a baseline for assessing these effects.

There are two broad trends in crop plant genetic engineering. The first is the rapid extension of the technology to a wider range of crops. In India for example important local crops such as millet, chickpea and pigeon-pea are all being genetically engineered. In Uganda biosafety laws are being established so genetic engineering for resistance of banana to wilt disease can be performed. Research programmes in China, Mexico, Australia and New Zealand and elsewhere are all working on developing local crops by genetic engineering.The second major trend is to find a far wider range of characteristics for genetic engineering. A wide variety of interesting leads is available, although many may not be taken up commercially due to the need to make a profit from seed sales to support development costs. Engineering crop plants with improved nutrition such as vitamin- and iron- enhanced rice is relatively well advanced, and a major research programme to combat human micronutrient deficiencies has begun.

Research aimed at producing antibodies and vaccines in plants is at an advanced stage, as plants can yield high quantities of specifically modified therapeutic proteins. Extensive research aimed at using genetically engineered plants to make commodities such as specialist starches and oils for industrial applications is also underway. Before these crops can be considered for release extensive precautions need to be established to prevent cross- pollination and to establish health risks due to ingestion of material from these crops. Current legislation provides a framework for these considerations but there are many issues related to the coexistence of industrial and food products that remain to be studied and debated before possible introduction.

REFERENCES

Anderson, P.K. et al. {Emerging infectious diseases of plants: pathogen pollution, climate change and agrotechnology drivers". *Trends Ecol. Evol.* 19, 535–544. 2004.

Banpot Napompeth, 1994. *Leucaena psyllid in the Asia-Pacific Region: Implications for its management in Africa.* FAO, Regional Office for Asia and the Pacific, Bangkok, Thailand, RAPA Publication: 1994.

Campbell, M.A. et al. "Engineering pathogen resistance in crop plants." *Transgenic Res.* 11, 599–613. 2002.

Ciesla, W.M. "IPM: New approaches to old problems". *American Forests* 88(2): 40-44, 51-52. 1982.

Dickinson, M . *Molecular Plant Pathology*. BIOS Scientific Publishers. 2003.

George N. Agrios. *Plant Pathology,* Academic Press. New York. 1997.

McDowell, J.M. and Woffenden, B.J. "Plant disease resistance genes: recent insights and potential applications". *Trends Biotechnol.* 21, 178–183. 2003.

Stuiver, M.H. and Custers, J.H.H.V. "Engineering disease resistance in plants". *Nature* 411, 865–868. 2001.

Wormald, T.J., *Mixed and pure forest plantations in the tropics and subtropics.* FAO, Rome, Italy, Forestry Paper 103, 152 pp. 1992.

8

Forests and Biodiversity

Until recently, efforts at assessing and monitoring forests have focused on the amount (area) of forest remaining and/or its standing timber volume. Increasingly the multiple benefits and functions of forests, including provision of non-wood forest products, hydrological functions, carbon sequestration and biodiversity protection have been recognised, and their consideration in forest assessment has been recommended. There is now a perception that not all forests are equal, whether it be for timber production or for conservation of biodiversity. The international community has placed a high priority on assessing and monitoring the capacity of forests to provide any given range of services and conveying this information to policy and decision-makers and the general public.

FORESTS AND BIODIVERSITY

Biological diversity, or biodiversity, is an imprecise term that is used to refer to the diversity of life at the levels of genes, species and ecosystems, in a single locality or over broad geographic areas, including the Earth as a whole. Biodiversity is used to refer to not only the numbers but the types of genes, species and ecosystems existing in the area in question. There is usually an implicit assumption that naturally occurring or original components of biodiversity are of more value than articficially introduced or degraded ones, and therefore biodiversity preservation or protection is frequently understood to refer to the preservation of these qualities and components.

Forests play a significant role in maintaining the world's biodiversity. They provide habitat for more than half of the world's species. A forest area contributes to global forest biodiversity by the number of species present and shared with other areas and the number of species it contains that are found nowhere else (endemics). The biodiversity complement of individual forests varies with forest type, and the climatic and soil factors that relate to it, as well as the biogeographic position and isolation of the forest. In general forests at low latitudes and low altitudes, with warmer and wetter climates, have higher species richness than those at high latitude or altitude, or with cooler and drier climates.

Natural perturbations, such as storms and wildfires can also be important influences on forest biodiversity.

Natural global patterns in biodiversity are also altered by human action. In the contemporary world, human activities may be the most important influence on forests' capacity to maintain their original biodiversity. Such activities as commercial and artisanal logging, large scale land conversion, fuelwood and charcoal production, slash and burn agriculture, harvesting of non-timber forest products, hunting and mining all affect forest biodiversity. Climate change resulting from modification of the atmosphere by anthropogenic emissions of carbon dioxide is also affecting the distribution and status of forest biodiversity.

Human Impacts on Forest Biodiversity

Each of these types of human influence affects forests differently, and the magnitude of the effects will depend strongly on the methods employed locally, the forest type, and on other factors within and around the ecosystem. For example, commercial logging in temperate forests is often by clear cutting, which entirely removes forest cover in some areas and fragments remaining forest cover. In contrast, commercial logging in tropical forests is usually by selective felling which disrupts canopies and forest structure and alters species composition, but may not alter total forest cover or its spatial configuration. Secondary effects of logging such as increased access and resultant hunting are also important determinants of the status of forest biodiversity and the prospects for its preservation. Small-scale timber extraction differs yet again in its effects on forest condition. The effects of climate change are less localised, and are only beginning to be recognised. Thus, many factors influencing forest biodiversity are affected in varying and complex fashions by human activity. In general human activities tend to affect any of three major aspects of forests:

1) The total area of forest remaining—many of man's activities remove forest cover either temporarily or permanently. Some forest types may disappear locally, and reduction in the total amount of habitat is a significant pressure on some forest species that can lead to local extinction.
2) The configuration of remaining forest cover—reduction in forest area is often accompanied by division of remaining forest cover into fragments, rather than continuous blocks. Forest biodiversity is affected by the consequent local reduction in habitat area, by the exposure of forest edges to new environmental and biotic influences and by isolation from other forest areas.
3) The structure and composition of remaining forest - some human activities alter canopy structure, or focus disproportionately on particular species and specific components of their populations.

Evaluating Forest Capacity

Thus, useful measures of forest capacity for biodiversity preservation are likely to address the amount of forest remaining, its configuration or integrity and its 'naturalness', or lack of anthropogenic disturbance. However, developing measures of disturbance and making them operational is not straightforward. Precise histories of forest disturbance and its intensity are rarely available, especially over broad geographical scales. The problem of describing and measuring human-induced change in ecosystems has been summed up by Groombridge in a review of habitat/ecosystem classification:

> . . . Just as it is impossible to define rigidly the limits of any given ecosystem or habitat, so it is impossible to determine how much a given area of ecosystem or habitat has to change before it can be considered destroyed or converted. The problem is compounded by the fact that the natural environment is not static but rather dynamic, sometimes highly so, on a time scale ranging from hours to millions of years. It is thus difficult even to define an undisturbed ecosystem or habitat as a standard against which to measure degree of disturbance.

There is a wide range of problems to be grappled with in successfully formulating methods for measuring the impact of humans in forest ecosystems. Difficulties arise from:

— *Different interpretations of the basic form and function of ecosystems*—Some concerns focus on ecosystem processes, while others emphasise composition. Such different conceptual approaches may lead to very different conclusions as to appropriate sets of information for the description and measurement of change.

— *Sensitivity to observation and process scale*—The scale at which ecological phenomena are observed and measured will have a major bearing on the conclusions drawn, and the uses to which they can be put.

— *Ambiguity in identifying benchmark conditions*—The notion of naturalness depends on a clear distinction between the presence and impact of human activity and natural ecological patterns and processes. This can be problematic if, as in much of Europe, there is little or no reference forest with little human influence to provide a basis for comparison

— Establishing whether human-induced change represents a fundamental shift in organisation or change within normal limits of forest dynamics and ecosystem processes;

— Uncertainty regarding the place of humans in the environment.

These difficulties pose significant technical problems and can give rise to potentially contradictory answers to questions concerning ecological change.

Two principal strategies can be adopted to address these issues and assess the naturalness and ecological integrity of forest ecosystems at broad geographic scales in biodiversity-relevant terms:

1. Use indicators to assess key aspects of forest structure and/or function. This strategy has the obvious advantage of directly addressing the primary concern, which is the state of the ecosystem, but is subject to the difficulties described above concerned with the defining and measuring human-induced change in ecosystems. Attention must focus on parameters that can be evaluated in a globally consistent manner at broad geographical scales and are clearly related to the status of component biodiversity within the forests. Fragmentation of forests and parameters that describe it are a promising avenue for developing such indicators.
2. Measure the driver of ecosystem change—human activity, rather than the response —ecological change. This approach avoids fundamental problems associated with describing and measuring ecosystem response to human activity. Separation of the driver (cause) and response (effect) components also allows for greater precision and flexibility in analysis. General indicators of naturalness or ecological integrity can be developed on the assumption that the greater the amount of ecosystem exposure to human activity the greater the potential for human intervention in these ecosystems.

Indicators are measurements that convey information about more than just themselves. They provide means for quantifying and simplifying information on complex issues. They are purpose-dependent, almost always open to various interpretations, and never tell the whole story. Indicators are needed because assessing and monitoring everything is impossible and because what is known needs to be conveyed to non-experts in policy-relevant form. Good indicators are:

- scientifically valid, i.e. they relate appropriately to what they are supposed to represent;
- based on easily available data;
- responsive to change;
- easily understandable;
- relevant to focal issues and users' needs;
- subject to target or threshold setting.

Assessing Human Impacts on Forest Biodiversity

Of the three types of human impacts on forests mentioned above, the only one for which a consistent effort at assessment has been made at regional or broader scales is the change in forest area. Since the 1980s FAO has provided statistical data on forest cover, based on forest inventory, national reporting and high resolution remote sensing. Since the mid-1990s a geographic overview of the current distribution of forest cover has been available . Thus it is possible to provide statistical and mapped data about the amount of forest cover remaining, and to a lesser degree, about the amounts of particular types of forests and trends in the amount of cover. The greater the level of detail about forest types available within such data sets, the more relevant to issues of the conservation of forest biodiversity they will become.

Forest configuration

The changes in forest configuration that accompany changes in land use and forest area can have substantial effects on the capacity of forest ecosystems to maintain their original biodiversity. As forest ecosystems are divided into smaller patches, there are numerous effects on their biota, and the responses may vary substantially among species and among forest types. The extensive literature on the effects of forest fragmentation suggests that the effects can be broken down into three major groups: area effects, edge effects and isolation effects. What follows is a brief summary of characteristic components of these effects.

Area effects

When large forest blocks are broken into smaller ones, not all species are included in the remaining patches, simply because of sampling effects. This is especially true for rare species and for non-mobile organisms, such as trees and many invertebrates, which may be sparsely or patchily distributed within the forest. Large animals and top carnivores are well known to require large areas of habitat. These species are especially vulnerable to the reduction in habitat area caused by forest fragmentation, and they may disappear entirely from forest patches because food or other resources are inadequate to support them.

Even smaller species are affected by the size of forest patches; amphibian species richness increased logarithmically with patch size in forest remnants in Madagascar. The disappearance of some species from forest fragments can profoundly affect the forest itself, as shown by the effects on tree communities of the disappearance of seed-eating rodents from forest islands in Gatun Lake in Panama. Other species persist, but in smaller populations, which may encompass less genetic diversity and lead over time to the vulnerability of those species to other ecological changes such as disease. Rare species and those that normally occur at low population densities are especially vulnerable to these effects. Smaller forest patches may also include less environmental variability and therefore fewer microhabitats than more extensive forest areas. This can contribute to the loss of individual species and may cause a reduction in the total species richness per area of forest. Fragmentation of forest cover may also alter the nature and proportional impact of natural disturbance regimes and regeneration processes.

Edge and gradient effects

Another important effect of forest fragmentation results from the creation of interfaces with non-forest environments. These interfaces are associated with environmental gradients resulting from the exposure of the forest edge to drying winds and increased solar radiation. The physical gradients affect ecological processes including canopy gap formation, biomass and nutrient cycling, regeneration and predation that can profoundly affect native species. Invasive species, both native and non-native, are often favoured by an increased incidence of forest edges within the landscape, so that substantive changes in species composition have been documented in forest fragments. Although some 'edge effects' have historically been regarded positively, principally because many game species make use of forest edges, they are generally regarded as detrimental to most native forest species. The magnitude of edge effects within forest fragments can be strongly affected by the land-cover characteristics in the surrounding landscape, the matrix harshness, and they are also dynamic, frequently increasing in magnitude and extent over time. Connections among habitat fragments are an important means of reducing genetic isolation and providing additional foraging and refuge areas.

Isolation effects

The other major group of effects of forest fragmentation results from the separation of the forest fragments from each other and from larger blocks of forest. This isolation reduces the movement of species that are reluctant or unable to cross non-forest areas and for those that depend on such species for dispersal. Reduced movement and dispersal also increases the chance of local extinction of individual species as a supply of colonisers or propagules is lacking. Isolation of fragments may also reduce the genetic neighbourhood of some trees, reducing the breadth of the local gene pool for cross fertilisation.

Responses to all of these effects vary among species, but a body of empirical evidence is accumulating that facilitates predictions about the likely effects of fragmentation on any particular forest ecosystem. Also from such evidence, a series of empirical generalisations concerning the spatial configuration of habitat with respect to biodiversity preservation can be paraphrased as follows.

i) Habitat that is more widely distributed across its original range is more likely to persist than habitat confined to small parts.

ii) Large blocks of habitat are superior to small blocks of habitat.

iii) Blocks of habitat close together are better than blocks far apart.

iv) Habitat in contiguous blocks is better than fragmented habitat.

v) Interconnected blocks of habitat are better than isolated blocks.

The long-term maintenance of forest integrity depends on promoting these characteristics in landscapes. The more that landscapes retain these characteristics, the less is their vulnerability to human-induced change. These generalisations provide a useful basis for assessment and communication of information about forest status in this respect.

Human Activity

Forest structure and composition and their implications for biodiversity are difficult to evaluate at broad geographic scales and may vary widely depending on the kinds and intensity of human activity and local ecological conditions. Therefore, assessing the amount of human activity may be a useful proxy for evaluating its impacts on biodiversity. It is well-documented, for example, that logging increases the probability of recurrent fire in Amazonian rain forests, and that this in turn will lead to long term changes in species composition. Logging also affects animal community composition and ecological relationships. Hunting activity near settlements substantially reduces the abundance of mammal species and the construction of roads facilitates both logging and hunting as well as land conversion and colonisation.

Measuring Forest Configuration

The evaluation of forest cover configuration depends on the availability of spatially referenced or mapped data. Most commonly, such data are derived from remote sensing, using either satellite-mounted or airborne sensors or aerial photography. For continental and global scale evaluation, satellite imagery is the most appropriate data source.

Satellite-mounted sensors measure radiation reflected from the Earth's surface in a variety of spectral bands; different land-cover types have different characteristic reflectances and spectral signatures. Satellite sensors vary in the frequency with which they return to a given portion of the earth's surface (temporal resolution), in the spatial resolution, or pixel size, of the data they provide, and in the spectral resolution or numbers and types of spectral bands in which data are recorded.

Satellite data are processed by adjusting them radiometrically to compensate for variations in atmospheric conditions, and then classifying the digital data by one of several methods. Classification can be achieved by plotting reflectance in particular spectral bands or band ratios such as NDVI (Normalised Difference Vegetation Index) and visually interpreting the results. Alternatively, satellite data can be classified digitally by grouping pixels with similar spectral characteristics, and by comparing their spectral responses to those of pixels from areas of known land cover (supervised classification). Additional power can be brought to the process by incorporating information on variation in spectral response of an area through time (seasonal and other changes) and ancillary data (on land forms, land use etc.). Expressing pixel composition in relation to mixtures of the spectral responses of possible component land covers (spectral mixture modelling) can also improve the resolution and accuracy of classification.

The processing of satellite data is expensive and time consuming, requiring sophisticated hardware and software to deal with the large volumes of digital data involved. Although very sophisticated and spatially detailed vegetation maps have been generated from satellite data, these have been confined to small areas because of the additional volume of data required for high spatial and spectral resolution processing.

The only currently available global land-cover data set derived from satellite data that have been processed in a consistent manner is the GLCCD, produced by the EROS Data Center and IGBP, from monthly averages of data from the AVHRR satellite during 1992-93. These data have a spatial resolution of 1 km, and their relatively low spectral resolution has been compensated by the large amount of data on temporal variation that is available and by incorporating large amounts of ancillary data into the classification process . At present, this is the only data set that could be used to evaluate forest fragmentation and regional and global scales.

Because of the coarse spatial resolution of the GLCCD, the data represent relatively coarse spatial mosaics and provide insufficient detail for certain kinds of analyses of forest configuration. There are also systematic errors in that result from the coarse spatial resolution: small non-forest patches in areas of high forest cover remain undetected as do small forest patches in landscapes with low forest cover. These phenomena lead respectively to over- and under- estimates of forest cover. Although these errors can be reduced by calibration against high-resolution data, it is not clear whether such a process has been applied to the GLCCD. Details of boundary configuration and interspersion of forest and non-forest land cover are lost at the lower spatial resolution. However, some consistent patterns that have ecological meaning for forest biodiversity do emerge. The following discussion focuses on the options for evaluating forest configuration that are appropriate for use with data like the GLCCD, but also includes examples of metrics that can be used productively with higher resolution data. The sensitivity of the different measures to data scale and resolution is discussed.

Other Constraints

Defining ecosystems is an issue in the extent to which the analysis attempts to focus on individual forest types and their configuration within the landscape. Ideally one would look at the fragmentation parameters of each forest type within a mixed matrix of forest and non-forest separately, but in fact it is unlikely that the source data on forest type distribution can support this. It will probably be necessary to look just at forest cover in relation to the non-forest matrix, or perhaps at the fragmentation properties of rather broad or regional forest types with a minimum of overlap.

Scale is of course an issue. Fragmentation is differently determined for different components of the biota of a forest ecosystem. A path or deforested strip of a few metres width may be a significant barrier to an invertebrate, whereas a deforested strip of several km presents little obstacle to a forest-dwelling bird of prey. The data available for doing a globally consistent evaluation of forest fragmentation are 1 km resolution satellite data. Therefore, any fragmentation metrics that are derived from these data will represent the distribution of forests only at this coarse scale, and indeed might be better said to represent the configuration of forested areas in the landscape than of individual patches of forest. The effects of coarse resolution data on forest area estimates discussed above also apply to the estimation of forest fragmentation. Because of the problems of detecting small patches within very high or very low forest cover landscapes, fragmentation will be underestimated in areas of both very high and very low forest cover.

The question of what is natural or a baseline condition is also an issue in evaluating fragmentation of forests. Some forest types and regions are naturally more continuous than others. For example, forest close to latitudinal or other limits of its distribution has a tendency to be naturally patchy, as in the taiga/tundra transition of the boreal regions and some forest-savanna boundaries. Similarly forests that exist on mountains in non-forested landscapes because of the greater humidity at higher altitudes are inherently restricted in area and isolated from each other. The fragmented nature of these systems has a great significance for biodiversity that is entirely distinct from that of the fragmentation of continuous forest cover by human activity.

Some level of forest fragmentation may also result from natural disturbance and dynamic processes within forests. However, the scales of these processes are usually such that they will be beneath the resolution of any global or regional scale analyses. Although some kinds of storm damage or wildfire impact may be at scales that could be detected by these analyses, the issue of whether changes in forest fragmentation parameters are within or outside the range of natural variation is unlikely to arise.

Analysis Tools

A number of packages for use with geographic information systems (GIS) permit the analysis and characterisation of landscapes in terms of their patch composition, spatial relations and dynamics. One such package, FRAGSTATS is widely used for the description and analysis of landscape configuration. It offers a wide range of measures of varying complexity.

Monitoring Forest Spatial Integrity

The choice of measures of forest fragmentation for use as indicators of forest capacity to retain biodiversity is dictated both by the source data and by the range of biological effects being targeted. Summary statistics of landscape metrics are of little use for predicting responses of individual species without more detailed information about both species requirements and environmental variation on the ground. However, to provide both an overview of forest status in relation to biodiversity and baselines to track changes that may affect forest biodiversity simple statistical expressions of forest configuration can be useful. It is also important that the metrics chosen be easily communicated and understood by the anticipated audience for the overview and monitoring, so conceptually complex indices are generally less useful.

Landscape ecological theory and GIS technology have generated a number of measures of the spatial distribution of habitat that express different aspects of its fragmentation in ways that relate to ecological processes. These have mostly been used in the evaluation of habitat and landscape processes at management unit scales. For example, Kramer used these analyses to describe landscape change as a result of management of two national parks in Costa Rica. She found that as forest cover increased, patch size increased and patch shapes became more compact for the shrinking pasture areas, but remained similar or became more complex over time for the expanding forest. Landscape diversity declined as the pasture areas shrank. In another example, Helmer used similar analyses to show that in the mountains of Costa Rica secondary forest is strongly associated with primary forest and occurs in smaller patches with more complex shapes. Logsdon et al.,. used FRAGSTATS to analyse and describe landscape composition, configuration and heterogeneity from remotely sensed data for a small part of central Amazonia. In one of the few studies that directly addresses the relationship between landscape parameters and potential management actions, Ranta et al.,. used landscape analysis to characterise the fragmentation of Atlantic rain forest in Brazil and to simulate the likely impacts of land use change or forest restoration. Still other studies have characterised forest fragmentation at broader scales. Skole & Tucker evaluated forest fragementation and potential edge impacts for the whole of the Amazon Basin, and substantial research has been done to characterise and describe patterns of fragmentation throughout the tropics.

As these examples show, spatial analysis of landscapes and the metrics it produces have been the subject of much research and have been used widely for descriptive purposes. However, they represent a complex suite of ecological processes and effects that are beyond the detailed understanding of many decision-makers. Also, they often focus at scales that are beyond the resolution of currently available classified satellite data on landcover at regional and global scales. Furthermore, they have rarely been used to identify the likely relative value of different patches of habitat in the context of fragmentation, as a basis for monitoring changes in that condition or for making policy and management decisions.

Thus, there is a need for an easily understood summary index that can be used both as a basis for visualising the relative biodiversity preservation capacity of different forest areas and to establish a baseline for tracking forest landscape change. Both of these uses support decision-making and evaluation of policy effectiveness. Such an index needs to reflect all three types of fragmentation effects outlined above:

— area or patch size;
— interface with non-forest, or edge effects
— isolation from, or interconnection with other patches.

Forest Patch Size

Area effects are most easily represented in terms of patch size. A GIS can be used to identify all patches of forest within the area of study, to measure their areas and assign them to a patch size class. If the data used are coarse resolution satellite data, this evaluation will identify contiguous blocks where forest is the predominant land cover, even if there may be some breaks in the forest cover at sub-pixel level.

The appropriate size class intervals can be selected empirically to provide a distribution with an easily analysed shape. The thresholds can also be adjusted to suit different regional characteristics or to address specific conservation issues and values. For example, selection of patch size thresholds might be related to average individual home range size or sustainable micro-population area of forest animals. Other criteria might be based on the spatial scale of natural regeneration and successional processes that provide continuity of forest ecosystems after disturbance. However, it is important to recognise that consistency of classes between sampling times is an essential component of any monitoring or comparative work.

For the evaluation of forest fragmentation based on global datasets (mostly 1 km^2 pixel size), a conservative analysis would assume that details of forest patches smaller than 10 km^2 cannot be analysed reliably because of the high impact of variation in shape and sub-pixel level structure. It can also be accepted that all forest patches larger than 300 km^2 (30,000 ha) can be regarded as continuous forest, and thus make up the largest patch size class.

In the present analysis, and potentially for future broad scale monitoring and data integration activities, logarithmic type scales are used, with the size class intervals delineated by rounded values in km^2. When higher resolution data are available, the scale might be extended to split the smaller patch size classes, while still keeping the established pattern of scaling. Equally, the upper end of scale could be expanded for broad scale or global level studies.

For the purpose of developing an integrated index of fragmentation, numeric ranks ranging from 1 (1-10 km^2 patch) to 10 (> 300 km^2) were assigned to classes, and each 1 km^2 cell classed as forest was assigned to one of these ranks according to the size of patch it belongs to. The result is a visualisation of where forest occurs in large patches and where it occurs in small ones that may be clearer to non-experts than simple maps of forest cover. Such an analysis can also generate a statistical distribution of forest area among patch size (P_s) classes that can be used as a baseline for assessment and monitoring of forest condition in relation to the capacity for biodiversity preservation.

Given the characteristics of coarse resolution data and the implications of connections or breaks between forest patches at this scale, additional components are needed to improve the strength of the evaluation. Furthermore, the patch size analysis alone does a poor job of distinguishing between the capacities to support biodiversity of outlying narrow branches of patches and core areas of forest patches within the same size class.

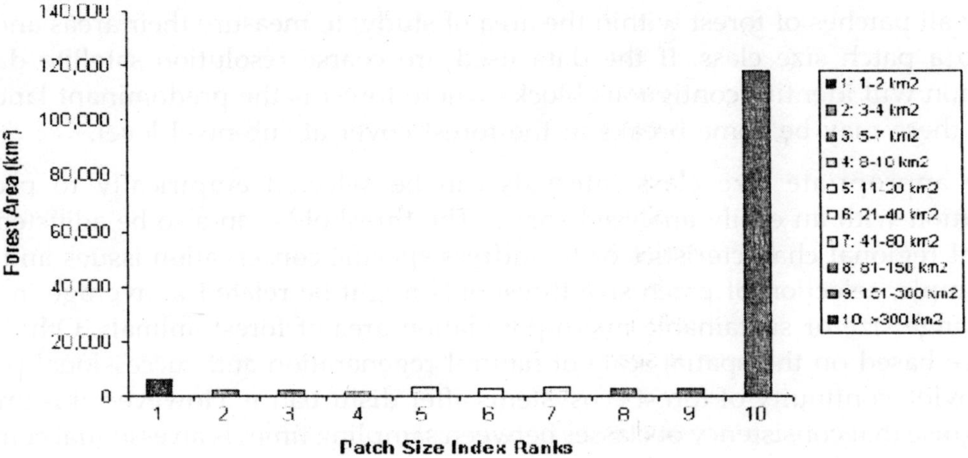

Figure 1. Statistical distribution of the forest area of Paraguay among different patch size classes.

Shape or Edge Influence

The effect of the interface between forest and non-forest can be addressed through the relatively commonly employed shape indices, such as perimeter to area ratios and edge-to-core ratios. However, these measures are more appropriate to ground level studies or high resolution data sets in which the forest cover is real and the extent of edge influence is well understood. For coarse resolution data in which patches may or may not represent actually contiguous forest, an alternative approach is to evaluate the percentage of the neighbouring cells that contain forest within a given radius of each cell. The radius can be chosen in the light of known scaling issues and concerns about specific ecological phenomena. In the current illustrative example a radius of 5 km was chosen as being consistent with the spatial accuracy of the data.

As can be seen in Figure 4, the value of S_F is high when the sample cell belongs to the interior of, or is near a dense or solid patch that is comparable in size or larger than the radius used for calculation of S_F. For areas at the periphery of, or distant from large continuous forest patches, SF is lower and mostly dependent upon patch shape and isolation. Small isolated patches have small amounts of forest in their immediate neighbourhoods; the extreme case is the single-pixel forest patch that is separated by more than the S_F evaluation radius from any other forest.

Points along patch edges, and in patches of complex shape or in narrow strips of forest, where edge effects may influence forest status, also have low proportions of forest cover in their neighbourhoods, while points within larger patches or continuous forest are entirely surrounded by forest. Therefore, spatially weighted forest density, S_F provides a basis for expressing the role of edge effects, without dependence on accurate definition of the edge or limit of a given patch.

Like the patch size analysis, this analytical approach can be presented in a way that provides a clear spatially referenced visualisation. This shows where forests are subject to edge and isolation effects as distinct from forest that is both part of a large patch and distant from the interface between forest and non-forest. Although the implications of these effects for forest biodiversity are strongly dependent on forest type and location, change in the amount and location of forest subject to such influences is likely to result in changing biodiversity status. Therefore, establishment of both visual and statistical baselines for this parameter is an essential step in the monitoring of forest biodiversity status.

Spatially weighted forest cover density, S_F, is measured directly in percentage units. In the present example, the scale is broken into ten equal intervals and the lower percentages are assigned to lower ranks. This reflects an implicit assumption that forest habitat that is less subject to the influences of edges and isolation is of greater value for forest biodiversity. Although this assumption may be invalid in some cases, it is a realistic view at the global scale.

As with the patch size distribution, appropriate breaks between classes can be selected on theoretical and empirical grounds, and a statistical summary of the forest area in each class can provide a basis for assessment and monitoring of forest condition in this context. However, consistency in approach between assessments is a critical component of any monitoring or comparative analysis. The use of GIS tools that retain the source data and intermediate parameters in separate grids, provides flexibility and ensures that a consistent approach can be applied across all data sets in a time series. Such a distribution is an initial assessment of forest condition with respect to fragmentation, which reflects proximity to non-forested areas and edges, and can be used as a baseline for monitoring purposes. If fragmentation of the forest increases, the amount of forest in the highest classes will decrease, and that in the smaller classes will increase. Distributions can be compared between times using non-parametric statistical tests.

A simplified version of the forest cover density analysis can be used to identify forest density zones, generalised outlines of the forest areas that are likely to have the highest integrity, those of least integrity and the intermediate values. This approach provides a useful way of defining "core" forest areas. It also provides a means of focussing attention on the forest areas of intermediate 'quality' or integrity. These are the areas that may appear least distinct to the non-expert observer, but are most likely to be immediately affected by policy and management decisions. Separate focus on this zone of intermediate forest density can enhance the visibility of pattern and change in the statistical data.

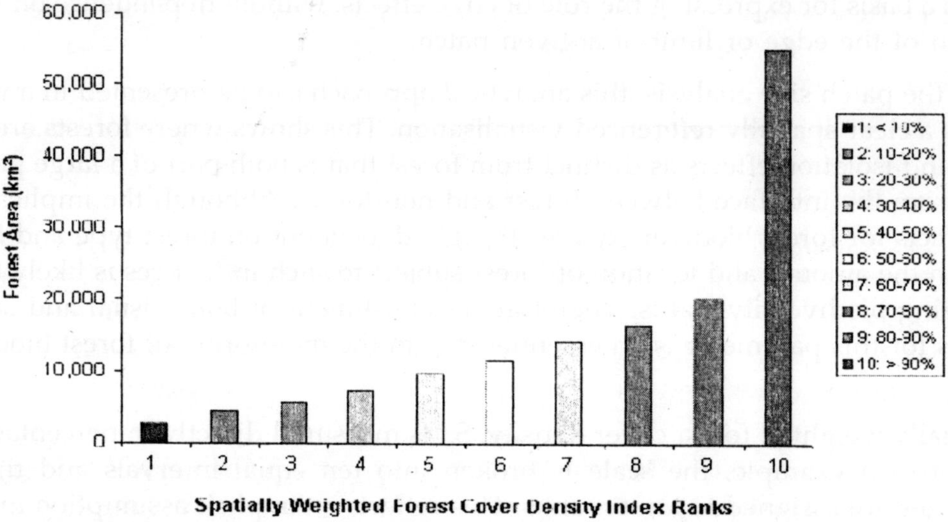

Figure 2. Statistical distribution of Paraguay's forest area among different classes of spatially weighted forest cover density, SF, forest occurrence within a 5 km radius.

Isolation and Inter-connection

Quantification of the third component the effects of fragmentation, isolation, requires some measure of distance to other forest areas. However, the degree of isolation and/ or the positive effects of interconnection are also dependent on the characteristics of the neighbouring forest. It is also true that many forest species are unable and/or reluctant to cross areas without forest cover, so forest areas that are directly connected to other forest areas are likely to be of greater value and more accessible to a greater range of forest species.

The possibilities of dispersal to and from a particular forest area depend on the species of interest and the forest stand characteristics. Furthermore, the overall sustainability of biological systems is increased by the presence of relatively intact "core" areas surrounded by peripheral areas that are important in buffering the system as a whole against external impac. These peripheral areas, in their turn, also benefit from connection to "core" areas, which provide necessary genetic resources that can be key to maintaining ecosystem function after natural or anthropogenic disturbance. Therefore, forest patches that are connected by forest to core forest areas may be viewed as more sustainable and of higher biological value that areas of similar forest cover density and overall shape that are not connected to core forest.

Core forest area, and connection to it, may be defined using thresholds appropriate to the particular components of biodiversity of interest, management considerations and properties of individual forest types. In the present study, for general illustration core forest area was defined using two criteria that are appropriate to the coarse scale of the data:

1. Core area is represented by continuous forest with density (SF) more than 90%
2. The size of an individual core area must be at least 100 km².

The distance to core forest areas, via cells containing forest cover, estimates the degree of connection of forest cells to core areas. The connectivity, C_F, is inversely proportional to distance from core areas, ranging from 10 (core area) to 1 (24-27 km), and in this example, forest cells at distances greater than a threshold of 27 km from core forest are regarded as effectively isolated and assigned a C_F value of 0. Forest cells that are connected to core forest by between one and 27 km of forest are assigned to intermediate classes of connection.

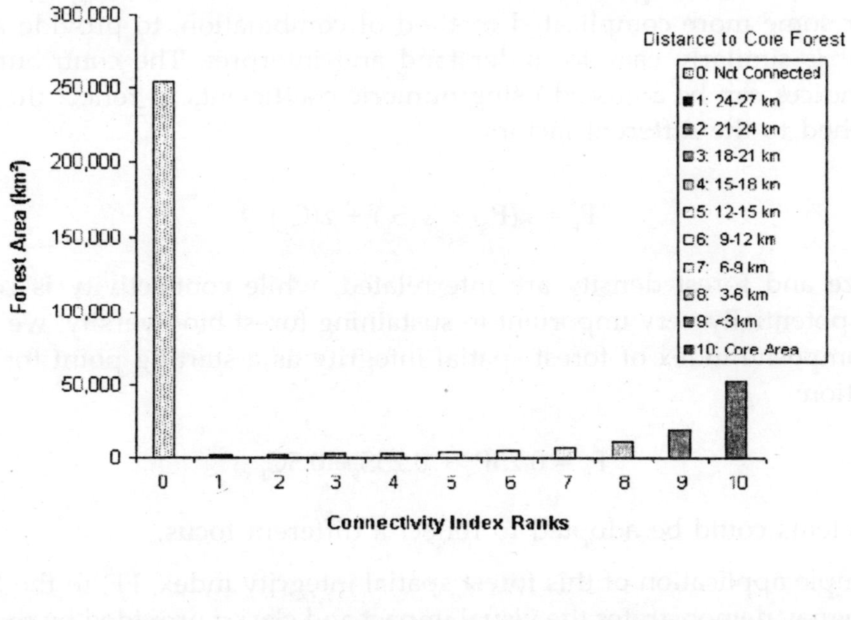

Figure 3. Statistical distribution of forest area in Paraguay among different classes of connectivity, C_F.

Like the previous two indicators, this approach can be used to provide both a spatially referenced visualisation and a statistical summary (Figure 3) of which forest areas are likely to be in the best condition for preserving forest species. In this case, the best condition refers to the closest connection to core forest, and therefore the greatest accessibility to the greatest range of forest species. Although this approach is proposed for the analysis of relatively low resolution (1 km) data, where individual pixels may include mosaic patterns, and species diversity and successional stage may not be incorporated, the same principles would apply at more detailed scales.

Forest Spatial Integrity Index

Despite the individual limitations of each of the above indices, between them they cover all three important aspects of forest fragmentation effects and make it possible to quantify most variations in the spatial distribution of forest cover. Patch size, P_s, facilitates comparison of forest stands larger than some minimal patch size threshold and below a size that can be regarded as continuous forest, regardless of variation. Spatially weighted forest cover density, S_F, provides a way of identifying both small, dispersed patches and areas subject to edge effects. However, it provides little detail at intermediate values and gives insufficient information about the interconnection of forest patches with forest patches of different status. The connectivity index, C_F, permits ranking patches of similar size in relation to their probable accessibility to forest species, and provides a way of distinguishing among forest areas of intermediate sizes and densities.

As all three indices range between 0 and 10, it is feasible to integrate them, using averaging or some more complicated method of combination, to provide a combined measure that is similarly easy to understand and interpret. The contributions of the individual indices can be adjusted using numeric coefficients to reflect the conceptual weight attached to the different factors:

$$F_F = x(P_S) + y(S_F) + z(C_F)$$

As patch size and forest density are interrelated, while connectivity is conceptually distinct and potentially very important in sustaining forest biodiversity, we suggest the following composite index of forest spatial integrity as a starting point for analysis of forest condition:

$$F_F = 0.25P_S + 0.25S_F + 0.5C_F$$

Other coefficients could be adopted to reflect a different focus.

The example application of this forest spatial integrity index, FF, to the forest cover data for Paraguay demonstrates the visual impact and clarity provided by this approach, which could ensure its utility for decision making. The statistical distribution of forest area among the different integrity index classes (Fig. 4) provides a baseline for monitoring forest spatial integrity over time. Such monitoring is an essential component of evaluating policy effectiveness and the impact of management decisions.

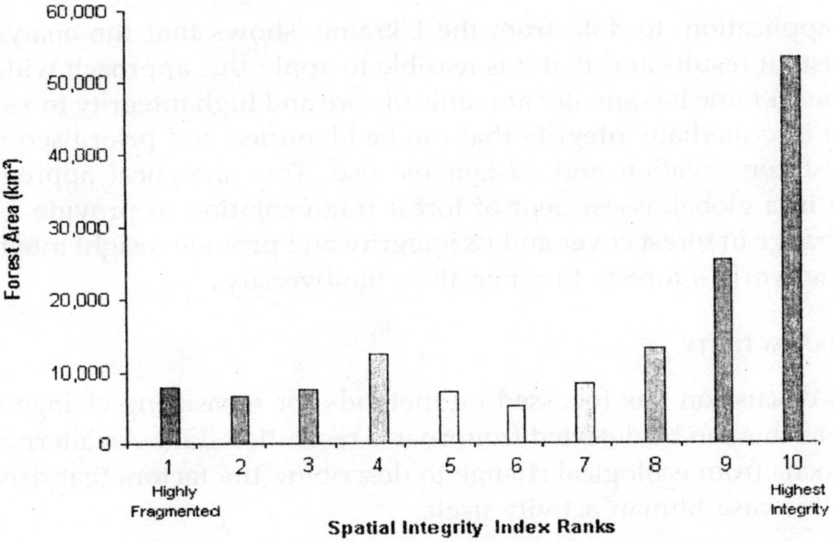

Figure 4. Statistical distribution of the forest area of Paraguay among different classes of spatial integrity as expressed by the Index of Forest Spatial Integrity, F_I, which combines information on patch size, shape and isolation.

Statistical data need to be presented in absolute areas rather than as percentages. Changes in percentages of forest cover in lower integrity categories may reflect either an improvement in the integrity of formerly low value areas, or simply a loss of forest area from those categories without an increase in others. Added clarity of both mapped and statistical presentation of data on forest spatial integrity may be obtained by delineating the three different forest density zones as derived from the simplified presentation of the Spatially Weighted Forest Density. This presentation might help the non-expert user to visualise the spatial relations among forest patches in relation to their overall spatial integrity index and to anticipate the effects of changes in the landscape more vividly. It also makes it possible to view the statistical distribution among intermediate integrity classes in more detail, as the scale can be expanded by excluding large areas of high and low integrity forest from the presentation.

Presentation of Results

Great care must be taken in the presentation of results of such assessment and monitoring. Data in a mapped context may be the most useful for supporting site-specific decision-making and, conceivably, for scenario testing. Data in statistical form are potentially more useful for monitoring and evaluation of policy effectiveness. It is crucial that consistent methods are applied for comparisons in space and time and that original data and analyses are retained to permit reanalysis in the event that changes in thresholds or approaches are deemed appropriate.

Another application, to data from the Ukraine, shows that the analysis produces visually consistent results and that it is feasible to apply this approach widely. It shows clearly that the Ukraine has smaller amounts of core and high integrity forest, but retains some areas of intermediate integrity that can be identified and prioritised in policies to promote forest conservation and sustainable use. This analytical approach could be implemented in a global assessment of forest fragmentation to provide a baseline for monitoring change in forest cover and its integrity and provide insight into the changing capacity of the world's forests to retain their biodiversity.

FOCUS ON HUMAN ACTIVITY

To this point discussion has focussed on methods for measuring change in the forest itself, insofar as this can be detected from coarse resolution data. An alternative strategy is to switch focus from ecological change to describing the factors that drive ecological response; in this case human activity itself.

Shifting emphasis in this way means that generalisations can be made about the level and extent of human intervention in ecosystems and, in turn, the exposure or vulnerability of these ecosystems. It also alleviates difficulties associated with identifying and measuring key ecosystem phenomena, facilitates an explicit treatment of scale issues, improves prospects of acquiring suitable data, and removes the necessity to ascertain whether particular ecological outcomes are attributable to human activity or natural processes.

Spatial Pattern

The analysis of spatial pattern in human activity in the landscape has a strong tradition in human geography, particularly from the 1930s to the 1970s, from which several key principles have emerged. Firstly, the spatial distribution of human activity reflects an ordered adjustment to distance. Of particular relevance is the notion of the attenuation of pattern or process with distance, as expressed in Tobler's 'first law of geography' which states that everything is related to everything else, but near things are more related than distant things. Secondly, human activity is generally located to minimise the 'frictional' effects of distance. A related principle is the notion of accessibility or functional centrality. Finally, human activity agglomerates in settlements.

Generalisations concerning the spatial configuration of habitat are encapsulated in the broader generalisation that the integrity of habitat is usually associated with spatial isolation from human activity. This broader principle also relates to more traditional geographical perspectives on the pattern of human activity in the landscape and the attenuation of pattern and process with distance. However, a reliance on spatial pattern to explain human interaction with ecosystems does have limitations. It involves, for instance, the necessary assumption that there is a direct relationship between the spatial location of human activity and its ecological effects. This precludes any satisfactory accounting for processes (e.g. hydrologic and atmospheric) where spatially distributed effects are non-linear or highly complex (e.g. multi-scale processes).

Measuring Isolation from Human Activity at Landscape Scales

Spatially explicit indicators of ecological integrity, which are consistent with a strategy based on measuring isolation from human activity in the landscape, should have the following characteristics:

- The index should be quantitative and the methodology should have the capacity to measure variation in exposure to human activity across the landscape.
- Index values should derive directly from primary attribute data in a systematic and repeatable manner, and they should reflect the scale characteristics of primary data inputs.
- The methodology should be transparent and as simple as possible. Notions of ecological integrity and naturalness (or isolation from human activity) are complex and difficult to deal with in a precise way. There is, therefore, little advantage gained from pursuing complex modelling techniques when these are likely to have deficiencies of a similar order of magnitude to simpler procedures. Complex modelling also has the disadvantage that it can become difficult to understand and interpret. Any modelling procedure in this area will have contentious aspects and it is far better for these to be explicit and well understood. Complex modelling also generally requires primary attribute data of accuracy and precision that is typically not available at global or regional scales. Global consistency and comparability is improved if the modelling procedure does not make demands for sophisticated primary attribute data.
- The methodology should be amenable to elaboration in a staged and systematic manner. This is essential to enable the model to benefit from new or improved attribute data and knowledge. A capacity for elaboration may also be useful in local or regional situations where its possible to take advantage of additional or enhanced local attribute data, or where particular local factors have a known and significant impact on ecological integrity.

Large Natural Areas

All other factors being equal, the core of a large natural area will be less exposed to (more isolated from) human activity than the core of a small natural area. The exposure of a natural area to human activity is therefore fundamentally related to its size.

Current interest in the size of natural areas, as a measure of relative isolation from human activity, is complemented by its long history in nature conservation science. Since the 19th century, for instance, it has been recognised that there is a link between the human-induced break-up of large natural areas and the extinction of species. The importance of size in nature conservation science has, however, been promoted most strongly by island biogeography theory and the species-area relationships of Preston, which hold that larger areas typically support a greater diversity of habitats and contain more species and larger populations of individual species than smaller areas. Size is today similarly recognised as important in process-functional terms. Larger natural areas have the capacity to better accommodate change in larger-scale processes and disturbance events.

The systematic evaluation of natural areas for conservation, based on size, is first evident in the US early in the 20th century with the emergence of interest in wilderness and roadless areas. During the 1930s the US Forest Service prepared an inventory of available wilderness areas within National Forest, rating undeveloped roadless areas on the basis of size. The US Wilderness Act of 1964 set up a framework for wilderness identification and protection in the US that is still in place today. It requires relevant Federal land-holding agencies to identify and assess roadless areas.

Interest in the identification of large natural areas that have limited exposure to modern technological society is not restricted to the US. It has been a mainstream aspect of nature conservation assessment in other 'frontier' regions of the world where there is a relatively clear distinction between the presence and impact of human activity and natural ecological patterns and processes. This is especially the case in Australia, Canada and New Zealand. Notably, the concept has also been pursued in parts of the world where this distinction is not so clear and where there is continuing indigenous habitation. This includes countries such as Finland, South Africa, and Italy. Identification and assessment methodologies typically involve criteria that specify minimum size and shape characteristics and require a 'primary' vegetation cover, no urban, agricultural or other commercial land use, minimal constructed access and no permanent settlement. Detailed specifications vary on a regional basis and from study to study.

Several assessments of this type have been conducted at the global level. The first global 'reconnaissance-level' assessment of large areas with minimal impact and proximity to modern technological society was completed by McCloskey and Spalding. Using 1:1,000,000 scale Jet and Operational Navigation Charts as a database, that study identified areas of at least 400,000 ha with no mapped human structures or roads. The global distribution of these areas is is heavily dominated by arctic and desert regions and includes little area in forested zones.

A second global-level evaluation of human activity mapped relatively undisturbed natural areas, and reduced the minimum size of mapped 'undisturbed' areas to 100,000 ha. It defined undisturbed and two classes of non-natural areas as follows:

— undisturbed - a record of primary vegetation and a very low population density (<10 persons per km^2 density or <1 per km^2 density in arid/semi-arid and tundra communities)

— partially disturbed - a record of shifting or extensive agriculture, or other record of human disturbance; and

— human dominated - a record of permanent agriculture or urban settlement, or where primary/potential vegetation is removed

Data were derived from a variety of information sources. Overall, the world was found to have around half of its total surface, but only 27% of its habitable surface undisturbed by man. A number of important forest areas, especially in the Indomalayan biogeographic realm, had no undisturbed area remaining and very little partially disturbed territory.

An evaluation of the status of the world's 'frontier forests' represents a third instance of a global-scale study involving the identification of relatively large natural areas. This assessment focuses solely on forest environments, distinguishing three classes of threat:

— frontier forests under low or no threat - large intact forest ecosystems that are relatively undisturbed and large enough to maintain their biodiversity

— frontier forests under medium or high threat - on-going or planned human activity (eg logging, agricultural clearing, mining)

— non-frontier forests - secondary forest, plantations, degraded forest, and patches of primary forest.

The threat classification in this mapping exercise was drawn largely from expert opinion. No specific size threshold was applied in identifying frontier forests, although it was required that areas be of sufficient size to maintain biodiversity and to absorb large-scale disturbances.

These three global-scale spatial studies represent simple and useful assessments of the naturalness or integrity of ecosystems. Each study primarily relies on the assumption that isolation from the impacts and influences of human activity is a reasonable indicator of ecological integrity, although some reference is made to certain biophysical conditions in the latter two instances. In each case the accuracy and precision of results is dependent upon the quality of available suitable data of global extent.

A greater concern, particularly in the latter two studies, is that the analyses are not explicitly scaled; nor are they systematic and repeatable. This means that the precision of the mapping and the accuracy of attribute class allocations are not transparent and direct expressions of the data and the analytical procedure. It also means that they can not form the basis for any kind of consistent assessment programme to monitor change through time.

Moreover, the selection of particular size thresholds to identify places that are isolated from the impacts and influences of human activity can be questioned. Exposure to human activity is not simply a matter of presence or absence, it is a matter of degree.

Measuring a Continuum: the Wilderness Index

Recognition of problems with the use of fixed size thresholds, qualitative terms and a lack of repeatability gave rise, in the early 1980s, to new quantitative indicator-based approaches to the identification of remote natural areas. For example, emphasis switched from the identification of wilderness areas based on qualitative thresholds to the 'continuum' concept of wilderness, involving the quantitative measurement of relative variation in remoteness from human activity across the landscape. This type of approach underpinned the Australian government's National Wilderness Inventory and the production of similar remote and natural lands databases elsewhere.

The Australian wilderness study places emphasis on measuring the extent to which points in the landscape are remote from, and undisturbed by, the influence of modern technological society. It does so by quantitatively measuring variation in remoteness and naturalness across the landscape using four indicators:

- remoteness from settlement (remoteness from places of permanent habitation);
- remoteness from access (remoteness from established access routes);
- apparent naturalness (the degree to which the landscape is free from the presence of permanent structures associated with modern technological society); and,
- biophysical naturalness (the degree to which the natural environment is free from biophysical disturbance caused by the influence of modern technological society).

The two remoteness indices and the apparent naturalness index are based on a measurement of Euclidean distance between each point in the landscape and ordered classes of settlement and infrastructure. Variations in exposure to different types of settlement and infrastructure features are accommodated through a weighting and distance-decay regime whereby more prominent feature types are accorded greater influence than less prominent type. In this way, distance-based measures represent spatial pattern that is specific to the location of individual landscape features, allowing for the attenuation of levels of technological activity according to the distance from the feature and its prominence or likely influence. The use of distance measures in this context is not based on any empirical relationship between distance and the flow of resources associated with types of technological activity.

The biophysical naturalness index is rated according to the intensity of land use in areas where the primary vegetation structure is essentially intact. Land use, in this context, refers to activity that is not confined to defined physical structures and includes a variety of forms of spatially distributed resource procurement activity, such as timber production and livestock grazing. Land use intensity is rated on the basis of historical records or of land use likelihood modelling.

This type of approach has a number of advantages over conventional mapping methods. Of particular relevance to the question of periodic assessments and monitoring is the fact the analysis is quantitative and repeatable. Estimates of isolation or exposure to human activity produced by the analysis are a direct expression of the data and the modelling that is applied. This means that the scale of the analysis can be explicitly matched to the accuracy and precision of data inputs. GIS-based application of the model, which effectively automates the analysis, also promotes flexibility so that new primary attribute data can readily be introduced and the analysis repeated or manipulated in a variety of ways.

Measures of Accessibility, Population Density and Resource Use

The Australian wilderness index measures environmental exposure or isolation from human activity in the landscape in terms of Euclidean distance from classes of settlement and infrastructure, along with a rating of land use intensity. No attempt is made to represent exposure or isolation in ways that are more exact or 'real'. The notion of exposure to, or isolation from human activity may be elaborated in two ways: 1) the refinement of the spatial, distance-based aspects, and 2) enhanced calibration of the intensity of human activity, taking account of data relating to appropriation and use of resources.

One obvious way in which the spatial, distance-based component of isolation may be enhanced is through use of more refined measures of accessibility. The accessibility of places in the landscape is not simply a function of Euclidean distance from access points and the quality of that access. Accessibility is also dependent, for example, on terrain. For instance, a forest that is located in rugged terrain at a given distance from a road is generally less accessible to timber harvesting than a forest that is located at the same distance from a road in flat or undulating terrain. The improving quality of digital elevation modelling and the increasing availability of elevation data sets means that it is now practicable to introduce terrain factors into accessibility modelling at region and global scale. A global digital elevation data set (DEM) modelled from satellite imagery is, now available at a grid resolution of approximately 1 km^2. The accessibility of forest environments has recently been modelled at regional scale using slope measures derived from elevation data at that scale.

Enhancements of this kind should, however, be treated with caution. There is no doubt that terrain-related factors have significant impact on accessibility in the real world, but terrain attributes are highly sensitive to DEM accuracy and precision, and estimations should be treated very cautiously. Typically, a DEM grid resolution in the order of 20 m is required if the tilt components of topographic variation are to be faithfully reflected in model output. It is therefore doubtful that terrain data at the scale presently available have the capacity to add meaningfully to accessibility analyses of the kind under discussion. However, the prospects for this type of analysis will certainly improve over time.

The second area of model elaboration involves calibration of the intensity of human activity. One simple way this may be developed is by factoring in statistical data on population and spatially relating this to settlement and infrastructure patterns. Indeed, patterns of population density have already been derived in this way on a global basis and show promise in this regard. The calibration of the intensity of human activity could potentially be further refined by combining population data with information about resource use. The intensity of demand pressure on forest ecosystems may, for example, be calibrated by inclusion of demand estimates for forest resources. A demand surface for fuelwood could be developed by using statistical data on fuelwood consumption, and distributing this spatially around urban and rural settlements on the basis of population density, forest cover, accessibility and known patterns of fuel wood consumption. A similar surface representing the pressures caused by industrial demand for timber resources could be developed and spatially distributed through settlement, infrastructure and land use components.

Developing Spatial Indicators of Naturalness

Discussion to this point reveals some key principles that guide the development of spatial indicators for assessing the naturalness of forests.

- Indicators based on ecological response to human perturbation can only provide limited, and possibly contradictory, answers to questions concerning ecological change at the ecosystem level.
- Indicators of naturalness that focus on human activity - the driver of human-induced ecosystem change - are highly promising in generic, landscape-scale applications. Such indicators rely on the assumption that the greater the exposure to human activity, the greater the probability of human interaction and intervention in ecosystems.
- Spatially explicit measures of relative environmental isolation from human activity can provide a generic foundation for describing and measuring the potential for human intervention in ecosystems.

— Methods for measuring relative environmental isolation should be quantitative, repeatable, transparent, appropriate to the input data, simple to interpret and amenable to elaboration.

Key Attributes

Human activity in the landscape can be described in terms of spatial patterns of land use and land occupation. Categories of landscape modification can be represented unambiguously in terms of (1) settlement, (2) infrastructure, and (3) land use.

1. Settlement can be defined as permanent human occupation. It is the focal point for human activity in the landscape where resources are transformed and used. Settlements range in scale from a single point of permanent occupation, such as a house, through to conurbations which may extend over thousands of square kilometres.

2. Infrastructure is the built fabric around which human activity concentrates. Infrastructure provides the physical means for accessing, distributing and transforming resources. Infrastructure includes all built structures, including those associated access and settlement.

3. Land use includes any resource procurement or transformation activity that can be spatially delimited on the land surface.

The representation of human activity using primary data sets comprising settlement and infrastructure features and land use has a number of advantages.

— These features represent fundamental elements through which the pattern of human activity in the landscape can be measured and described at both small and large scale, encapsulating complex resources procurement and transformation processes.

— These features are unambiguous landscape phenomena, which are amenable to classification and measurement. This provides for some control over accuracy and precision in spatial (and temporal) representation. It also facilitates analyses that use spatial information technologies.

— These features provide flexibility in relating human activity in the landscape to ecological effects, allowing for either aggregated or disaggregated analyses.

Using settlement, infrastructure and land use features for representing human activity does, however, have the important limitation of excluding resource manipulation techniques that rely on naturally occurring physical or biotic phenomena, such as fire or specific plants and animals, commonly associated with indigenous societies. In modern societies at least, there may be some association between the distribution of these factors and discernible patterns of settlement, infrastructure and land use.

AN EFFECTIVE SPATIAL MODEL: BASIC REQUIREMENTS

Spatial analytical technologies such as GIS provide powerful tools for modelling patterns of human activity in the landscape. Established theory connected with the spatial structure of settlement and urban structure provides also some useful principles to guide development of a modelling framework for representing spatial isolation from technological activity in the landscape. A set of spatial indicators that represent a gradient of exposure to technological activity can be developed using distance-decay functions to rate locations across the landscape on the basis of their relative exposure to settlement, exposure to infrastructure, and land use intensity. Relatively high levels of exposure to technological activity are encountered in places that are close to urban areas. Relatively low exposure is associated with places that are isolated from the settlement, infrastructure and land use practices associated with modern technological society.

Validation

Any attempt to link ecological integrity or naturalness with measures of exposure to human activity raises the issue of validation. It has already been stated that an emphasis on human activity alone means that specific ecological impacts cannot be inferred. However, some form of validation is critical in order to establish the extent to which there is agreement between the model and the real world. The validation process consists of showing that a model accords with facts as known, with what is accepted as true or reasonable, or is justifiable and appropriate for a stated set of purposes.

One attempt at validating the AWI approach, which was carried out at WCMC, focused on forest reserves in Uganda and Sri Lanka. The study compared a number of measures, including average wilderness scores and surrounding human population density, with expert on-the-ground evaluations of the relative condition or naturalness of each reserve. The AWI averages were the measure most closely correlated with the expert evaluations, and were more effective than, for example, human population density in predicting forest condition. This suggests that the approach may well be appropriate as an indicator of forest naturalness at broader scales.

REFERENCES

Dallmeier, F. 1992. *Long-Term Monitoring of Biological Diversity in Tropical Forest Areas*. MAB Digest, UNESCO.

Gould. S.J. 1977. Ever Since Darwin. Norton Press.

MacIver, D.C., 1998. *Atmospheric Change and Biodiversity. Environmental Monitoring and Assessment* 49:177-189.

Smit, B. I. Burton and R.J.T. Klein. 1999. The Science of Adaptation: A Framework for Assessment. Mitigation and Adaptation Strategies to Global Change 4.

Stankey, G.H. 2001. Too Early to Tell, or Too Late to Rescue? Adaptive management under Scrutiny. US Forest Service, PNW, Science Findings, April 2001.

Watson, R., M. Zinyowera, and R. Moss (editors). 1998. The Regional Impacts of Climate Change. A Special Report of IPCC Working Group II. Cambridge University Press, Cambridge.

9

Computerised Forest Management

There has been a rapid increase in the use of the Internet in developing countries, although this expansion is still largely an urban phenomenon. Rural communities represent the "last mile of connectivity" in both developing and developed countries with regard to access to Internet services and the telecommunication connections that help transmit those services. People in rural areas are generally unable to take advantage of the services available to their urban peers.

Internet initiatives for rural development need to be approached with a degree of caution. One cannot expect less privileged farmers and food-insecure residents of rural communities to list computers and digital telecommunication services as high-priority items for improving their lives. However, there are various intermediaries that serve these populations which, together with small and medium enterprises (SMEs) in rural areas, could take advantage of these technologies to improve their work, improve communication capacity, gain efficiencies and reduce telecommunication costs. With SMEs, intermediary organizations such as extension field offices' rural NGOs, health clinics, government satellite offices and church organizations can offer benefits in numerous ways. Strategies for improving access to the Internet and use for rural development will necessarily involve the full participation of intermediary organizations and other rural stakeholders.

Today we truly live in a global village, but it is a village with privileged "information haves" and many "information have-nots". To face the unprecedented challenges brought on by the changing global economy, dynamic political contexts, environmental degradation and demographic pressures, and to make critical decisions, people at all levels of society - especially the food-insecure and the organizations that serve and represent them - *must be able to access critical information and to communicate*, improved communication and information access are directly related to social and economic development.

POSSIBLE RANGE OF PROJECT ACTIVITIES AND OUTPUTS

In the case of forest-based rural development activities, depending on local circumstances, the following range of outputs can be achieved through the use of Internet applications:

indigenously developed learning tools, learning games, learning networks

- interactive extension information networks
- local, regional and national market information networks
- distance learning programmes and networks
- participatory research and action networks
- interactive expert systems networks
- early warning system information dissemination and data gathering networks
- shared curriculum databases and curriculum development networks
- research sharing and information dissemination systems
- training tools
- small and medium enterprise development networks and marketing mechanisms
- rural media networks (e.g. rural radio, newsletters)
- indigenous knowledge networks
- disease and pest monitoring action networks (plant and animal)

The Internet is a multipurpose tool, a medium of communication and perhaps the most flexible medium currently available. It has the potential to be integrated within a wide variety of efforts that have objectives such as local participation, training, education, research (especially participatory research), technical support and institutional strengthening. Thus, endeavours that might find a role For Internet applications could range from: training in forest harvesting to community forestry or the development of criteria and indicators for sustainable forest management.

A DECENTRALIZED "PEOPLE'S NETWORK"

The Internet was conceived and designed in 1963 by Larry Roberts, working for the Advanced Research Projects Agency (ARPA) with funding from the United States Department of Defense. ARPAnet, as it was then called, emerged as a communication tool in the late 1960s for a handful of Department of Defense workers and contractors. It was designed to be a fail-safe communication system because it would be fundamentally *a decentralized* network. People could send packets information from one computer to another, across the United States, and those packets could travel by a variety of routes to reach their destination. If one or several routes were destroyed or malfunctioning, the packets would find alternative routes and eventually reach their destination. ARPAnet grew during the late 1960s and 1970s because new "nodes" and routes were added to include university researchers. The many benefits of electronic mail (e-mail) were attractive to users and, as a wider community of faculty and students began to use the tool, it quickly gained popularity as a communication tool in North America. Because it was a decentralized network, there were few means to control its popular expansion, and it soon transcended its Department of Defense mentors.

The Internet today is a people's network. Anyone with basic computer equipment and a telephone line can connect to it, communicate through it, host information on it and look through it. Unlike many other media such as television and radio, every user of the medium can be an information producer and knowledge sharer. No one knows for sure how many people are using the Internet today, but estimates range from 40 million to 100 million people.

Communicating over the Internet costs a fraction of what it does to use traditional telephones or fax because information flows in discrete packets of digital bits that are able to share telecommunication lines with hundreds of other packets. Where a traditional transatlantic telephone call will tie up a single telephone line for only two people, an e-mail message can travel along a telephone line with hundreds and even thousands of other messages. Thus, the Internet is cheap, powerful, decentralized and in the hands of civil society.

A fact that is well known to veteran Internet users is that the Internet has the power to cut across social and geographic distance and help people find new ways of facilitating the flow of information and knowledge. Within bureaucratic organizations it has a way of levelling hierarchies, facilitating new communication patterns and helping enable activities that might not otherwise occur.

With regard to Internet use in support of rural and agricultural development, applications fall into five main areas: economic development for agricultural producers, community development, research/education, SME development and media networks. The following sections explore these areas and, where possible, highlight their actual or potential application to forestry.

APPLICATIONS IN PLANNING AND MARKET INFORMATION FOR AGRICULTURAL PRODUCERS

Rural communities and small-scale agricultural producers, including those involved in forest-based activities, are deeply affected by global economic, environmental and political forces. The idea that communities of small-scale agricultural producers are isolated and living in closed, self-sufficient societies is a myth. In fact, with the appropriate knowledge, small-scale producers can even have a competitive edge over larger operations. When knowledge is harnessed by strong organizations of small producers, strategic planning can be used to provide members with lower-cost inputs, better storage facilities, improved transportation links and collective negotiations with buyers. The Internet is one tool that can enhance this flow of information. It is an inexpensive way to communicate and access global information. Local Internet services can be easily managed by well-organized local user groups and farmers' organizations. Information and analyses can be tailored to local, regional and national knowledge and communication needs and realities. When combined with national and global market information systems, and with the ability to communicate quickly with potential buyers and brokers, local Internet systems become valuable strategic planning and decision-making tools.

Community information centres and farmers' organizations can also gather information from the Internet and disseminate it via local radio stations, newspapers and other local information-sharing networks and tools. For example, daily market prices and agricultural news can be posted at cooperatives, local stores, transportation hubs, agricultural supply outlets and social gathering points. Simple newsletters can be developed using Internet information and distributed to members of farmers" organizations. When integrated with other media tools, the Internet can be a powerful information resource and research tool.

COMMUNITY DEVELOPMENT APPLICATIONS

"Modern communication technologies, when systematically applied and adapted to conditions in rural areas of developing countries, can be used for rural communication to increase participation, disseminate information and share knowledge and skills The establishment of new institutional frameworks, including all stakeholders, which are autonomous and income-generating, can lead to sustainable and cost-effective efforts, as opposed to working only with government agencies."

Internet services are also valuable when placed at the service of rural development-oriented organizations which act as local communication conduits or intermediaries. Along with providing improved market knowledge, they can also:

- develop locally appropriate applications and creative services;
- provide knowledge about successful development strategies;
- enable efficient regional, national and global organizational efforts (the use of the Internet as a global communication and organizing tool in Chiapas, Mexico, is an excellent example of the latter);
- improve access to a huge variety of information, training, research and educational resources (including distance education services) which are typically unavailable in rural and remote areas because of the costs associated with printed materials and books;

— enable rural young people to learn about computers and to have access to the technologies and information available to their urban peers;

— provide access to critical technical information for rural professionals such as physicians, health care workers, technicians and engineers, thereby providing further encouragement for these professionals to continue practising in rural and remote communities;

— be used as marketing tools to promote rural tourism and market the products of small secondary industries and home-based businesses;

— enable local NGOs to gain a global presence and make better contact with potential donors and supporters through the on-line publication of resources and information and through the use of electronic mail; and

— sensitize urban policy-makers to the realities and needs of rural populations.

RESEARCH/EDUCATION APPLICATIONS

Toolnet is a network for small-scale development projects that fosters exchange of information, experiences, expertise and solutions to technical problems. It provides multifunctional electronic mail to link field workers, local organizations, technological institutions, international development organizations anti individuals... directed toward technology transfer among developing countries... Points are operating or planned in about 25 countries worldwide.

Within national, regional and international research communities, increased attention has been directed towards "participatory research" strategies. These strategies place farmers and rural residents at the centre of the research process and enable them to enrich their knowledge base and share it with one another as well as with field workers, researchers and decision-makers. Internet use among intermediary organizations and leaders involved in participatory research can provide a cost-effective method for documenting and sharing lessons learned and research results.

Internet use also has the potential to strengthen linkages between and among farmers' organizations, extension workers, researchers and policy-makers.

The cost of accessing printed academic materials within developing countries is usually so high that students and faculty members have great difficulty acquiring books and journals. Furthermore, the time required to obtain printed materials from overseas can be long enough to render some information outdated by the time it arrives. Via the Internet, any information published on-line can be accessed almost instantly and at a fraction of the cost of obtaining printed materials. Information on the Internet is easy to access and archival lists of resources can be easily reviewed and assessed in remote locations.

Electronic distance education services are already in use in North America, Australia and Europe (particularly among people in rural areas), and with the continued growth of Internet access in developing countries there is a very good chance that similar services will develop a significant demand. Distance education (as well as traditional education) partnerships between universities in the North and the South (such as the University of Guelph's partnerships with universities in Cameroon and India to develop distance education extension worker training programmes) have proved to be beneficial to the institutions involved. With the assistance of Internet tools, these partnerships can be further strengthened and Internet learning resources can be cooperatively developed across oceans to be utilized by participants in developing nations. Of course, this process can work in the other direction too, to enable students in the North to learn more about the conditions, challenges, potentials and knowledge development of the South.

Overall, the Internet holds significant potential to enhance learning and research relationships among researchers, academics and students wherever they may be located. The list of potential applications is infinite and thousands of informal linkages of this sort take place every day in Internet discussion groups. Development agencies such as FAO can play a role in helping to formalize and provide credentials and diplomas for people who participate in specific electronic learning initiatives delivered via the Internet.

SMALL AND MEDIUM ENTERPRISE DEVELOPMENT

The removal of international trade barriers has brought quickly changing global markets. Large international corporations can now compete for the SMEs' market, but SMEs traditionally have not had the infrastructure or necessary resources to fight back. Our Mission [is] to provide a productive and professional Internet/WWW-based network to help SMEs communicate about business needs, share their resources and expand their markets.

Private sector businesses, large and small, are using the Internet to reach new markets, promote products and services globally and access critical business and financial information.

In many developing countries, the Internet is being used to promote national timber production activities, for example those of the Malaysian Timber Council and the Timber Export Development Board in Ghana.

The tourism sector has been quick to recognize the benefits of the Internet for advertising destinations, tours and holiday services. Of particular interest are the World Wide Web sites for "ecotourism", game parks and adventure tours in areas of southern Africa where rural tourism is a growing industry. Tourism operators in rural and remote areas have a difficult time marketing their destinations through traditional media owing to production and distribution costs. The Internet now represents a very inexpensive way for them to showcase their sites to the world and interact directly with potential tourists.

NEWS MEDIA NETWORKS

The news media in developing countries have also been at the forefront of developing Internet applications. For example, in Zambia, both national daily newspapers mirror their daily copy on the World Wide Web (http://www.zamnet.zm), making the local news accessible to expatriate Zambians around the world. E-mail discussion groups provide these expatriates with an opportunity to discuss the daily news with one another and with their Internet-connected peers in Zambia. A discussion group joined by the author generated a minimum of 30 e-mail messages per day! Such e-mail discussion groups for expatriates and nationals exist for virtually every developing country in the world and represent a relatively untapped resource for accessing the views, ideas and creativity of members of civil society with regard to development policy and initiatives.

In addition to the latter news and information applications, organizations such as Inter Press Service (IPS) Third World News Agency (http://www.link.no/IPS/eng/intro.html) use the Internet to source news stories from local writers in developing countries and share those stories with international wire services such as Associated Press. IPS is also able to provide Internet feeds that enable African news media to have access to African news from around the continent. This is particularly relevant to rural radio stations and other rural newspaper and newsletter producers that would otherwise be unable to obtain the same news from other sources. IPS can also provide an outlet for rural news writers to share their stories regionally, nationally and globally. Similar Internet strategies for rural radio networks, which might also incorporate digital audio transmissions, may well emerge in the near future.

The Internet is not a panacea for rural development, but it does bring new information resources and can open new communication channels for rural communities. It offers a means for bridging the gaps between development professionals and rural people by initiating interaction and dialogue alliances, interpersonal networks and cross-sectoral links between organizations. It can create mechanisms that enable the bottom-up articulation and sharing of local knowledge. Benefits include increased efficiency in the use of development resources, less duplication of activities, reduced communication costs and global access to information and human resources.

There are ten common elements among successful rural and agricultural Internet communication and information systems:

- preliminary participatory communication and information needs assessments with intended users
- awareness-building campaigns designed to sensitize decision-makers to the possible uses of Internet services
- executing agencies committed to participatory rural and agricultural development

- local private sector or not-for-profit (university or NGO) Internet Service Providers with a social service orientation
- open participation of user community in design, implementation and management of communication and information services
- institutional and user commitment to manage and sustain Internet services
- ongoing provision for technical training, user support and outreach within the user community
- combination of centralized and decentralized information production, analysis and distribution
- ongoing provision for technical support and system maintenance and upgrading
- user community with a financial stake in the communication and information system (e.g. ownership of hardware, user fees' salaries, infrastructure)

The Internet may help in meeting peoples' information and communication objectives in order to attain their development goals and objectives, but it must be integrated within human contexts and seen as a communication process tool, not simply a static "information technology" or unidirectional broadcast medium. Otherwise. Internet tools will be: relegated to the junk heaps of inappropriate development technologies or dismissed because of previous failures to make the medium locally relevant and useful. If, for example, the information outputs (in accessible form) derived from highly technical forest resource assessment systems are not made available to the people who live in and around forest areas(in addition to policy-makers at - national and international levels), then we are failing to ensure full leverage of the large infrastructural investments involved and assist people in making appropriate decisions based on such valuable information. Of course, the Internet is not the only communications tool that may be used, and radio and television may have equal or even greater potential, at least for the moment) We must avoid contributing to the gap between the information haves (experts, academics, researchers, policy-makers, etc.) and the information have-nots (usually the ultimate beneficiaries of development work) that can emerge when we create Internet applications to serve only privileged researchers and bureaucrats. In particular, we must strive to find ways to bring knowledge producers, such as researchers and policy-makers, closer (in the social as well as the geographical sense) to the other less recognized knowledge producers: the people who are the ultimate beneficiaries of their development programmes.

Early Internet users in developing countries have proved that they can develop excellent local services and locally appropriate knowledge resources. However, without support from development agencies, there is a risk that such efforts will never meet the needs of people in rural communities.

Given FAO's mandate to help improve the lives of rural populations and foster sustainable agricultural development, FAO has an important and historic role to play in insuring that the benefits of Internet and information and communication technologies reach rural and agricultural stakeholders. Several development agencies are currently assisting in the expansion of indigenously managed Internet services in developing countries. FAO pilot projects, linked to indigenous rural and agriculture organizations, can help ensure that rural communities remain part of regional and national Internet initiatives.

Adopting a proactive strategy and acting to bring the Internet to rural and agricultural communities in developing countries will help enable rural people to face the unprecedented challenges brought on by the changing global economy, political changes, environmental degradation and demographic pressures. To deal with these challenges and to make critical decisions people at all levels of society, and especially the food-insecure and the organizations that serve and represent them, must be able to access critical information and communicate. Improved communication and access to information are directly related to social and economic development (Tallero and Gaudette, 1995).

The time to act in support of Internet knowledge and communication systems in developing countries is now. Today we truly live in a global village, but it is a village with privileged "information haves" and many "information have-nots". With the new technologies available to us, we have an opportunity to change this.

INFORMATION SYSTEMS IN FORESTRY

Like all ages in the development of human civilization, in today's world the power of information drives our governments, makes our industries productive and competitive and enriches our personal lives. Humans today are better informed than at any previous time in history, but they are also assaulted daily with a barrage of information which is of no direct relevance to them, cannot be used because of inadequate technology or understanding, or is actually misleading and potentially harmful to them or to their organizations. In today's world it is not enough for managers to understand the business or organization for which they are responsible; they must also be able to manage the relentless flow of information, selecting what is relevant to the proper functioning of their business or organization and ignoring what is irrelevant or misleading for them. They must also recognize information that is needed but is not currently available or able to be used, and they must put in motion the processes that will enable them eventually to obtain and make use of this information.

This is easy to say but difficult to practice. Few foresters have had training in the information sciences. Even for specialists in information systems, the half-life of knowledge in this field is estimated to be about two years. This means that, after two years, half of what is currently considered the state-of-the-art in information management will have become obsolete. This is due principally to the rapid pace of technological development in information systems.

To some extent, of course, this has always been the dilemma of managers; as time passes their technical knowledge risks becoming outdated. In today's world this occurs faster than ever before, as effective management requires an understanding of information technology, and how to make it function on behalf of the manager's organization, as well as technical skills.

Information has always been an important "stock in trade" of forest managers. Knowledge of tree species, stand age, growth characteristics, site quality, soil and terrain classes, logging costs and product markets have all been essential for profitable forest management for many decades. What has changed is the technology for collecting, analysing and presenting this information, and particularly the speed at which this can be done. Whether they like it or not, forest managers must also be information managers. In particular, they must understand the implications for forest management of new types of information and new information technologies and they must understand how information needs will affect both the personnel and structure of their organizations.

Most authorities on forestry information systems recognize three types of information needs: i) information for strategic planning and analysis; ii) information for tactical (medium and short-term) planning; and iii) information for operational management and control. Each of these has its own requirements, although to some extent the technologies overlap.

Strategic Planning and Analysis

For any organization, the most fundamental strategic question is "What business should we be in?" In some cases, such as state forestry agencies, this question has a more or less obvious answer, although some of the details may change from time to time (e.g. some assets may be divested through privatization, or the organization's relative emphasis on production forestry versus amenity forestry may change). In the case of private enterprises, the question has no obvious or fixed answer. As conditions change, enterprises must change also or they will cease to be competitive. Thus, forestry enterprises should always be asking this question and they should always be open to the possibility of new answers.

In the early 1970s two forest companies in the State of Oregon were both largely dependent on the United States Forest Service for their timber supplies. In response to the above question one company's answer was: "We are a producer of lumber and plywood from timber purchased from the United States national forests. " The other company's answer was: "We are a producer of lumber and plywood, currently from timber purchased from the United States national forests hut, by 1985, we intend to be entirely self-sufficient for raw material supplies." The latter company undertook a difficult and expensive programme of timberland acquisition, both in Oregon and in the southeastern United States. During the 1970s and 1980s, its profits were generally lower than those of the first company. Then in 1991, a United States District Court suspended nearly all timber harvesting from national forests in the Pacific Northwest until the Forest Service could prepare a satisfactory plan for protecting the northern spotted owl and other old growth-dependent wildlife and plant species. This plan has now been delivered and it will reduce timber harvests on the Oregon national forests by two-thirds as compared with the harvest levels of the mid-1980s. The first company has now essentially collapsed and it seems likely that it will eventually file for bankruptcy. The second company is more profitable than ever before and its timberlands have increased in value enormously.

What does this have to do with strategic information? Both of these companies had access to the same basic information about the way society was changing in the United States. One company concluded that the environmental movement would have no fundamental effect on the management of the national forests by the United States Forest Service. This company benefited in the short term because its raw material costs were low. The other company concluded that the national forests would eventually become an unreliable source of timber. In the short term this conclusion was costly but in the long term it has enabled the company to thrive in what would otherwise be an almost impossible situation.

One major difference between these two companies may have been their use of strategic planning. The first company relied on its board of directors to make strategic policies, virtually without formal inputs other than balance sheets and income statements. The second company employed a full-time planning team to develop a series of simulation models aimed at exploring the consequences of various raw material and production strategies. The planning team also brought in outside consultants who specialized in anticipating societal changes. As a result of both quantitative and qualitative inputs from this planning team, the company's board of directors made the difficult decision to embark on a programme that would eventually lead to raw material self-sufficiency.

What kinds of information and information technology can help in making strategic decisions? The following paragraphs provide a list which probably applies to nearly all forestry organizations.

Simulation Models

These are useful not so much for forecasting what is likely to happen, but rather for making projections of the potential consequences of futures that might happen. If a company can see that bankruptcy is a probable outcome of its current practices, then it might well consider this question more seriously than otherwise. The availability of general-purpose software such as spreadsheet programs makes the development of computer simulation models much easier now than it was even a few years ago. What has not changed is that an organization that is serious about strategic planning must maintain a full-time effort in this area and ensure it has full support from the highest levels of management.

Forestry Data

A forest company or agency that does not have up-to-date, accurate information about its forests cannot make effective strategic decisions. This includes detailed information on forest boundaries, site quality, age classes, conditions of topography and soils and the location and condition of infrastructure. Increasingly, non-timber resource information such as recreation potential and the condition of wildlife habitat is also becoming important for strategic planning. Continuous timber inventories are of course essential, and models should also be developed for projecting growth and yield into the future so that strategy alternatives can be tested using computer simulation models.

From a strategic perspective, forestry organizations require information not only for their own forests but also for the forests of competing and client organizations, or those on which they depend for raw materials.

Market Information

Private enterprises, and to some extent also government agencies, rely on markets to set the levels of demand and prices for the goods and services they provide. Information about these markets and how they are likely to change in the future is essential for strategic decision making. Forest companies and agencies should continually ask "What kinds of products and serviced should be providing?". The answer to this question will change as society changes. Market information and analysis is required to anticipate such possibilities.

Information Related to Policy and Socio-economic Considerations

Forestry organizations, whether in the private or public sector, are in the "business" of providing goods and services that society values. For proper strategic planning it is thus essential for all such organizations to utilize experts who will help identify trends in the relative values society places on these products and services, and also questions of land tenure, legal status, organizational aspects (unions, associations, etc.), demographics, decentralization, democratization, local governance, etc. In today's fast-changing world, it is becoming clear that forestry organizations that fail to assess this kind of information on a regular basis are increasingly at risk.

Tactical (medium- and short-term) Planning

Most of what we normally think of as forest planning is really tactical planning. Even though foresters are accustomed to thinking in terms of one or more timber rotation periods, their plans typically concentrate on five or ten years into the immediate future, with less-detailed projections beyond that point to prevent surprises.

In its tactical plans, the forestry organization provides a tentative answer to the question "What kinds of operations will we undertake during this planning period and where will they be carried out?". For the most part, this is the type of question that makes the forestry profession attractive in the first place, and professionals spend a major part of their careers working to provide answers to it. The answers are always tentative because unseen factors such as changing markets and natural disasters may force fundamental changes in the plan. This does not negate the importance of the plan's quality. The best tactical plans, in fact, provide for such possibilities and make it easier to accommodate changes.

What kinds of information technology can contribute to better tactical planning in forestry? The easy answer is that most of the "traditional" information sources, such as maps and inventories, as well as nearly all the new crop of information systems, are focused squarely on tactical planning. However, providing a detailed, comprehensive answer is more difficult because the field is changing so rapidly and specialized knowledge is required to keep pace with it. Moreover, it is clearly necessary to figure in socio-economic considerations, not all of which lend themselves to information technology "solutions". Nevertheless, nearly all forestry organizations would probably benefit, most of them in substantial ways, from the information technologies described in the following paragraphs.

Remote Sensing Imagery

Aerial photography remains, as it has been for more than a generation, a fundamental tool for capturing forestry data remotely. Aerial photographs can be used with relatively inexpensive equipment, together with field sampling, to produce quickly and with satisfactory accuracy maps showing vegetative types, soils, streams and lakes, infrastructure and topography. The technology is well understood and has been fully integrated into many forestry organizations.

Orthophotographs are reproductions made from aerial photographs in which image displacements caused by the camera tilt and topographic relief have been corrected. When orthophotographs are assembled into a mosaic, the result is a photographic map with both scale and planimetric detail of a high reliability. Orthophotomosaics are often overprinted with topographic contour lines. The production of these orthophotograph maps requires special photomechanical equipment and is usually beyond the capabilities of individual forestry organizations; in most countries there are government or private sector organizations which can produce them.

Digital orthoimagery is a relatively new development in mapping technology. It is equivalent to orthophotography except that the imagery exists in digital form and can therefore be manipulated by software or used in combination with a geographic information system.

Satellite imagery used in forestry generally comes from one of two satellite sources: Landsat (United States National Aeronautics and Space Administration) or SPOT (French satellite system). SPOT panchromatic imagery has a higher resolution but Landsat Thematic Mapper imagery provides more wavelengths of data. Satellite images are well suited to the classification of digital data into broad land cover classes and to the detection of changes in forest area or condition over time. They are not replacements for aerial photographs because their resolution is so much lower. Furthermore, the analysis of satellite imagery is a relatively specialized field, requiring a considerable investment in computer hardware and software and specially trained personnel. Like aerial photograph interpretation, it requires field verification.

Geographic Information Systems

Computerized geographic information systems (GIS) represent an important development in information technology. A GIS is a computer hardware and software system designed to organize, store, retrieve, analyse, display and publish spatially referenced data. Although most GIS are capable of producing high-quality maps, their capabilities go far beyond this. The key to the importance of GIS, in fact, lies in their advanced analytical capabilities which enable managers to analyse complex situations rapidly, sometimes in entirely new ways (Berry, 1986). As an example, with GIS it is possible to "overlay" map layers representing timber classes, soil types, and topography so that a map can be produced which shows all areas of mature timber on fragile soils where the ground slope exceeds 40 percent. Because this "derived" map layer resides in the computer system, its area can be quickly computed. Inventory information can then be combined with the map data to estimate the reduction in the allowable annual harvest volume that would result if this area were removed from the allowable cut base.

GIS represents a tremendously powerful tool that has the potential to enhance greatly the capabilities of forestry organizations in tactical planning. To achieve this potential, the organization must be prepared to commit substantial resources in terms of both personnel and funding.

GIS will be useful for forestry analysis, only if the foresters use it. This means that all professional forestry staff (including top-level managers) should receive at least "awareness" training in GIS, and those who will use the system to extract information and carry out analyses should receive additional, more technical training. Sufficient hardware must also be provided so that each professional who will be using the GIS has convenient access to it. In most cases, the GIS will reside on a high-performance computer workstation, and individual users wills connect to it through a network accessed ' from their desktop personal computers.

Global Positioning Systems

A recent development in information technology that has managed to capture the imagination of foresters around world is the Navstar Global Positioning System (GPS), an all-weather navigation and position-fixing system based on radio signals transmitted by satellites launched and maintained by the United States Department of Defense. This network comprises a fleet of satellites which, together, provide position-fixing capabilities almost anywhere on earth, day or night. One important application of GPS is to provide data for GIS and, in fact, the system makes it possible to collect certain kinds of spatially referenced data that it would otherwise probably not be economically feasible to obtain.

Like all such technologies, acquiring skill in the use of GPS requires practice, and organizations which adopt the system must recognize that data accuracy will be relatively low at first but will improve over time as field crews become proficient both with the equipment and with the procedures for its correct use.

Operational Management and Control

As important as they are for strategic and tactical planning, advances in information technology are also changing the way forestry organizations manage their day-today operations. As with planning technologies, information systems for operational management and control are of many different types and they are used in different ways by different organizations. Some of the more common information systems currently in use or under development include the following:

Electronic Data Recorders

Electronic data recorders (EDRs) are basically hand-held computers which have been developed for recording field electronically. Many are available with software written specifically to enable the collection and-preliminary analysis of forestry inventory data. Those used in forestry must be rugged so that they can withstand exposúre to rain, cold and hot temperatures and rough handling. Besides a keypad, screen and memory for recording and storing data. EDRs must generally have some way (typically a serial port) for transferring data to a computer for incorporation into the organization's permanent database (which might, for example, be a GIS).

On-board Computers

With the advent of timber processors and cut-to-length harvesting machines, it is becoming more common for such equipment to include on-board computers which can help decide on the optimal crosscutting strategy for individual stems. In the future, it is likely that on-board computers will be connected through radio transmission links to a central computer at the mill. This will permit crosscutting to satisfy order requirements, rather than simply considering each stem independently, and will also permit crosscutting rules to be changed at intervals during the day as new information on product orders becomes available.

Manual Crosscutting

In places where trees are large and the topography is rugged, such as the Pacific Northwest of the United States, manual felling and crosscutting still predominates. Even here, through hand-held computers, technology is now available that permits crosscutting decisions to be rapidly optimized (Garland, Sessions and Olsen, 1989). Such practices can routinely increase log values by 10 to 15 percent as compared with conventional crosscutting methods.

Truck-mounted GPS

Some forestry organizations are experimenting with the use of GPS to help improve the efficiency of truck scheduling and routing. Trucks equipped with both GPS and automatic radio-transmitting equipment to send the GPS data to a central receiving point make it possible for transportation equipment to be tracked in real time on a display map of the operating area. The feasibility of such systems depends to a considerable: extent on the availability of reliable radio communications. Such systems have perhaps been implemented most extensively in the Nordic countries.

Real Time Data Transmissions

In an effort to reduce log inventories and improve order responsiveness, some forestry organizations are experimenting with the transmission of production and transport data from the production site to the mill and the transmission of order requirements from the mill to the production site. This is usually done by radio at intervals during the day. The intention eventually is to use automated transmission systems which would require little or no human intervention. Theoretically, it should be possible to maintain almost no log inventory at the mill, and to harvest and crosscut trees in response to specific orders. This will require more reliable radio transmission capabilities than are currently available in most countries.

COMPUTERIZATION IN ANDHRA PRADESH FOREST DEPARTMENT

The FAO Country Capacity Building in Forest Resources Assessment Project(GCP/INT/621/SWE), funded by Sweden, is providing technical assistance to India's Andhra Pradesh Forest Department in the introduction of computers and, in particular; of a geographic information system (GIS) to replace the current traditional method of storage, retrieval and processing of forestry information.

The State of Andhra Pradesh has a forest area of 6.3 million ha, managed by about 300 professional and 3000 technical staff. The Forest Department has more than 100 years of forest management tradition. However, recently it was realized that the department needed to increase its capacity for planning, implementing and monitoring large-scale forestry investment programmes. The baseline information about the state's forest resources was considered weak; information on the quantity and quality of forest resources as well as on supply of forest produce and on commercial and noncommercial demand for timber and non-timber forest products was lacking. Existing work plans were not realistic and often not implemented in their entirety.

In 1994, with financial assistance from a World Bank project, the state's Forest Department initiated a major effort to improve its forestry programme. The project includes components focusing on forest management; plant propagation; research and training; joint forest management; biological diversity conservation; fodder development, and tribal development. The project is providing incremental inputs in equipment and materials, physical facilities, staff training, technical services and technical expertise to carry out basic surveys, formulate strategic investment plans, develop timely and adequate working plans and monitor their implementation. It is supporting the development within the forest administration of a broad base of basic skills in computer data processing and in the use of satellite and other imagery for survey and mapping needed for the formulation of strategic plans and local operational plans. It provides financial assistance for improving the facilities for data collection, management and analysis and is also funding special surveys related to high-priority production and conservation areas within Andhra Pradesh's forest estate as well as to forest production on non-forest areas. All of these activities are being developed within the framework of an integrated management information system.

In October 1994, within the context of the larger effort, the Andhra Pradesh Forest Department signed a special agreement with FAO to provide the following services through the capacity-building project:

— conduct a relatively small-scale and low-cost, statewide forest survey based on remote sensing and limited field data;

— establish a GIS with the specific capacity to guide activities in joint forest management;

— improve planning capacity and computer infrastructure in order to permit the implementation of a new forest management system.

Scope of Computerization

The main areas of computerization being undertaken through the project are listed in Table 1 together with the name of the agency responsible for implementation.

Table 1. Computerization modules and responsibility for their implementation

Modules	Implementing agency
Office automation/local area network	A local company
Management information system (MIS)	A local company
Forest inventory system	FAO
Monitoring based on remote sensing and forest cover mapping at 1:50000	FAO in cooperation with forest survey of India
Geographic information system (GIS)	FAO with a local contract for digitization of input maps
Computers in planning&decision making	FAO with full involvement of Andhra Pradesh Forest Department staff

Statewide Forest Inventory

A statistically sound procedure has been developed for collection and analysis of the forest inventory data. The forest inventory data processing system (FIDAPS), an FAO software package for processing inventory data, was adapted to fit the needs of the state and data from two districts has been processed. This user-friendly software runs on a PC under Windows 3.1 and has now been installed in Hyderabad where it is being used by local staff.

GIS Planning and Implementation

An information needs analysis has been conducted to find out what kind of spatial information is used and by whom, the scales used, what base data are available and so on. As a part of the development study, the forest inventory data were combined with the georeferenced databases derived from remote sensing. With the combination of the two more precise estimates of forest cover, biomass and other characteristics were obtained.

In cooperation with FAO, the Forest Department organized a workshop with the participation of all levels of its staff to make the staff aware of problems and potentialities associated with GIS. A methodology for the priority zoning of silvicultural treatment using GIS data has also been implemented and this identifies joint forest management areas by treatment type.

Office Automation and Networking

In the first instance, computers are being provided only at the head offices and circle and divisional offices, which are located in cities and big towns. The range offices, generally located in small towns, will not be provided with computers owing to uncertainties of power supply but also because of a lack of proper maintenance support. The range officers will use computer-coded forms and periodically enter the data at the divisional offices. Although not an ideal solution, it is the only one considered feasible today.

New Planning System

Work on the planning system is currently in progress. Two types of planning situation are being distinguished development planning and forest management planning. Development planning will be carried out at the district and state level with the objective of defining broad goals and identifying required means without going into prescriptions or other specific details. The planning horizon will be five years for the medium term and 15 years for the long term.

Forest management planning will be carried out at the divisional level. The divisional forest officer will prepare a five-year and an annual management plan of the division consistent with the larger development plans.

A statewide continuous forest inventory and mapping exercise is proposed so that all districts and, therefore, the whole state will be covered by specialized inventory teams using a systematic sampling design. In the fifth year, the idea is to revisit the plots inventoried in the first year to monitor changes in the stand and site. The remote sensing and mapping is proposed to be done using data from the Indian remote sensing satellite LISS-III (23 m resolution) for the whole state in a uniform manner and establishing a division-, district- and state-level GIS database. At the village level, PAN data with a resolution of 5.8 m will be utilized.

Silviculturists, using the comprehensive georeferenced data and their specialized knowledge, will classify state-owned forests into working circles (treatment types) on a uniform basis and provide prescriptions for their optimum handling. The prescriptions by working circles will be revised as and when new silvicultural knowledge on treatment of various types of forests and their natural and artificial regeneration becomes available. Silvicultural prescriptions will be written in local languages so that local foresters (section officers) can understand and implement them.

Human resources development, not only to run the computers but also to make productive use of them, is a critical factor. Computerization should not be seen as the introduction of a new tool into the old system, but rather as the opportunity to make a quantum jump in management efficiency. The old system must change to cater for new needs. Selection of hardware and software and the customization of MIS and GIS are already demanding tasks in view of fast-changing technology, but they do not present an unsurmountable problem. There is enough expertise and knowledge available in the department or commercially in Hyderabad, if needed. The real problem lies in operationalization or institutionalization of the technology.

Information-based planning is not an extension of the traditional system. It requires a review of the entire decision-making and control process and the establishment of a new system based on the enhanced use of information and knowledge. To receive the full benefit of computer technology, the system change must affect all levels of an institution and its functions. Jobs need to be redefined and responsibilities placed in a new context. A reorganization of the institutional structure must involve a reallocation of posts.

In Andhra Pradesh, the computerization process has now reached precisely this level: progress is: slow and there is not only resistance but active opposition to change. However, there are signs of hope. Hard decisions are called for and being made to get the full benefits of a new technology requiring a new way of thinking and decision-making. With wise computerization will come a new planning system and way of working which wilt not only enhance the state's forestry sector but, more important, make a lasting impact on the life of the people, including the least privileged of the poor who live in and around the forest of Andhra Pradesh.

FAO COMMUNITY FORESTRY UNIT ELECTRONIC CONFERENCE ON CONFLICT MANAGEMENT

FAO's Community Forestry Unit is now conducting Phase II of FTPP. The aim of the programme is to improve the livelihoods of rural communities through the sustainable management of tree and forest resources. In recent years, the institutions working in partnership with the FTPP have identified conflict management (also known as dispute resolution) as a priority activity at both the global and regional levels. In 1991, in response to requests from various countries, a research project was conducted to examine the relevance of conflict resolution literature to community forestry. In October 1992 the Community Forestry Unit formally launched a programme on Conflict Management and Community Forestry as part of the FTPP and has since carried out a number of national activities in Latin America, Africa and Asia.

Recently, the development of the topic reached a stage where it was necessary to:

- assess the current understanding of the topic and identify gaps in the knowledge; gather information at the global level for the development of a conceptual document;
- strengthen the institutional capacity in the regions; and
- work at both the global and regional levels to define the role and objectives of both the unit and the programme in conflict management, incorporating the specific needs and strategies in the regions.

Choosing the E-conference Medium

E-conferencing was selected as a means for achieving the objectives in conflict management set out above because it was felt that it would be an excellent opportunity to:

- reach a wide cross-section of individuals and institutions in the different regions, particularly those involved in conflict management and natural resource management;
- develop global knowledge based on comprehensive and widespread regional inputs;
- propose regional activities based on a global understanding;
- advance the institutionalization of the topic (by opening direct communications with numerous individuals within an organization or institution); and
- generate outputs on the topic more quickly than through other media.

An electronic conference (e-conference) is a process which allows a group of individuals to discuss a given topic through the medium of electronic mail An e-conference uses listserver software to create a list of e-maid addresses with one collective address for the whole list. Conferees follow a subscription process to add their e-mail address to the list. Once subscribed to the list, conferees can send messages to the collective e-mail address from where the messages are relayed to all the other addresses on the list. An e-conference differs from electronic mailing lists, newsletters and news groups (which often use the same listserver software) in that discussions follow a set agenda for a defined period of time, discussing specific issues or sending out papers and asking conferees for comments. In addition to proceedings, e-conferences can be geared towards developing specific outputs such as bibliographies, conceptual documents, collections of case-studies, position papers and action plans. One of the key distinctions between an e-conference and other kinds of lists is that an e-conference is usually moderated by a subject matter expert and/or facilitator who lead(s) the discussions while other kinds of lists may not be moderated.

Preparation for the e-conference

E-conferencing is a relatively new medium, and the e-conference on conflict management was one of the first such events undertaken at FAO. When, in March 1995, initial research and planning for this event was undertaken, information on e-conferencing was not widely available, particularly with respect to possible models for structuring and operating such conferences.

E-conferences vary tremendously in scale, complexity and organization, desired outputs and, therefore, in their preparation time and staffing requirements. It is important to bear in mind that the e-conference on conflict management represented only one approach to e-conferencing.

This particular e-conference was organized by a central team or secretariat working from the Community Forestry Unit at FAO headquarters in Rome. It used a total of eight electronic conference rooms (listservers), provided by FAO/UNDPO's SARD-FORUM conferencing facilities. SARD-FORUM is an electronic, venue for the global exchange of information among people involved in sustainable agriculture and rural development (SARD).

It is a cooperative initiative of FAO's Sustainable Development Department (SDD); the Sustainable Energy and Environment Division (SEED) of the United Nations Development Programme (UNDP); and IN FORUM, an international non-governmental organization. (A training manual on how to organize and operate different types e-conference is currently in preparation by INFORUM in conjunction with FAO's Staff Development Group of the Human Resources Planning Service.)

E-mail connectivity

To help overcome limited e-mail connectivity in some regions (particularly in Asia and West and East Africa), the regional focal points were encouraged to identify individuals and institutions with access to email in the different regions, and to set up working groups to participate collectively through these connections in the e-conference discussions. Alternatively, national and regional working groups without e-mail access could fax or mail comments and outputs from their meetings to the regional focal point, which served as a liaison with the e-conference secretariat in Rome. The focal points would then distribute material from the e-conference in printed format for discussion in face-to-face working group meetings.

Conference discussants and participants

To oversee the relevance and quantity of contributions circulated in the e-conference, the secretariat decided to create two categories of conferees, discussants and participants, and to assign each group to separate e-conference rooms (listservers).

Discussants were actively recruited by the e-conference secretariat: more than 100 invitations were sent to individuals and institutions working in community forestry, conflict management and related areas several months prior to the start of the e-conference.

Discussants were requested to follow the proceedings closely and to provide comments and insights related to their particular area of expertise. The aim was to create a core group that could be relied on to provide regular and informed participation throughout the three months of discussions.

During the initial planning and preparation phase (June-August 1995), the secretariat consisted of a coordinator and a logistical assistant working part-time. For the intensive preparatory phase (September-December 1995), a community forestry intern assisted the secretariat, with all three members working full-time on the e-conference.

Operation

Once the eight e-conference rooms (listservers) had been set up by the technical staff at SARD-FORUM, the operation of the rooms was largely overseen by the logistical assistants at FAO headquarters in Rome. Occasional technical support was required from SARD and FAO e-mail administration staff to resolve software problems. In total, 95 discussants and 368 participants (463 conferees altogether) from 55 countries subscribed to the e-conference, with another three countries represented by non-electronic working groups. Included were practitioners, researchers and students from a range of disciplines (including the social and technical sciences) and institutions. From the field of conflict management there were mediators and trainers in conflict resolution (both from the field and from academic institutions), while natural resource management was covered by people working at the policy level, in the legal system and on project planning as well as by field project personnel.

Of the 55 countries which directly subscribed to the e-conference, 31 were developing countries (with non-electronic working groups in a further three developing countries participating through the regional e-conference focal points). While there were fewer e-mail addresses subscribed from the regions, in many instances there were several people participating through one e-mail connection. Approximately 20 percent of the conferees sent in at least one e-mail message. Contributions were less frequent than originally anticipated, but the discussions did generate a large volume of material (more than 1.5 Mb). Contributions sometimes consisted of well-researched and referenced pieces of up to three pages in length.

In addition to the discussions taking place in the e-conference rooms, a number of parallel meetings and events were held in the regions. The e-conference focal points coordinated the activities of 17 national and regional working groups in Asia (Nepal), East Africa (Ethiopia, Kenya and the United Republic of Tanzania), West Africa (Benin, Burkina Faso, Cameroon, Mali, the Niger and Senegal) and Latin America (Bolivia, Chile, Colombia, Costa Rica, Ecuador, Nicaragua and Venezuela). Each of the groups developed their own strategies for participating in the e-conference discussions. Some of the working groups circulated material from the e-conference to their own electronic networks. Others, without access to e-mail, held satellite meetings where printed material from the e-conference was distributed for discussion.

Costs

The costs involved in e-conferencing can vary tremendously depending on the duration of the event, the number of conferees involved, the objectives and desired outputs of the organizers and staffing requirements. E-conferences can cost as little as US$ 5000 (when, for example, the event is run from an organizer's computer as an "extra" or part-time activity involving no staffing costs and a limited number of conferees). A more average cost (based on e-conferences held through INFORUM and SARD-FORUM, involving a moderator, subject matter expert/facilitator, information manager and conference coordinator) would be between US$ 40000 and $ 50000. The cost of organizing and operating the Community Forestry Unit's e-conference, which was particularly large and complex, was almost US$ 250000, of which the main budget of some US$ 173000 was supplemented by donations in equipment and staff time valued at an additional US$ 66000.

Lessons learned

This section addresses some of the lessons learned from the e-conference, based on reflections of the organizers and feedback received from conferees.

The logistical aspects of the e-conference were not complicated. Initial technical support was required to set up the eight e-conference rooms (listservers) for the e-conference on conflict management, but the operational tasks (e.g. overseeing subscriptions, loading electronic files in the SARD-Library, managing the circulation of messages) were quickly picked up by the two logistical assistants, neither of whom had been involved in e-conferencing before.

The number of conferees was manageable. It would have made little difference, in terms of workload, if an even larger number of conferees had subscribed to the e-conference. Thus, depending on the objectives of the organizers, and the administrative resources available, it is not necessary to set a limit on the number of conferees.

It was extremely helpful to have a mechanism for monitoring the circulation of messages. Contributions sent to the participant e-conference room were first channelled to the secretariat for review, which allowed "junk mail" to be filtered out and lengthy contributions to be edited or summarized. Also, in all eight e-conference rooms, messages longer than 40000 bytes or containing attachments were automatically diverted to the secretariat. It was important to control both the volume and quality of the contributions circulated to conferees. People quickly lose interest if they are "flooded" with messages. Also, costs for participating in an e-conference can otherwise be prohibitive for many people (who have to pay per page/transmission to receive e-mail), while others have accounts that can only receive a set amount of e-mail.

The e-conference focal points were an effective means for stimulating participation and disseminating information in the regions. A large number of people in the developing regions, who would otherwise have been excluded because they did not have e-mail access, were able to participate in the discussions through national and regional working groups coordinated by the e-conference focal points. As the global e-conference discussions were held in English, the regional working groups also provided people with the opportunity to work together in their own languages.

Preprepared materials were helpful for focusing the discussions and were widely appreciated by the conferees. Of the participants who responded to the final evaluation, two-thirds indicated that they had circulated material (the discussion papers in particular) to colleagues in their institutions, government officials, officers in forestry and natural resource departments and directors and field coordinators in NGOs. Some of the respondents also indicated that they were using discussion papers and virtual conflict cases from the e-mail conference in working groups, student seminars, for project planning and implementation or to develop research projects. Whether discussion papers, opening statements, case-studies and/or virtual cases are used, it is helpful to develop some sort of input for an e-conference, both to help focus the discussions and as an incentive ("carrot") for people to subscribe and contribute to the e-conference.

The schedule for the e-conference was (overly) intensive. The e-conference ran for a total of 15 weeks and was divided into several one- or two-week sessions, including five main two-week sessions in which two topics or papers were discussed simultaneously.

The final conferee evaluation indicated that many people had difficulty in keeping up with the discussions and reading all of the material that was circulated. In the developing regions, the two-week sessions did not allow enough time for people to download and read the material from the e-conference, translate and/or circulate this material in printed format to the working groups as well as organize working group meetings to formulate responses.

With hindsight, it might have been better to have included fewer items on the agenda, or to circulate shorter or fewer discussion papers. Alternatively, the e-conference could have been run over a longer period (most e-conferences run for six months or longer); however, this might have posed the problem of maintaining the conferees' interest over an extended time.

The logistical aspects of the e-conference were time-consuming. The logistical assistants processed thousands of "service messages" from the software supporting the e-conference rooms (such failed mail notices) and replied to more than 500 administrative or technical inquiries from conferees. In addition, more than half the conferees joined the e-conference once it was already underway and needed to be updated.

The "open subscription" process for the participant room was problematic. Undoubtedly, circulating the subscription instructions (for the participant room) in promotional material meant that a large number of people quickly learned about the e-conference and subscribed to it. However, apart from a list of e-mail addresses, it was difficult to collect information about the 463 individuals and institutions that had subscribed to the e-conference. There was a period for introductions at the beginning of the e-conference, but only one-third of conferees sent in an introduction. Consequently, it would have been better to use a registration system whereby conferee data (name, address, institutional affiliation, nationality, etc.) were collected prior to giving people subscription instructions.

It was not necessary to have two different categories of conferees, nor separate e-conference rooms for the two groups. By directly inviting principal discussants to participate in the e-conference, the secretariat had hoped to create a core group committed to following the discussions closely and offering comments and insights related to their area of expertise. However, it was not possible to "guarantee" a certain level of participation in advance. Owing to time constraints and other commitments (e.g. travel), discussants did not participate as actively as anticipated. In fact, it was the participant group who proved to be the most "talkative". In retrospect, all conferees could have been assigned to the same e-conference room for the plenary discussions.

Assessing the medium

In addition to lessons-learned from the organizational and operational aspects of the e-conference on conflict management, several insights have also been gamed into the strengths and weaknesses of the e-conference medium in general.

The first obvious strength of an e-conference is that it can include and reach far more people from more places for less financial cost than can any conference attended in person. A second obvious strength is that organizers and conferees can immediately and easily "network" with a large number of people, and quickly disseminate and share a large amount of information.

From the point of view of substance, the e-conference actively contributed to the integration of the fields of community forestry and conflict management and helped define the interface between the two fields. It accelerated the conceptualization of natural resource conflict management as an important issue and helped to increase the visibility of institutions working on the topic. The e-conference significantly advanced the Community Forestry Unit's own development of the topic: a concept note, annotated bibliography and collection of case-studies are planned for publication in 1997.

Another strength is that the electronic medium works entirely in the written word, generating a verbatim transcript/permanent record. This had the effect of eliciting very thoughtful and organized discussion pieces or responses to others' messages. A further advantage is that people who join the e-conference late or miss part of the discussion are in a position to "catch up" in the reading and feel comfortable contributing to the discussion, secure in the knowledge that they are not repeating what others have said or speaking out of context.

As to the issue of cost-effectiveness, e-conferencing is not as expensive as one might imagine: the money saved by not holding a "real" conference (e.g. no travel or accommodation costs), is spent in other ways, including creating capacity at local and regional levels and building "info-structure".

An e-conference could also be valuable when used in conjunction with other communication or conferencing forms. For example, e-conferencing could be used prior to or as a follow-up to a face-to-face conference, workshop or consultation in developing an agenda or position paper, writing the proceedings, etc.

Weaknesses of the e-conference medium

Even if e-mail access is now available throughout the world, and even as its use expands, the cost of computers and the medium's technology does exclude large numbers of people, especially those removed from national and regional capitals in the developing world and people outside academic institutions or development organizations. An unfortunate irony of the medium is that people from universities in North America and Europe are typically provided free access to e-mail, while many who were participating in this e-conference from developing regions had to pay to send or receive messages.

There are tradeoffs in how people spend their time in an e-conference compared with a face-to-face conference. E-conferences often last months, while face-to-face conferences typically last a week or less. Attendees of real conferences, if they are held in a place other than where they are working, are usually in a position, if they wish, to devote full time to the discussions. Conferees in an e-conference participate at their regular workstation, they do not get "release time" from their normal work activities and are expected to participate as an "extra activity" with neither compensation nor professional recognition for having participated. Finally, people involved in an e-conference feel that "virtual relationships" are no substitute for the in-person networking that occurs in other conferences.

Some of the medium's strengths are also its weaknesses. People contributing to an e-conference might find the process both inhibiting and intimidating, knowing that their written statement is part of a permanent record. The time and care that people take, writing and rewriting their messages, make the exchanges less spontaneous. A related weakness has to do with the "ownership" of ideas: even if conferees are given a recommended citation format for referencing others' papers and contributions, conferees might feel reluctant to discuss their original research, fearing that others might "steal" their material. Finally, when ideas are written down, not only are they more tangible but they also can be seen as quite voluminous; without careful facilitation, readers can feel overwhelmed with the volume of e-mail messages.

E-conferencing has several other weaknesses as a means of communication and interaction. First, however quickly email messages are transmitted, face-to-face (or telephone) exchanges are much faster and less time-consuming e-conferencing does not appear to be as effective as face-to-face communication as a means of building a dialogue or coming to a consensus (apparently the silence represented by an empty "in box" is less uncomfortable in "cyberspace" than silence in a real conference or meeting).

In conclusion, e-conferencing proved an efficient and cost-effective means of networking and disseminating large amounts of information quickly. It provided the Community Forestry Unit [through FTPP) with the opportunity to involve both a large number and a diverse range of conferees in its discussions e-conferencing would appear to be particularly effective when used in combination with other communication or conferencing forms (e.g. as a preliminary or follow-on activity for face-to-face workshops, conferences and consultations). However, e-conferencing also has several limitations. It can be time-consuming both for organizers and conferees. It can be difficult to encourage regular and widespread participation in an e-conference and to involve some categories of actors (e.g. policy-makers and grassroots organizations), either owing to time constraints or limited e-mail connectivity in many parts of the world (particularly in developing countries).

ELECTRONIC OPPORTUNITIES FOR FORESTRY EDUCATION NETWORKING

A search of the Internet in October 1996 identified about 30000 documents containing the keyword "forestry education", about 80000 documents containing the keyword "forestry", about one million documents containing the keyword "education" and about three million documents containing the keyword "university". It can be imagined that, within this large body of information, there are many different types of applications. In addition, Internet software, servers and connections are being improved so rapidly that new instruments and options are appearing almost on a daily basis. The five following examples of Internet's electronic applications for higher education, and particularly forestry, emphasize the potential of the Internet to deliver solutions to problems that could not previously have been addressed by most universities of developing countries at a feasible cost.

There are a large number of universities and private companies working to develop Internet-based distance learning services and applications. Studying a course via the Internet means that, in addition to the features of a distance learning course carried out through mail correspondence, the student can also communicate with his or her instructor and fellow classmates via electronic mail (e-mail) and electronic conferences (e-conferences), submit assignments in real time via e-mail and participate in electronic tutorials from home. These aspects make studying via the Internet particularly suitable for those who live far away from university centres or for those whose employment or domestic situation makes it difficult to travel to a study centre. Courses currently presented via the Internet normally have the paper-based materials, any audio or video tapes and home experiment kits sent via conventional surface mail or courier before the course begins, although some course material may also be available in electronic form.

On-line distance learning could become a very powerful tool to promote cooperation and exchange among forestry educators from different countries, and, in particular, to facilitate the exchange of educational expertise and technology. Course syllabuses, texts, exercises, exams, lecture notes and other such materials could be exchanged at a very low cost through the Internet among universities participating in networks or other cooperation agreements.

A common constraint for many universities in developing countries is the difficulty in maintaining up-to-date libraries and other forms of data and information banks. Without access to these resources, it is very difficult to improve the quality of higher education. In this area, the Internet is already having a vast impact.

Today, the Internet provides access to the catalogues of almost all the main libraries throughout the world, either free (if the cost of the Internet connection itself is discounted) or at a very low cost. All of the main bibliographic databases, such as FAO's AGRIS and CARIS (www.fao.org/library/default.htm) or CAB International's Abstracts (www.cabi.org), are now searchable online. In addition, several of the most important libraries have now set up services to provide low-cost copies of the actual documents to overseas users.

There are several large projects under way to improve the situation further by creating "virtual libraries", in which it will be possible not only to obtain catalogue information online but also to download the actual documents in digital format. An example is FAO's Virtual Library Project which may eventually place the entire FAO collection on-line, including several million forestry-relevant documents.

Accessing basic information about universities

The Internet makes it possible for universities to make information about their departments, the undergraduate and graduate degrees they offer, the curricula of each degree, the faculty and their current research and recent publications, upcoming events, scholarships and other resources for students available to an incredibly wide audience.

Examples of universities that are already using this potential in the field of forestry include the School of Forestry and Resource Conservation of the University of Melbourne (www.agfor.unimelb.edu.au/agroforestry/agfor.html) and the Department of Forestry of Aberdeen University (www.abdn.ac.uk/~for257/index.htm). Universities can also share information about the syllabus of their courses.

CollegeNET administers a searchable database with information on 2500 colleges and universities (www.collegenet.com). Another database with thousands of universities listed can be found at www.mit.edu:8001/people/cdemello/univ-full.html

The FAO Directory of Forestry Education and Training Institutions, available on-line at (http://www.fao.org/WAICENT/faoinfo/forestry/free/foedu/foedu.htm) provides concise information on about 810 forestry education and training institutions in more than 110 countries. Also included is a Survey Form on Forestry Education Organizations to allow institutions to enter or update inputs on-line. FAO also maintains on-line directories of forestry research institutions, and directories of forestry extension organizations and/or arid lands research organizations are under development. There are many databases about recent publications and ongoing research being made available, free or at a low cost, through the Internet. These can be very important resources for the improvement of higher education programmes.

E-conferences are "meetings" that take place over the Internet, using two combined electronic services: e-mail and listservers. A listserver is a software that administers a list of e-mail addresses (of the participants in the conference) and facilitates the exchange of messages between them; it usually also maintains an archive of the messages exchanged. Electronic conferencing can be used for higher education purposes by itself or as part of an on-line distance learning programme.

Setting up Electronic Networks, Services and Applications

In the preceding section it was suggested that a body such as FAO's Advisory Committee on Forestry Education could play a strong role in promoting the use of the new electronic applications in advancing forestry education. To be effective, it would need to organize and set up an electronic network.

Outlined below are the basic steps that should be followed when setting up an electronic network parallel with an institutional network such as the FAO Advisory Committee on Forestry Education. The section is based on the experience gained by the authors in the development of FIDAMERICA, an Internet-based network of Latin American rural and agricultural development projects supported by the International Fund for Agricultural Development (IFAD).

1. Analyse the current information exchange needs of the institutional network. It is important to start by understanding the characteristics of the network or association of institutions that is considering setting up an electronic network. The key questions are: how is information currently being exchanged or shared? What (non-electronic) applications have been set up? Typically, these will include face-to-face meetings, newsletters, technical reports exchanged by air mail, telephone and fax to communicate between individuals. Are these non-electronic applications cost-effective? If they are working well, does it make sense to set up electronic applications, services and networks?
2. Design an appropriate network, services and applications. Start by identifying potential high-priority applications, not by designing a network and services and

then trying to identify applications that could use services. For example, begin by identifying what functions people and institutions need in order to share information and knowledge more effectively. Identify the services that need to be set up. Design a network with one or more server computers that could provide these services. If at all possible, design your network in such a way that you make use of the servers already made available (normally at a cost) by specialized organizations; not only is this option likely to be less costly but it will allow you to concentrate on exchanging information and knowledge and not on solving computer problems or upgrading your systems every time they become obsolete (which is about once a year at a minimum!). Analyse the current capabilities of the institutions where the client computers will be set up and the types of telecommunication networks that are available in the region and in each country. Use whatever telecommunication services are available; in most cases it is likely they will tend to improve over time.

3. Compare the cost-effectiveness of current vis-à-vis electronic applications. This can be a difficult analysis. The cost of setting up an electronic network and services and of learning how to use the electronic applications is often difficult to estimate. It is usually not clear how many of the non-electronic applications should be replaced. It is very important to analyse personnel costs carefully since they are always much higher than the cost of computers, software or communications.

4. Obtain commitment from member institutions. While setting up an electronic network sounds like a computer and software issue that should be left in the hands of engineers, it is actually a personnel issue that will affect everyone. It is very important to demonstrate to senior staff how the applications will work and how the plan will be implemented.

5. Train staff on how to implement electronic applications and use services. A key activity is to organize training workshops for staff from the institutions. These workshops will have two objectives: to teach people how to use the new tools and to motivate them to invest the time and money necessary to set up client computers with client software with access to telecommunications networks so that they can begin to use the applications. For example, of the 35 institutions trained at four FIDAMERICA workshops, only three had not taken these steps after four months; perhaps a key element in this successful "adoption rate" was the incorporation in these workshops of simulated electronic workshops and other applications that clearly showed the potential contribution of the new tools to improving the work of the participants.

6. Set up a first service on a server and organize pilot applications. The first electronic services that are set up to facilitate the exchange of information will depend on the demand from the institutions. In most cases, the first electronic service that networks will set up is a Listserver; this will allow the development of several electronic

services, such as electronic bulletin boards, e-conferences, electronic document archives, electronic newsletters, and simplified communication and information exchange procedures among the members of the institutional network.

7. Reinforce linkages between institutional and electronic networks. It is very important to set up other applications quickly to follow up on the enthusiasm that is usually generated by the first exchange. At this point, the link between the institutional and the electronic networks must be strengthened or there is a danger that they will separate. For example, managers of the institutional network may continue to use their old non-electronic communication procedures, while the staff assigned to manage the new tools will want to get involved with applications that are not set up by the institutional network. The best way to reinforce the link is to set up electronic applications that are directly relevant to the needs of the institutional network.

8. Expand services and applications in response to demand. Once the institutions are actively using electronic applications, staff will begin to see the potential of organizing other applications that will require other electronic services. For example, researchers may decide they want to share data, and distributed databases may be set up to respond to this demand. In the case of FIDAMERICA, an Electronic Farmers' Market was established to respond to the demand of the members of the institutional network to take advantage of the new tools to improve access to market information and marketing opportunities. Mechanisms must be set up to ensure that member institutions are aware of the potential applications, and also so that the specialized staff develop new applications in response to the members' needs and demands and not their own technical preferences. In FIDAMERICA all the members were encouraged to send in written suggestions to develop the first annual work plan, and over 40 such recommendations were received.

InterCambio is the Virtual Centre for Information and Knowledge Exchange in Support of Rural Development in Latin America. It was established by the Red Internacional de Metodologías de Investigación en Sistemas de Producción (RIMISP), a Latin American Production Systems Methodology Network, and INFORUM (an international NGO that supports electronic networking).

InterCambio is a virtual electronic centre that uses various types of electronic services, such as electronic conferencing, electronic libraries, electronic marketplaces, on-line databases, on-line technical assistance services, etc. Its mission is to work with national and regional development, research and policy institutions to facilitate their access to information and knowledge through modern electronic communication tools in order to enhance the impact and efficiency of their efforts to promote changes in rural life through improved agriculture and natural resource management.

The first institution to become a founding member of InterCambio was IFAD. RIMISP and INFORUM are actively looking for other institutions that would be interested in joining IFAD as founding member.

InterCambio's supporting members are institutions that want to take advantage of the electronic structure that has been set up to make available specific electronic services such as databases, e-conferences, etc. Supporting members benefit from the synergy derived from being associated with members providing other services, much as an individual store benefits from being located in a shopping mall. Supporting members have three options: i) they can contract InterCambio to develop, set up and manage an electronic service; ii) they can develop the electronic service on their own computers and then contract InterCambio to move the service to the InterCambio Center; or iii) they can set up the electronic service on their own computers and then ask InterCambio to set up a direct connection so clients accessing the InterCambio Center can link directly to their electronic service.

InterCambio's first supporting member was FIDAMERICA, a network linking IFAD-funded institutions in Latin America. FIDAMERICA has announced plans to convene at least five moderated electronic conferences during the next two years, and it will develop an electronic library of publications on rural and agricultural development in Latin America, a database on human resources for rural development and, in partnership with other institutions, an electronic marketplace for agricultural and nonagricultural products from Latin American rural communities.

InterCambio will provide various types of electronic services that are relevant to the purposes of a network or association of institutions and individuals involved in forestry education, including: electronic newsletters; electronic bulletin boards; e-conferences; an electronic library; on-line databases; and guidelines to access to information in the World Wide Web. With time electronic services and applications will change the nature of forestry education for the same basic reasons that they will affect many other knowledge- and information-based human activities: they will reduce the constraining impacts that distance and, consequently, time and costs have had until now on the flow of information and knowledge.

Forestry educators need to ask themselves the following question: What could be done if distance, time and cost were not significant factors in the flow of information?

Internet technology is developing at an incredible pace. Five years ago the only practical use for most people was sending short electronic messages; today, all sorts of digital data move rapidly and cheaply in the Internet, and these services are becoming available to an increasing cross-section of the world's population. These technologies will not replace the classroom, at least for a long time, but they hold promise of allowing a revolution in education and, in particular, of allowing students and professors in developing countries access to vast information and knowledge resources.

REFERENCES

Ahituv, Niv & Newmann, Seev (1990). *Principles of information systems for management*. Third edition. Won C Brown Publishers, 2460 Kerper Boulevard, Dubugue 145 2001

Anthony, Robert N., Deavden, John & Bedford, Norton. M. (1989). *Management control systems*. Sixth edition, Richard D Irwin, Inc, Homewood, Illinois 60430

Hennessy, Nik & Lawlor, Damian (2000). *Computer software requirements in forest management*. National Council for Forest Research and Development (COMFORD)

IIFM (2002). *Manual for forestry management control system (FORMACS)*. Indian Institute of Forest Management, Bhopal, India.

Kroeber, Donald W. & Watson, Hugh J. (1984). *Computer based information systems: a management approach*. Macmillan Publishing Company, Inc.

Lucas, Henry C. (1986). *Information systems concepts for management*. McGraw-Hill series in management information system. McGraw-Hill Inc.

Maciariello, Joseph A. (1984). *Management control systems*. Prentice-Hall International, Inc., Englewood Cliffs, New Jersey 07632.

Murdick, Robert G. (1986). *Management information systems concept and design*. Prentice-Hall International Inc., Englewood Cliffs, New Jersey 07632

Newman, William H. (1975). *Constructive control: design and use of control systems*. Prentice-Hall International Inc., Englewood Cliffs, New Jersey.

10

Medicinal Plants and Biodiversity Issues

Since time immemorial, people have gathered plant and animal resources for their needs. Examples include edible nuts, mushrooms, fruits, herbs, spices, gums, game, fodder, fibres used for construction of shelter and housing, clothing or utensils, and plant or animal products for medicinal, cosmetic or cultural uses. Even today, hundreds of millions of people, mostly in developing countries, derive a significant part of their subsistence needs and income from gathered plant and animal products. Gathering of high value products such as mushrooms (morels, matsutake, truffles), medicinal plants (ginseng, black cohosh, goldenseal) also continues in developed countries for cultural and economic reasons.

Among these uses, medicinal plants play a central role, not only as traditional medicines used in many cultures, but also trade commodities which meet the demand of often distant markets. Demand for a wide variety of wild species is increasing with growth in human needs, numbers and commercial trade. With the increased realisation that some wild species are being over-exploited, a number of agencies are recommending that wild species be brought into cultivation systems. Cultivation can also have conservation impacts, however, and these need to be better understood. Medicinal plant production through cultivation, for example, can reduce the extent to which wild populations are harvested, but it also may lead to environmental degradation and loss of genetic diversity as well as loss of incentives to conserve wild populations. The relationship between *in-situ* and *ex-situ* conservation of species is an interesting topic with implications for local communities, public and private land owners and managers, entire industries and, of course, wild species. Identifying the conservation benefits and costs of the different production systems for Medicinal and Aromatic Plants (MAP) should help guide policies as to whether species conservation should take place in nature or the nursery, or both.

CBD AND THE ECOSYSTEM APPROACH

Since its adoption in 1992, the Convention on Biological Diversity (CBD) has strived to implement its three major goals: the conservation of biological diversity, the sustainable use of its components, and the fair and equitable sharing of the benefits from the use of genetic resources. Although MAP have not been explicitly on the agenda of the various CBD meetings, all three goals of the Convention are fully applicable to MAP resources.

CONCEPT OF SUSTAINABILITY

As a base line element of the ecosystem approach it has to be recognised that humans, with their cultural diversity, are an integral component of ecosystems. In conceptual terms, the essence of sustainable development is expressed by the relationship between people and the ecosystem around it. This implies that ultimately one is entirely dependent upon the other. Human and ecosystem well-being need to be assessed together. A society is thought to be sustainable when both the human condition and the condition of the ecosystem are satisfactory or improving. The system improves only when both the condition of the ecosystem and the human condition improve.

The number of plant species which have at one time or another been used in some culture for medicinal purposes can only be estimated. An enumeration of the WHO from the late 1970s listed 21 000 medicinal species. However, in China alone 4 941 of 26 092 native species are used as drugs in Chinese traditional medicine, an astonishing 18.9 percent. If this proportion is calculated for other well-known medicinal floras and then applied to the global total of 422 000 flowering plant species, it can be estimated that the number of plant species used for medicinal purposes is more than 50 000 (Table 1).

Table 1. Medicinal plants used world-wide

Country	*Plant species*	*Medicinal plant species*	%
China	26 092	4 941	18.9
India	15 000	3 000	20.0
Indonesia	22 500	1 000	4.4
Malaysia	15 500	1 200	7.7
Nepal	6 973	700	10.0
Pakistan	4 950	300	6.1
Philippines	8 931	850	9.5
Sri Lanka	3 314	550	16.6
Thailand	11 625	1 800	15.5
USA	21 641	2 564	11.8
Viet Nam	10 500	1 800	17.1
Average	13 366	1 700	12.5
World	422 000	52 885	

We recognise, however, that certain plant families have higher proportions of medicinal plants than others. Good examples are the Apocynaceae, Araliaceae, Apiaceae, Asclepiadaceae, Canellaceae, Guttiferae and Menispermaceae. In addition, these families are not distributed uniformly across the world. As a consequence, not only do some floras have higher proportions of medicinal plants than others, but also have certain plant families a higher proportion of threatened species than others.

It is difficult to assess how many MAP are commercially traded, either on a national or even an international level. The bulk of the plant material is exported from developing countries while major markets are in the developed countries. An analysis of UNCTAD trade figures for 1981-1998 reflects this almost universal feature of MAP trade (Table 2). Adding the volumes for the five European countries in this list (94 300 tonnes) marks the dominance of Europe as an import region. Germany ranks fourth and third as importer and exporter, expressing the country's major role as a turntable for medicinal plant raw materials world-wide.

Table 2. The 12 leading countries of import and export of medicinal and aromatic plant material from 1991-1998

Country of import	*Volume [tonnes]*	*Value [1000 US$]*	*Country of export*	*Volume [tonnes]*	*Value [1000 US$]*
Hong Kong	73 650	314 000	China	139 750	298 650
Japan56 750	146 650	India	36 750	57 400	
USA	56 000	133 350	Germany	15 050	72 400
Germany	45 850	113 900	USA	11 950	114 450
Rep. Korea	31 400	52 550	Chile	11 850	29 100
France	20 800	50 400	Egypt	11 350	13 700
China	12 400	41 750	Singapore	11 250	59 850
Italy	11 450	42 250	Mexico	10 600	10 050
Pakistan	11 350	11 850	Bulgaria	10 150	14 850
Spain	8 600	27 450	Pakistan	8 100	5 300
UK	7 600	25 550	Albania	7 350	14 050
Singapore	6 550	55 500	Morocco	7 250	13 200
Total	342 550	1 015 200	Total	281 550	643 200

Iqbal estimates that about "4 000 to 6 000 botanicals are of commercial importance", another source refers to 5-6 000 "botanicals entering the world market". A thorough investigation of the German medicinal plant trade identified a total of 1 543 MAP being traded or offered on the German market. An extension of this survey to Europe as a whole arrived at 2 000 species in trade for medicinal purposes. Recognising the role of Europe as a sink for MAP traded from all regions of the world, it is a qualified guess that the total number of MAP in international trade will be around 2 500 species world-wide.

Threatened MAP

To satisfy the regional and international markets, the plant sources for expanding local, regional and international markets are harvested in increasing volumes and largely from wild populations. Supplies of wild plants in general are increasingly limited by deforestation from logging and conversion to plantations, pasture and agriculture.

In many cases, the impact through direct off-take goes hand-in-hand with decline owing to changes in land use. Species favoured by extensive agricultural management like *Arnica montanain* central Europe go into decline with changes in farming practices towards higher nutrient input on the meadows. This requires habitat management as the key factor in managing species populations.

One of the goals of the IUCN Medicinal Plant Specialist Group is to identify the species that have become threatened by non-sustainable harvest and other factors. The enormity of this task is illustrated by the following estimate: According to Walter and Gillett, 34 000 species or 8 percent of the world's flora are threatened with extinction. If this is applied to our earlier estimate that 52 000 plant species are used medicinally, it leads us to estimate that 4 160 MAP species are threatened (Table 3).

Table 3. How many medicinal plant species are threatened?

Number of flowering plant species worldwide	422 000
12.5% of them are used medicinally	52 000
8% are threatened	4 160

Cultivated MAP

Many medicinal plants, especially the aromatic herbs, are grown in home gardens, some are cultivated as field crops, either in sole cropping or in intercropping systems and rarely as plantation crops.

In a survey carried out for the Rainforest Alliance, companies involved in trade and production of herbal remedies and other botanical products were asked what percentage of their material is from cultivated sources and what percentage from the wild. On average, companies reported that 60-90 percent of material was cultivated, with the remaining wild harvested. However, when asked about species numbers rather than volume of material, the figures are generally inverted. Lange and Schippmann state that of the 1 543 species traded in Germany, only 50-100 species (3-6 percent) are exclusively sourced from cultivation.

Of more than 400 plants species used for production of medicine by the Indian herbal industry, fewer than 20 species are currently under cultivation in different parts of the country. In China, about 5 000 medicinal plants have been identified and about 1 000 are more commonly used, but only 100-250 species are cultivated. In Hungary, a country with a long tradition of MAP cultivation, only 40 species are cultivated for commercial production. In Europe as a whole, only 130-140 MAP species are cultivated.

Based on these figures, we assume that the number of MAP species currently in formal cultivation for commercial production does not exceed a few hundred world-wide. A global survey on the extent of MAP cultivation in terms of species, volumes and values would be highly desirable. On the other hand, however, we recognise that many more MAP species are cultivated on a small-scale in home gardens, either as home remedies or by herbalist or cultivation by local people can take place as enrichment planting.

Given the demand for a continuous and uniform supply of medicinal plants and the accelerating depletion of forest resources, increasing the number of medicinal plants species in cultivation would appear to be an important strategy for meeting a growing demand. But why are so few species cultivated and why are some species cultivated and so many others not?

One explanation may be found in the observation that cultivated plants are sometimes considered qualitatively inferior when compared with wild gathered specimens. For instance, wild ginseng roots are 5-10 times more valuable than roots produced by artificial propagation. The reason is primarily cultural, as the Chinese community, which is the largest consumer group of wild ginseng, believes that the similarity in appearance of gnarled wild roots to the human body symbolises the vitality and potency of the root. Cultivated roots lack the characteristic shape of wild roots and are therefore not as highly coveted by consumers. In Botswana, traditional medicinal practioners said that cultivated material was unacceptable, as cultivated plants did not have the power of material collected from the wild.

Scientific studies partly support this. Medicinal properties in plants are mainly due to the presence of secondary metabolites which the plants need in their natural environments under particular conditions of stress and competition and which perhaps would not be expressed under mono-culture conditions. Active ingredient levels can be much lower in fast growing cultivated stocks, whereas wild populations can be older due to slow growth rates and can have higher levels of active ingredients. While it can be presumed that cultivated plants are likely to be somewhat different in their properties from those gathered from their natural habitats it is also clear that certain values in plants can be deliberately enhanced under controlled conditions of cultivation.

In general, in all countries, the trend is towards a greater proportion of cultivated material. The majority of companies, the mass-market, over-the-counter pharmaceutical companies as well as the larger herb companies, prefer cultivated material, particularly since cultivated material can be certified biodynamic or organic.

From the perspective of the market, domestication and cultivation provide a number of advantages over wild harvest for production of plant-based medicines:

(i) While wild collection often offers material adulterated with unwanted, sometimes harmful other plant species, cultivation provides reliable botanical identification.

(ii) Wild harvest volumes are dependent on many factors that cannot be controlled and the irregularity of supply is a common feature. Cultivation guarantees a steady source of raw material.

(iii) Wholesalers and pharmaceutical companies can agree on volumes and prices over time with the grower.

(iv) The selection and development of genotypes with commercially desirable traits from the wild or managed populations may offer opportunities for the economic development of the medicinal plant species as a crop.

(v) Cultivation allows controlled post-harvest handling and therefore

(vi) Quality controls can be assured and

(vii) Product standards can be adjusted to regulations and consumer preferences.

(viii) Cultivated material can be easily certified organic or biodynamic although certifiers are also presently developing wildcrafting standards.

However, domestication of the resource through farming is not always technically possible. Many species are difficult to cultivate because of certain biological features or ecological requirements (slow growth rate, special soil requirements, low germination rates, susceptibility to pests, etc.).

Economical feasibility is the main rationale for a decision to bring a species in cultivation but it also is a substantial limitation as long as sufficient volumes of material can still be obtained at a lower price from wild harvest. Cultivated material will be competing with material harvested from the wild that is supplied onto the market by commercial gatherers who have incurred no input costs for cultivation. Low prices, whether for local use or the international pharmaceutical trade, ensure that few species can be marketed at a high enough price to make cultivation profitable. Domestication of a previously wild collected species does not only require substantial investment of capital but also requires several years of investigations.

On a time scale of sometimes many decades, the transition from wild harvesting to possible cultivation goes through various phases (Figure 1):

(i) *Discovery phase:* At this point the demand can be met by wild harvest. Extractivism is done for local use or for barter with others.

(ii) *Expansion phase:* It is clear that the product is potentially useful and that demand is likely to increase. Harvest is done for local or regional sale and eventually for international markets. In general, species with naturally low densities are unlikely to become important sources of commercially large quantities.

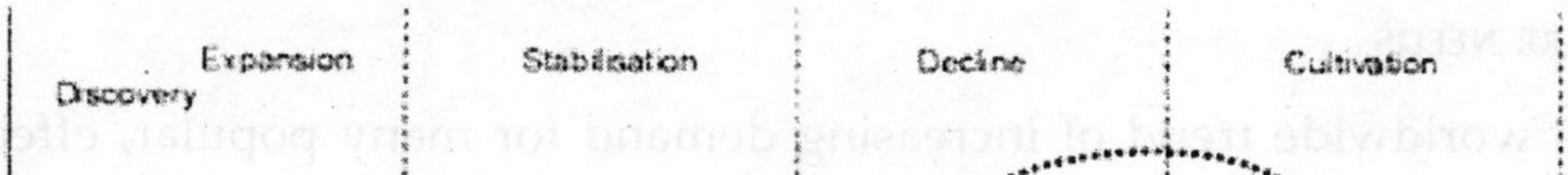

Figure 1. Transition phases from wild harvesting to cultivation: after wild resources decline with over-harvesting, raw material prices increase and cultivation becomes economically feasible; more resilient species can recover.

(iii) *Stabilisation phase:* The species is unlikely to be attractive to growers unless prices are high enough and wild-harvested resources are scarce enough. However, desirable species may be grown on farm land and planted around settlements.

(iv) *Decline phase:* Prices increase with scarcity due to transport costs, search time and the long-distance trade. Wild populations will have to decline further before cultivation is a viable option. The trade is characterised by fluctuations in supplies, often to the extent of disrupting the trade balance. For slow growing species, if controls on collection are not strictly enforced, wild populations will be more seriously eroded before cultivated material is available.

(v) *Cultivation phase:* Now, formal cultivation systems are developed and instituted. The plants are domesticated and incorporated in agroforestry systems sometimes for the benefit of small-scale farmers. If international market opportunities exist, commercial plantations are created with substantial investment and genetic selection, cloning, breeding and biotechnology may be applied. More resilient species may recover in their wild populations (Figure 1).

HEALTH CARE NEEDS

There is a worldwide trend of increasing demand for many popular, effective species in Europe, North America and Asia, growing between 8-15 percent per year. Rapid urbanisation and the importance of herbal medicines in African health care systems stimulated a growing national and regional trade in Africa. Demand for medicinal plants also reflects distinct cultural preferences. In the USA, for example, only 3 percent of people surveyed had used herbal medicine in the past year, whereas in Germany, with a strong tradition of medicinal plant use, 31 percent of the over-the-counter products in pharmacies in 2001 were phytopharmaceutical preparations.

The level of herbal medicine use in most developing countries is much higher than this. While most traditional medicinal plants are gathered from the wild, these are not static health care systems, and introduced species are commonly adopted into the repertoire of plants used by African or South American herbalists. In many cases, herbal medicines can also be cheaper than western medicines, particularly where access to traditional healers is easier. Demand for traditional medicine continues in the urban environment even if western biomedicine is available.

INCOME GENERATION

Wild harvesting of medicinal plants is a chance for the poorest to make at least some cash income. Especially those people who do not have access to farm land at all depend on gathering MAP to earn at least some money. However, local people generally get a low price for unprocessed plant material. Although income from *Prunus* african abark sales is an important source of revenue to villagers in Madagascar, in some cases generating >30 percent of village revenue, the price paid to collectors is negligible compared to Madagascan middlemen. In Mexico, Hersch-Martinez found that medicinal plant collectors only received an average 6.17 percent of the medicinal plant consumer price.Whether fruits, roots, bark or whole plants are involved, the potential yield from wild stocks of many species is frequently over-estimated, particularly if the effects of stochastic events is taken into account. As a result, commercial harvesting ventures based on wild populations can be characterised by a "boom and bust" situation where initial harvests are followed by declining resource availability.

Secondly, the susceptibility or resilience to collection pressure varies among species owing to biological characters such as different growth rates (slow growing vs. fast growing), reproductive systems (vegetative or generative propagation; germination rates; dormance; apomixis) and life forms (annual; perennial; tree).

SMALL-SCALE CULTIVATION AND HOME GARDENS

Small-scale cultivation, which requires low economic inputs, can be a response to declining local stocks, generating income and supplying regional markets. This can be a more secure income than from wild harvest which is notoriously inconsistent. For farmers that integrate MAP into agroforestry or small-scale farming systems, these species can provide a diversified and additional source of income to the family. Home gardens are increasingly a focus of medicinal plant propagation and introduction programmes intended to encourage the use of traditional remedies for common ailments by making the plant sources more accessible.

LARGE-SCALE CULTIVATION

Large-scale cultivation has a number of socio-economic impacts on rural people: "Commercialisation is both necessary and potentially harmful to farmers. It is necessary in that without it the market for products is small and the opportunity does not exist for rural people to generate income. A degree of product domestication is therefore desirable. On the other hand, commercialisation is potentially harmful to rural people if it expands to the point that outsiders with capital to invest come in and develop large-scale monocultural plantations for export markets. Rural people may benefit from plantations as a result of available employment and hence off-farm income [...]. However, plantations may also distort market forces to their advantage, for example, by imposing low wages which will restrict the social and economic development of local people. The major beneficiaries of large-scale exports will probably be the country's elite and, perhaps, the national economy".

Also, those socially disadvantaged groups who actually depend on gathering MAP for their survival and cash income, may not have access to farm land at all and are therefore not able to compete with large-scale production of MAP by well-established farmers. Other limitations to the domestication approach include boom-bust and fickle markets that let farmers down when consumers turn their attention elsewhere.

Cultivation of medicinal plants is widely viewed not only as a means for meeting current and future demands for large volume production of plant-based drugs and herbal remedies, but also as a means for relieving harvest pressure on wild populations. Booming markets with rapidly rising demands often have devastating effects on wild collected species. A closer look reveals that not all species are affected in the same way by harvesting pressures. The seven forms of rarity described by Rabinovitz make clear that a species which (i) has a narrow geographic distribution, (ii) is habitat specific, and (iii) has small population sizes everywhere, is more easily over-harvested than species of any other pattern (Table 4).

Species can be distinguished quite well in their susceptibility to over-collection if their life form and the plant parts collected are viewed together (Table 5). Harvesting fruits from a long-lived tree presents a far lower threat to the long-term survival of the species than does collecting seeds from an annual plant. In the latter case, if the seed is gone the plant is gone. In some cases the harvest impacts are more complex, e.g. with slow growing trees which reproduce from seed but only produce few, large fruits (example: *Araucaria araucana,* monkey puzzle tree). This will increase their susceptibility to over-harvest from low to medium or even high. A thorough summary of predictors of resilience or vulnerability to harvesting wild populations is presented by Cunningham.

Table 4. Seven forms of rarity

Geographic distribution	Habitat specificity	Local population size	
wide	broad	somewhere large	least concern
		everywhere small	
	restricted	somewhere large	
		everywhere small	
narrow	broad	somewhere large	
		everywhere small	
	restricted	somewhere large	
		everywhere small	highly susceptible

In summary, we can state that species most susceptible to over-harvest are habitat specific, slow growing and destructively harvested for their bark, roots or the whole plant. These species suffer most from harvesting and many of them have been seriously depleted, for example *Prunus africana* in West Africa, *Warburgia salutaris* in southern Africa and *Saussurea costus* in the Himalayas.

For threatened medicinal plant species cultivation is a conservation option because the constant drain of material from their populations is much higher than the annual sustained yield. If the demand for these species can be met from cultivated sources the pressure on the wild populations will be relieved. In these cases, the need for strict conservation of remaining populations, improved security of germplasm ex-situ and investment in selection and improvement programmes is extremely urgent as the example of Jaborandi (Pilocarpus jaborandi) in Brazil shows.

However, among the species that can be marketed at a high enough price to make cultivation profitable, only few are in the highest threat categories. Examples for threatened but cultivated species are *Garcinia afzelii, Panax quinquefolius, Saussurea costus* and *Warburgia salutaris*. With respect to economic viability many highly endangered MAP do not qualify for cultivation. This group of plants will enter cultivation only with the help of public domestication programmes.

Table 5. Susceptibility of species to overcollection as a function of Life Form and Plant Parts Used

	Wood	*Bark*	*Root*	*Leave*	*Flower*	*Fruit / Seed*
Annual		—	high	medium	medium	high
Biannual	—	—	high	medium	medium	high
Perennial	—	medium	high	low	low	low
Shrubmedium	medium?	medium ?	low	low	low	
Tree	medium	medium?	medium?	low	low	low

For all other harvested MAP species the priority conservation option is sustainable harvest from wild populations, for a variety of reasons.

Let's imagine that a valuable medicinal plant is exploited by local collectors. A pharmaceutical company has domesticated and begun to cultivate the plant on a commercial scale. When the company no longer needs the wild-harvested material, local harvesters have to abandon the harvest and any incentive the local collectors might have had to protect the wild populations is gone. The domestication of MAP species has an environmental implication in the sense that it reduces the economic incentives for forest dependent people to conserve the ecosystems in which the MAP species occur.

If collectors and collecting communities can be involved in the development of propagation and management methods, the likelihood of their having an interest in protecting the wild populations from over-exploitation, particularly if these are understood to be the genetic resource "bank" for the domestic enterprises, will be greater.

Another aspect to consider is the genetic diversity of the species which is in demand. Long before non-sustainable harvest practices lead to extermination of a whole species, selection of favoured growth forms and concentration on certain harvesting areas which may hold certain ecotypes will lead to a degradation of genetic diversity of the wild populations. The same is true under domestication: Industry requirements for standardisation encourage a narrow genetic range of material in cultivation. Domestication will not achieve conservation of genetic diversity because a narrow group of high yielding individuals will be selected for planting.

Challenges of Harvesting from Wild

Sustainable harvest is increasingly seen to be the most important conservation strategy for most wild-harvested species and their habitats, given their current and potential contributions to local economies and their greater value to harvesters over the long term. The basic idea is that non-destructive harvests and local benefits will maintain population, species and ecosystem diversity.

Besides poverty and the break-down of traditional controls, the major challenges for sustainable wild-collection include: lack of knowledge about sustainable harvest rates and practices, undefined land use rights and lack of legislative and policy guidance.

The most important ingredient required to achieve a truly sustainable form of resource use is information. In reality, resource managers are always confronted with the lack of adequate information about the plants used, their distribution, the genetic diversity of wild populations and relatives and, above all, the annual sustained yield that can be harvested without damaging the populations. Research on the conservation and sustainable use of medicinal plants and their habitats has fallen far behind the demand for this globally important resource. Each species has unique ecological, socio-economic, health and cultural associations that must be understood. Model research approaches are feasible, model solutions are not.

In many cases, access to the resource is open to everybody, rather than a limited access or private ownership. To make a living, commercial medicinal plant gatherers therefore "mine" rather than manage these resources. Open access schemes to harvestable plant population prevent rational and cautious use and make it difficult to adhere to quotas and closed seasons. Information on trade in MAP is scarce and data are rarely collected or published at a national level. Much production and consumption is at subsistence level and as a consequence the economic importance of these activities is largely under-estimated in government decision making regarding rural development, natural resource management planning and in government budget allocations. Therefore, national legislation and policies mostly fail to provide frameworks for a rational and sustainable use of wild resource.

Opportunities for governments to develop legislation to control and monitor harvest and trade of medicinal plant species and to consider conservation and sustainable use of medicinal plants as a priority in establishing protected areas have been greatly enhanced by two recent developments in international legislation: the addition of medicinal plant species to the Convention on International Trade of Endangered Species (CITES) and the entry into force of the CBD.

FUTURE TRENDS AND SOLUTIONS

How will the market demand develop in the future? People in developing countries are already and will increasingly depend on medicinal plants as sources for their primary health care. An estimate by the World Health Organisation that more than 80 percent of the world's population relies solely or largely on traditional remedies for health care is frequently cited. Also in the northern countries, use of medicinal plants is expected to rise globally, both in allopathic and herbal medicine. This upward trend is predicted not only because of population explosion, but also due to increasing popularity for natural-based, environmentally friendly products. In general, the demand for medicinal plants and herbal remedies and especially its renaissance in the developed countries is driven by the following factors:

- increasing costs of institutional, pharmaceutical-based health care;
- interest of individuals, communities and national governments in greater self-reliance in health care;
- interest of communities and national governments in small and large-scale industrial development based on local/national biodiversity resources;
- increasing success in validating the safety and efficacy of herbal remedies;
- legislation improving the status of herbal medicine industry;
- renewed interest of companies in isolating useful compounds from plants;
- search for new drugs and treatments of serious and drug-resistant diseases;
- marketing strategies by the companies dealing in herbal medicine.

The limitations of cultivation as an alternative to wild harvest have been examined by Sheldonet al. in several case studies. We share their conclusion that, notwithstanding the level of interest in cultivation as a means for enhanced production and in a few cases as an effort to contribute to conservation of the resource, most medicinal and aromatic plant species will continue to be harvested wild to some extent. There is therefore a need to recognise and strengthen the role of local people in forest inventory, monitoring and impact assessment processes and to integrate non-timber product uses into forest management

Need for Implementation of Management Plans

Limiting the harvest to a sustainable level requires an effective management system and sound scientific information (Table 6). The management system must include annual harvest quotas, consider seasonal or geographical restrictions and restriction of harvest to particular plant parts or size classes. In addition, clarification of the access and user rights to the resources providing MAP is part of the essential baseline information. Continuous monitoring and evaluation of the success is necessary to adapt the management strategy.

In many cases harvesting techniques need to be improved as the extraction of the roots or bark is often negatively affecting the recovery of the species or may even kill it. Collecting methods are often crude and wasteful, resulting in loss of quality and reduction in price. Field-based methods have already been developed for sustainable harvest assessment and monitoring of non-wood forest products, resulting in the publication of research guidelines and predictive models.

Table 6. Steps and standard elements of a management plan for MAP utilisation

- resource inventory of population abundance and distribution
- assessment of regional and global threat based on all available knowledge and expertise
- biological studies (growth and regeneration rates, pollination system, seed dispersal, potential for confusion with similar species, etc.) and assessment of harvest impact on viability of individuals
- assessment of annual sustained yield
- review of local knowledge and harvest practices
- review of harvest and trade levels in the past and evaluation of market trends
- revision of national regulations for the utilisation in source country
- assessment of tenure and access
- design and implementation of management scheme: annual harvesting quota, seasonal or regional restriction and on certain plant parts or size classes, domestication programme
- installation of continuing monitoring and re-evalutaion (adaptive management)

Eco-labelling and Certification

Given that sustainable harvesting from the wild is difficult to achieve, certification standards can play a role to assure that a product meets certain standards of sustainability. Certification programmes related to natural resource use have mainly been developed for timber and agricultural products but they are presently being adapted for wild harvest of non-timber plants. Various schemes focus on different areas along the supply chain: production, processing, trade, manufacturing, marketing. Four categories of certification schemes have been identified to be of relevance for MAP products: (i) forest management certification, (ii) social certification, (iii) organic certification, and (iv) product quality certification.

The latter include parameters such as product identity, purity, safety and efficacy. The Good Harvesting Practices (GHP) developed for medicinal plants cover to some degree ecological aspects but need to be more clearly focused on this aspect before they can make a meaningful contribution to ensuring sustainability. Dürbeck, Walter and most comprehensively Pierceet al. present overviews of certification programmes and their activities.

RECOMMENDATIONS

1) To overcome significant knowledge deficits, a global MAP cultivation survey should be commissioned by an international organisation. Aims are to identify species cultivated, in which countries they are grown, volumes produced and their market values. This survey should also assess public domestication programmes, as well as in-situ and ex-situ conservation efforts for wild populations of species in cultivation (e.g., in protected areas, in genebanks and botanic gardens).

2) Wild harvesting of MAP will continue to prevail owing to economic reasons outlined above. Sustainable wild harvest management schemes need to be supported by governments and authorities. Management plans need to be installed as a standard prerequisite for any such harvesting in the wild. There is a need to monitor and audit the harvesting process to determine whether it is sustainable.

3) Primary producers need help to improve returns from sustainable harvesting of MAP. Community based small scale cultivation enterprises need to be strengthened to enable them to compete with large-scale high-tech cultivation.

4) Secure ex-situ field gene banks need to be developed, particularly for habitat specific, slow-growing species with high susceptibility of being over-harvested.

5) Medicinal plant domestication programmes need to be expanded, taking fuller advantage of the genetic and chemical diversity within species over wide geographical areas.

6) Capacity to assess and monitor the conservation status of MAP and to manage harvest within the limits of sustainability is extremely limited worldwide and needs to be developed through training courses and curriculum development in ethnobotany and applied ecology . Research to investigate the sustainability of production systems is lacking and needs to be stimulated for a better understanding of the biological dynamics of the resource in the wild and in domestication.

7) Management planning has to take the diversity of tenure systems which apply to medicinal plants into account to a far greater degree. Clarification of user rights over the resource and access to it, particularly where it is considered common property, needs to be recognised as a crucial factor enabling or preventing a sustainable harvest from wild populations.

8) Eco-labelling and other social and economic incentives to strengthen market credibility and competitiveness of biodiversity-friendly products need to be promoted. The efforts of certifiers to develop certification standards for wild harvested plant material need to be supported as well as the approaches of industry to set up self-binding product quality standards. The private sector should be encouraged to consider local livelihoods and biodiversity when setting up ethical and environmental standards.

9) Conservation of medicinal plants currently lacks priority in policy and law. There are opportunities to change this within the implementation of legal instruments such as the CBD and CITES. Government policies and legislation need to be adapted and implemented to recognise the value of and need for sustainable wild harvesting management regimes, to implement national and/or regional permit systems and make medicinal plant conservation a priority for national health and economic policy.

10) The Global Environment Facility (GEF) needs to consider medicinal plant conservation as a programme priority worthy of funding.

11) Medicinal plants warrant priority in national efforts to implement the Global Strategy for Plant Conservation of the CBD.

12) Local communities can take more responsibility for sustainable harvest of medicinal plants only if they have the choices afforded by adequate income, control over the resource and the knowledge and skills required. On the issue of intellectual property rights it needs to be elaborated how the country, the local user or other entity can be adequately compensated for use of the resource by outsiders.

REFERENCES

Agelet, A., M.A. Bonet & J. Valles2000. Homegardens and their role as a main source of medicinal plants in mountain regions of Catalonia (Iberian Peninsula). *-Economic Botany*54: 295-309.

Bannerman, R.H.1982. Traditional medicine in modern health care. *-World Health Forum*3(1): 8-13.

Cunningham, A.B.1993.*African medicinal plants. Setting priorities at the interface between conservation and primary healthcare*. - Paris, UNESCO (People and Plant Working Paper 1).

Duke, J.A. & E.S. Ayensu1985.*Medicinal plants of China*. Vol. 1 & 2. - Algonac, USA, Reference Publications (Medicinal Plants of the World 4).

Eisenberg, D.M., R.C. Kessler, C. Foster, F.E. Norlock, D.R. Calkins & T.L. Delbanco1993. Unconventional medicines in the United States. *-The New England Journal of Medicine*328: 246-252.

FAO1995.*Non-wood forest products for rural income and sustainable development*. - Rome (Non-wood Forest Products 7).

Groombridge, B. & M. Jenkins1994.*Biodiversity data sourcebook*. - Cambridge, UK, World Conservation Press (WCMC Biodiversity Series 1).

Iqbal, M.1993.*International trade in non-wood forest products. An overview*. - Rome, FAO.

11

Biodiversity Conservation Services

As natural habitats have come to be increasingly restricted and degraded, increasing attention has been paid to conserving biodiversity in agricultural landscapes. This can be both an end in itself, driven by the realisation that agricultural landscapes can have high levels of biodiversity, and a means of complementing conservation in protected areas. Classical approaches to conservation, attempting to preserve pristine habitats within protected areas, are necessary but insufficient in the face of growing pressure on land.

Efforts to enhance biodiversity in agricultural landscapes need to consider the incentives faced by individual land users, who decide what practices to use on their land, generally without considering what biodiversity benefits different land use practices may have. When biodiversity-friendly agricultural practices are the most profitable, there is a happy convergence of private and social interests. This is the case of jungle rubber in Indonesia, for example. But biodiversity-friendly agricultural practices are not necessarily the most profitable from the perspective of individual land users. In some cases, the profitability of biodiversity-friendly practices can be boosted by inducing consumers to pay a premium for their outputs, as in the case of shade-grown coffee. But this approach requires complex certification schemes and is not always feasible.

A further approach, which has received increasing attention in recent years, is to provide direct payments for the provision of biodiversity services. This approach internalises what had been an externality, ensuring that it is taken into consideration in decisionmaking.

This is the approach taken by the Regional Integrated Silvopastoral Ecosystem Management Project (RISEMP), which is being implemented with financing from the Global Environment Facility (GEF). The project is piloting the use of payments for environmental services as a means of generating biodiversity conservation and carbon sequestration services in watersheds at three sites in Colombia, Costa Rica, and Nicaragua.

SILVOPASTORAL PRACTICES

Cattle production has long been associated with deforestation in Latin America, and as such has been an important cause of the loss of natural habitat and biodiversity in the region. In most countries, the prevailing policy framework encouraged deforestation for timber extraction and conversion of forest areas to pastures and crops, which were encouraged by subsidised credit, guaranteed prices, and other incentives. The extent of these policy distortions has been substantially reduced in recent years but pressure from poor landholders and-in some areas-large scale ranches continues to result in large-scale deforestation in many areas. In many countries, the legal framework encourages this process, by granting titles to land that is deemed to be 'improved'.

Table 1 summarises changes in pasture and forest area in Colombia, Costa Rica, and Nicaragua. Forest cover has been in retreat throughout the region. The area under annual crops has fallen in many countries. There has been some expansion of permanent crops, although this trend has reversed in recent years, due to low coffee prices. Permanent pasture, on the other hand, has expanded steadily in all countries for which data are available, although at different rates.

Table 1. Changes in pasture land and forest area in Colombia Costa Rica, and Nicaragua

	Colombia		*Costa Rica*		*Nicaragua*	
	Area, 2000 ('000 ha)	*Change 1990-2000 (%)*	*Area, 2000 ('000 ha)*	*Change 1990-2000 (%)*	*Area, 1995 ('000 ha)*	*Change 1990-1995 (%)*
Annual crops	2,818	-14.7	225	-13.5	2,457	25.2
Permanent crops	1,766	6.2	281	12.2	291	14.3
Permanent pasture	40,925	2.1	2,339	0.4	4,820	—
Natural forest area	49,650	-3.6	1,966	-7.5	3,278	-26.4

In addition to the environmental problems caused by the initial loss of forest, traditional approaches to pasture are often unsustainable. After an initial period of high yields, soil fertility is depleted and grass cover diminishes, resulting in soil erosion, contamination of water supplies, air pollution, further loss of biodiversity, and degradation of landscapes. Lower income for producers results in continuing poverty and in pressure to clear additional areas.

Silvopastoral systems, which combine trees with pasture, offer an alternative to prevalent cattle production systems in Latin America. They provide a deeply rooting, perennial vegetation which is persistently growing and has a dense but uneven canopy. These systems can be grouped in four major categories:

- Systems in which high densities of trees and shrubs are planted in pastures, providing shade and diet supplements while protecting the soil from packing and erosion.

- Cut and carry systems, which replace grazing in open pasturelands with stables in which livestock is fed with the foliage of different trees and shrubs specifically planted in areas formerly used for other agricultural practices. Cut and carry systems have been particularly successful in Central America and in Colombia.
- Use of fast-growing trees and shrubs for fencing and wind screens. This system, widely used in some countries of tropical America, provides an inexpensive alternative for fencing and supplements livestock diets.
- Livestock grazing in forest plantations. In this system, grazing is used to control the invasion by native and exotic grasses, thus reducing the management costs of the plantations.

On-site Benefits

Silvopastoral systems can provide a range of on-site benefits. The introduction of trees in pasture areas can improve pasture productivity. Silvopastoral systems tend to increase nutrient re-cycling across a deep portion of the soil profile occupied by the root systems of a wide variety of plants associated with silvopastoral systems. Depending on the species of trees being used and on local climate characteristics, trees extract water and nutrients from soil horizons inaccessible to grasses, and deposit the nutrients on the ground with the natural fall of foliage, twigs, and fruits. The biomass and amount of nutrients released by pruning the trees of the agroforestry systems varies depending on the kind of management in use. As much as 18 tons/ha of dry matter can be deposited on the ground annually, and the amount of nitrogen flowing through the system can reach values of up to 380 kg/ha/year. In addition, the trees can provide direct benefits in the form of products such as fruit, fuelwood, fodder, and timber. From the farmers' perspective, the benefits of silvopastoral systems derive from (a) additional production from the tree component; (b) maintaining and/or improving pasture productivity; (c) diversification of production; and (d) contribution to the overall farming system. The shade provided by trees may also enhance livestock productivity, especially milk production.

Biodiversity Benefits

The increased complexity of silvopastoral systems relative to traditional pastures means they often bring important biodiversity benefits. These take two main forms. First, they tend to support much higher species diversity than traditional pastures. Second, they help connect protected areas.

Silvopastoral systems have been shown to play a major role in the survival of wildlife species by providing scarce resources and refuge; to have a higher propagation rate of native forest plants under these scattered trees; and to provide shade for grazing animals, and shelter for wild birds. Food availability for wild birds is high in silvopastoral systems, and the complex structure of the vegetation provides a more adequate nesting substrate and better protection against predators than other agroecosystems. Silvopastures and other agroforestry systems also harbor a larger and more complex assemblage of invertebrates than monoculture pastures. By providing alternative sources of fuelwood and other wood products, silvopastoral systems can also help reduce pressure on remaining natural habitats.

In agricultural landscapes characterised by the fragmentation of the natural habitats, silvopastoral systems can serve as biological corridors, helping to connect remaining habitats. At the regional level, silvopastoral systems may play an important role in the implementation of the Mesoamerican Biological Corridor, given the vast area of pasturelands in Central America and Colombia. It is expected that these corridors would provide adequate habitat for wildlife while facilitating seed dispersal and the regeneration of the native vegetation.

Other Benefits

Silvopastoral systems are capable of fixing significant amounts of carbon in the soil under the improved pastures and in the standing tree biomass. Research in Colombia, Panama, and Costa Rica has shown that soils under silvopastoral systems have higher carbon content. Additional carbon is sequestered by the trees found in such systems. Moreover, grass-based pastures tend to sequester most of the carbon in the deeper part of the soil profile (between 40 and 100 cm depth), thus making it less prone to oxidation, and hence loss.

Silvopastoral systems are also likely to affect water services, though the specific impact is likely to be site specific. Infiltration generally increases with the presence of trees, reducing superficial runoff with its attendant soil erosion. Improved livestock management can help reduce compaction, thus further reducing surface runoff. The presence of trees also leads to increased evapotranspiration, however, thus tending to decrease water yield.

In hilly areas, trees have an additional protective role in the ecosystem, that of preventing landslides. Not only is the presence of trees essential for soil protection on slopes, but also the variety of species is important. Trees of different root depths are required for effective soil anchorage, in particular during torrential rain events accompanying tropical storms.

BARRIERS TO ADOPTION

Despite their many benefits, silvopastoral systems have only been adopted to a limited extent. National-level data typically do not distinguish land uses with a sufficient level of detail, but Table 2 shows average land use in the Quindío, Colombia, RISEMP project site. Pasture with no or low tree density dominates all other land uses. Overall tree cover is low, although there is a significant amount of forest remnants, most of which is riparian forest. Permanent crops, mostly coffee, account for about 10 percent of the area. Coffee was once the dominant land use in this area, but it has been replaced by pasture in the last decade due to low coffee prices. Fodder banks are practically non-existent: only 7 in 110 farms surveyed had any, with an average of less than 1 ha each.

Table 2. Average land use in farms in Quindío, Colombia

	(ha)	*(%)*
Annual crops	0.9	2.6
Permanent and semi-permanent crops	3.7	10.8
Pasture with no trees	21.9	64.1
Pasture with low tree density	0.6	1.8
Pasture with high tree density	0.0	0.1
Fodder banks	0.0	0.1
Plantations, riparian forest, and forest remnants	7.0	20.5
Total	34.2	100.0

An important constraint to the adoption of silvopastoral practices is their limited profitability from the perspective of individual land users. Establishing silvopastures can entail high initial costs, as shown in Table 3. Increasing the livestock herd to take advantage of the increased fodder production entails additional costs. In addition, there are opportunity costs resulting from the time lags before the systems become productive.

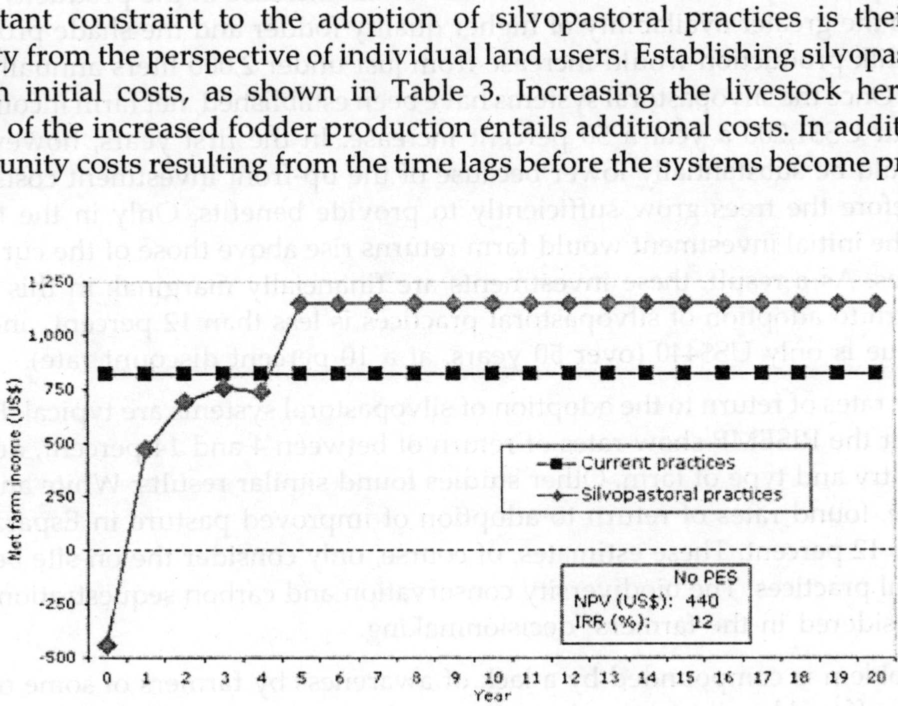

Figure 1. Typical time profile of benefits of silvopastoral systems

Table 3. Initial investment costs for selected silvopastoral practices

		Quindío, Colombia	*Esparza, Costa Rica*	*Matiguás-Río Blanco, Nicaragua*
Improved pasture	(US$/ha)	375	250	}265
Planting 100 trees in improved pasture	(US$/ha)	55	50	
Planting 1,000 *leuceana* trees	(US$/ha)	1,000		
Protein bank	(US$/ha)	960	660	475
Live fencing	(US$/km)	700	610	390

Figure 1 illustrates the typical time profile of returns to adoption of silvopastoral practices. The example shown is of a 20 ha farm in Nicaragua raising livestock for milk and meat. About 15 ha are used for unimproved pasture, with 2 ha devoted to cultivating basic grains for the household's own consumption and 3 ha in brush. Ccurrent land uses generate a net farm income of about US$800 a year. The proposed investment involves switching 3 ha of pasture to improved pasture with low tree density, and establishing a 0.75 ha fodder bank. This would allow an increase in the livestock herd from 14 to 15, but more importantly it would result in a substantial increase in the productivity of the herd due to the greater availability of higher quality fodder and the shade provided by the trees. Milk production would increase from just under 2,000 liters annually to over 4,000 liters. Once the silvopastoral systems have been established, net farm income would rise to about US$1,200 a year-a 50 percent increase. In the first years, however, farm income would be substantially lower because of the up-front investment costs and the time lag before the trees grow sufficiently to provide benefits. Only in the fifth year following the initial investment would farm returns rise above those of the current land use practices. As a result, these investments are financially marginal: in this case, the rate of return to adoption of silvopastoral practices is less than 12 percent, and the net present value is only US$440 (over 50 years, at a 10 percent discount rate).

The low rates of return to the adoption of silvopastoral systems are typical. Estimates prepared for the RISEMP show rates of return of between 4 and 14 percent, depending on the country and type of farm. Other studies found similar results; White and others, for example, found rates of return to adoption of improved pasture in Esparza, Costa Rica, of 9 to 12 percent. These estimates, of course, only consider the on-site benefits of silvopastoral practices. The biodiversity conservation and carbon sequestration benefits are not considered in the farmers' decisionmaking.

This problem is compounded by a lack of awareness by farmers of some of the on-site benefits offered by silvopastoral systems, such as reduced dependency on chemical fertilisers and pesticides, savings in water for irrigation, soil protection and enhanced fertility, and the potential for additional incomes from harvesting fruit, fuelwood, and timber. Limited knowledge of these on-site benefits further reduces the perceived benefits to land users.

Even if silvopastoral practices are financially viable, the high initial investment costs required pose problems for credit-constrained land users. In the Quindío project site, only 25 percent of households had access to credit in the past five years. Access to credit is higher in the Matiguás-Río Blanco area in Nicaragua, thanks to presence of several NGOs that offer credit. About 50-75 percent of households in this area report having used credit in the past five years. However, credit is often only available for specific purposes and with collateral requirements that are difficult for farmers to meet.

The long-term nature of investments in most silvopastoral practices means that tenure security is an important factor in their adoption. Tenure is not a constraint in the three study sites, however. In the Costa Rica and Colombia project sites, all farmers have formal ownership of the land. In the Nicaragua project site, most ranchers occupy public land, but long-term occupancy gives them secure tenure.

PAYMENTS FOR ENVIRONMENTAL SERVICES

From the land users' perspective, the biodiversity conservation and carbon sequestration benefits are externalities. As such, they do not take them into consideration in making their land use decisions, thus reducing the likelihood that they will adopt practices that generate such benefits, including silvopastoral systems. Recognition of this problem and of the failure of past approaches to dealing with it has led to efforts to develop systems in which land users are paid for the environmental services they generate, thus aligning their incentives with those of society as a whole. The simple logic of Payments for Environmental Services (PES) is that compensating land users for the environmental services a given land use provides makes them more likely to choose that land use rather than another.

There has been considerable experimentation with PES and other market-based approaches in recent years. Latin America has been a particularly fertile ground for such experimentation. Costa Rica has developed an elaborate, nation-wide PES program, the Pago por Servicios Ambientales (PSA). Under the 1997 Forestry Law, land users can receive payments for specified land uses, including new plantations and conservation of natural forests. The PSA program is now being supported by a World Bank loan and GEF grant under the Ecomarkets Project. The town of Heredia has established an 'environmentally adjusted water tariff', the proceeds of which are used to pay landholders to maintain and reforest watershed areas. In a separate initiative, hydropower producer La Manguera SA is paying the Monteverde Conservation League to maintain under forest cover the watershed from which its plant draws its water. In Colombia, irrigation water user groups and municipalities in the Cauca valley are paying to conserve the watersheds that supply them with water. In 2003, Mexico created the Payment for Hydrological Environmental Services program, which pays for the conservation of forests in hydrologically critical watersheds using revenue from water charges. In southern Mexico, the Scolel Té project is paying farmers to provide carbon sequestration services. In Ecuador, the city of Quito has created a water fund with contributions from the water utility and the electric power company to pay for conservation in the protected areas from which it draws its water.

The bulk of PES programs to date have focused on water services, reflecting both the urgency of addressing water issues in many developing countries and the relative ease with which the beneficiaries of water services can be identified. The approach has been used for biodiversity benefits in a few cases, mostly with GEF support as in the case of the Costa Rica Ecomarkets Project. Environmental NGO Conservation International has also used the approach in several cases, including Guyana and Peru. Many of these efforts have focused on relatively untouched areas, however, rather than on agricultural areas.

From Theory to Practice

The RISEMP, which began implementation in July 2002, is seeking to pilot the use of the payment for environmental services approach to encourage the adoption of silvopastoral practices in degraded pastures areas in Central and South America. The project is being implemented in three microwatersheds: Quindío, in Colombia; Esparza, in Costa Rica; and Matiguás-Río Blanco, in Nicaragua. Participating land users enter into contracts under which they receive a payment for the environmental services that they generate. They receive annual payments over a two- or four-year period, based on the increment in environmental services provided relative to the baseline situation for that particular farm. Through this mechanism, the project aims to establish silvopastoral systems on 3,500 ha, thus enhancing the environmental benefits generated in watersheds covering about 12,000 ha.

The project was prepared with support of the multi-donor Livestock, Environment and Development Initiative (LEAD), hosted by the Food and Agriculture Organisation (FAO). It is financed by a US$4.5 million GEF grant, with the World Bank acting as implementing agency. In each country, field activities are being undertaken by local non-governmental organisations (NGOs): the Centre for Research on Sustainable Agricultural Production Systems in Colombia, the Center for Teaching and Research on Tropical Agronomy in Costa Rica, and Nitlapan in Nicaragua, with CATIE coordinating the work. The American Bird Conservancy (ABC) is providing technical assistance for the development of a common and consistent methodology for the monitoring of biodiversity at the three project sites.

Contracting for land users to provide biodiversity benefits is all very well in theory, but in practice it is clearly unrealistic to ask them to deliver biodiversity. A way is needed to communicate what is desired to potential participants in ways that they can understand. The typical solution has been to offer to pay not for biodiversity itself, but for land uses that are hospitable to biodiversity.

But land use alone can be a relatively blunt instrument. In Costa Rica's PSA program, for example, most contracts call for conservation of existing forest, and pay all participants the same amount. While this approach has the virtue of simplicity, it fails to recognise the very different levels of services that different land uses can provide. The biodiversity-friendliness of agricultural practices is not a binary, yes/no proposition. On the contrary, there is a spectrum of effects, ranging from relatively inhospitable systems such as monocultures with heavy agrochemical use to relatively hospitable systems such as organic coffee grown under a diverse shade canopy of native species. Location also matters: biodiversity-friendly practices in proximity to protected areas, for example, might be more valuable by helping to buffer and protect them. Failing to take these differences into account risks either under-paying for desirable land uses, or over-paying for relatively less desirable ones.

The solution adopted in the RISEMP was to prepare a list of land uses and associate each with a point system upon which payments are based. This approach is similar to that of the Environmental Benefits Index (EBI) used in US Conservation Reserve Program (CRP). Separate indices were developed for the biodiversity conservation and carbon sequestration benefits of each land use. These two indices were then aggregated to form an environmental service index to be employed as the basis for calculating payments to participants. A similar index for water benefits was not included, partly because of the lack of data needed to develop it, and partly because improved water flows would be national benefits, and thus are not eligible for GEF funding. The biodiversity conservation and carbon sequestration indices are presented in Table 4.

Table 4. Environmental service indices used in the RISEMP

Land use	*Biodiversity index*	*Carbon sequestration index*	*Environmental service index*
Annual crops (annual, grains, and tubers)	0.0	0.0	0.0
Degraded pasture	0.0	0.0	0.0
Natural pasture without trees	0.1	0.1	0.2
Improved pasture without trees	0.4	0.1	0.5
Semi-permanent crops (plantain, sun coffee)	0.3	0.2	0.5
Natural pasture with low tree density (<30/ha)	0.3	0.3	0.6
Natural pasture with recently-planted trees (> 200/ha)	0.3	0.3	0.6
Improved pasture with recently-planted trees (> 200/ha)	0.3	0.4	0.7
Monoculture fruit crops	0.3	0.4	0.7
Fodder bank	0.3	0.5	0.8
Improved pasture with low tree density (< 30/ha)	0.3	0.6	0.9
Fodder bank with woody species	0.4	0.5	0.9

Natural pasture with high tree density (> 30/ha>	0.5	0.5	1.0
Diversified fruit crops	0.6	0.5	1.1
Diversified fodder bank	0.6	0.6	1.2
Monoculture timber plantation	0.4	0.8	1.2
Shade-grown coffee	0.6	0.7	1.3
Improved pasture with high tree density (> 30/ha)	0.6	0.7	1.3
Bamboo (*guadua*) forest	0.5	0.8	1.3
Diversified timber plantation	0.7	0.7	1.4
Scrub habitats (*tacotales*)	0.6	0.8	1.4
Riparian forest	0.8	0.7	1.5
Intensive silvopastoral system (>5,000 trees/ha)	0.6	1.0	1.6
Disturbed secondary forest (> 10 m2 basal area)	0.8	0.9	1.7
Secondary forest (> 10 m^2 basal area)	0.9	1.0	1.9
Primary forest	1.0	1.0	2.0
New live fence or established live fence with frequent pruning< (per km)	0.3	0.3	0.6
Wind breaks (per km)	0.6	0.5	1.1

The biodiversity conservation index was scaled with the most biodiversity-poor land use (annual crops) set at 0.0 and the most biodiversity-rich land use (primary forest) set at 1.0. Within this spectrum, the points given to each specific land use were set by a panel of experts, taking into consideration factors such as the number of species (of plants, birds, small mammals, and insects), their spatial arrangement, stratification, plot size, and fruit production. Higher scores were given to land uses that have greater potential to maintain the original biodiversity of the region. Note that the index estimates the environmental benefits of all land uses, and not only silvopastoral practices.

This approach can take into consideration the different impact that different land uses are likely to have on biodiversity. There are, of course, limitations. The biodiversity impact depends not only on the characteristics of the land use, but also on its location, its extent, and its relationship to other land uses. At the pilot scale of the RISEMP, issues of location are not significant, as all three pilot areas were specifically chosen for their proximity to protected areas or to corridors between them. All three project areas were selected in part because of their location in ecologically-sensitive areas. The Quindío project site is in one of the most severely degraded regions of Colombia, with few, mostly unconnected remnants of natural habitats. Restoring a degree of habitat heterogeneity and connectivity would increase the chances of survival of species requiring large home ranges in an area considered as a priority for bird conservation.

The Esparza area in Costa Rica is in the vicinity of conservation areas such as La Fortuna, the Monteverde Reserve complex, and the Alberto Brenes Biological Reserve.

More biodiversity-friendly land use practices would help the chances of survival of several species occurring in these protected areas. The Matiguás-Río Blanco watershed in Nicaragua is part of the buffer zone of the Cerro Musún Natural Reserve, and is very close to one of the priority areas for bird conservation in the country. If this approach were to be scaled up and applied on a broader scale, location effects could be incorporated either by varying the points for activities in different locations or by varying the payment per incremental point. Issues of scale and contiguity are harder to address. Some biodiversity benefits may be obtained only after appropriate land uses cover a minimum area, or if the areas covered are contiguous rather than scattered. To an extent, these effects might be addressed by adding bonus points if the area covered by a given land use passes a threshold. Such an approach could quickly result in an excessively complex point system, however. Another approach would be to set a minimum participation threshold for the PES program to take effect; this approach was followed by New York City, for example.

A similar procedure was used to establish the carbon sequestration index, with different land uses given points according to their capacity to sequester stable carbon in the soil and in hard wood through the years. Recent studies indicate that secondary forest can fix an average of 10 tonnes of carbon per year in wood and in the soil. As secondary forest has a value of 1.0 in the index, 0.1 points correspond to an estimated sequestration of 1 tonne of carbon. Data from studies conducted by CATIE were used to calibrate the carbon sequestration index.

As data were insufficient to derive country-specific indices, the same index is being used in all three countries. Data from the monitoring efforts will be used to improve the indices, and it is expected that these will differ from country to country. Should downstream water users be willing to pay for hydrological services, the approach could also be extended by adding an index denoting the contribution of each land use type to the desired water services, though developing such an index would certainly prove difficult.

Note that under RISEMP, biodiversity and carbon sequestration benefits are given equal weight in calculating payments. The two indices could easily be de-coupled, however, with separate payment levels for each kind of environmental service. Alternatively, different weighting schemes could be used to give proportionally more weight to one or the other, depending on the interest of those making the payments.This index approach was tested with potential participants, and is proving quite intelligible to them in practice. Dissemination materials such as posters and handbooks have been prepared showing precisely what the payments would be for specific land uses.

HOW SHOULD PAYMENTS BE MADE

The second challenge in developing an appropriate contract is the need to understand the economics of the farming system, so that the appropriate amount and form of payment can be determined. Payments for environmental services will have the desired effect only if they reach the land users in ways that influence their decisions on how to use the land.

Analysis of the time path of benefits generated by silvopastoral systems showed that they are unattractive to land users primarily because of their substantial initial investment, and because of the time lag between investment and returns. This leads to the hypothesis that a relatively small payment provided in the early period of adoption would be sufficient to 'tip the balance' between current and silvopastoral systems. This effect works by increasing the net present value of investments in silvopastoral practices, but also by reducing the initial period in which adoption of these systems imposes net costs on land users. By the time payments end, the silvopastoral practices themselves are ready to begin generating income for land users. The payments also alleviate the liquidity problems faced by many land users and help them finance the required investments.

Based on this analysis, it was decided to provide a relatively small, up-front payment to participating land users. This payment is of US$75 per incremental point, per year over a four-year period, up to a maximum of US$4,500 per farm (US$6,000 in Colombia, where input prices are higher). Both of these aspects deserve further discussion.

In principle, the amount should be no less than the land users' opportunity cost, and no more than the value of the benefit provided. In practice, the actual value of the benefit provided is extremely difficult to estimate, and particularly so for benefits such as biodiversity conservation. In contrast, the farmers' opportunity cost can usually be estimated relatively easily. For this reason, as well as to limit the budgetary requirements of the payment, payment levels are usually set at slightly more than the opportunity cost of the main alternative land uses. All the existing systems of payments for environmental services implicitly or explicitly use this approach. Costa Rica's PSA program, for example, currently pays US$45/ha/year for forest conservation. This payment has proven to be quite attractive, with far more applications for this contract than the program has been able to finance. In Mexico, a specific study was commissioned of the opportunity cost of land to provide a basis for payments levels under the PSAH; no study was made of the magnitude of benefits. Zelek and Shively propose a scheme to pay the opportunity costs of Philipino farmers who adopt practices that sequester carbon. Paying the opportunity cost of adopting the desired practice also accords well with GEF's policy of paying for the incremental costs of generating global environmental benefits.

In terms of payments for carbon emissions reductions, the US$75/point/year payment level is equivalent to paying US$7.5 per tonne of carbon sequestered, or US$2 per tonne of CO_2 equivalent. This compares favorably to current world prices for carbon emissions reduction of US$3-5 per tonne of CO_2 equivalent—although these payments typically require a higher degree of assurance of the permanence of the emissions reduction and a more intensive monitoring regime than the RISEMP offers. No similar comparison is possible for payments for biodiversity conservation. In general, emerging guidelines for payments for environmental services indicate that payments should be on-going rather than finite. In Costa Rica's PSA program, for example, payments for forest conservation contracts are for 5 years, but they are renewable indefinitely by mutual agreement. The logic for this is simple: if environmental services are to be generated over a long period of time, then payments for these services should also be made over a similarly long period. Ending payments sooner creates the risk that land users will revert to their previous land use practices. This is a risk that has been observed time and time again in projects that attempted to change land use practices, such as soil conservation or reforestation projects. This risk was thought to be relatively low in this instance, as the silvopastoral practices, once established, are privately more profitable (Figure 1).

Table 5 illustrates the application of this contract for the 20 ha farm in Matiguás, Nicaragua, used in the previous example. In the baseline year, the farm has 2.5 ha under annual crops, 14.5 ha under natural pasture without trees, and 3 ha under brush (tacotal). Motivated by the project, it converts 3 ha of its pasture to higher tree densities: 1 ha a year for the first three years. It also plants an 0.5 ha fodder bank, and fences off the scrub areas so that secondary forest can regenerate. Finally, it plants trees along 1.5 km of its fence lines. Using the environmental service index in Table 4, the resulting scores can be calculated for the baseline and for each subsequent year. These scores are then used to compute the payments due to the farmer, including the initial baseline payment for existing services and the main payment for incremental services provided under the project. Figure 2 shows the impact of these payments on the time profile of benefits to adopting silvopastoral practices, and the resulting impact on the profitability of the investment. What had been a marginally viable investment now becomes more attractive.

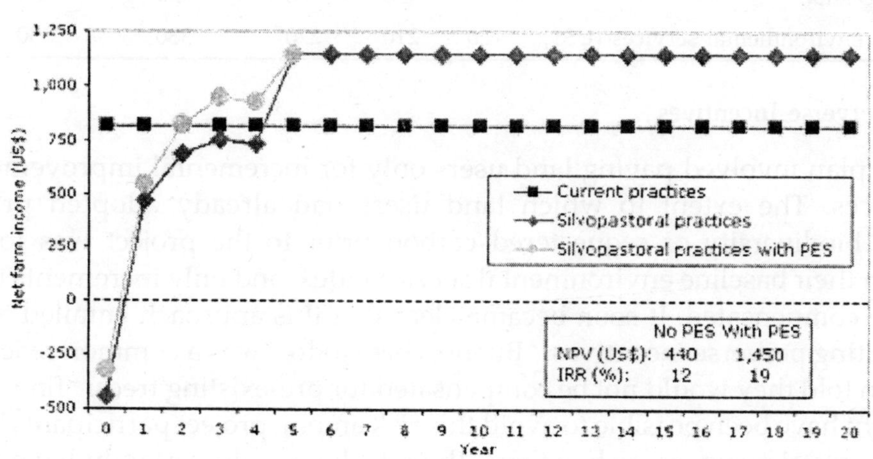

Figure 2. Effects of PES on the profitability of silvopastoral systems

Table 5. Example of payment computation

		Years from contract signing				
		0	*1*	*2*	*3*	*4*
Land use						
Crops (annual, grains, tubers)	(ha)	2.5	2.5	2.5	2.5	2.5
Natural pasture with low tree density	(ha)	14.5	13.0	12.0	11.0	11.0
Natural pasture with high tree density	(ha)	0.0	1.0	2.0	3.0	3.0
Fodder bank	(ha)	0.0	0.5	0.5	0.5	0.5
Scrub habitat	(ha)	3.0	0.0	0.0	0.0	0.0
Secondary forest	(ha)	0.0	3.0	3.0	3.0	3.0
Total area	(ha)	20.0	20.0	20.0	20.0	20.0
Wire fences with trees	(km)	0.0	0.5	1.0	1.5	1.5
Environmental service score (points)						
Crops (annual, grains, tubers)	0.0	0.0	0.0	0.0	0.0	
Natural pasture with low tree density	8.7	7.8	7.2	6.6	6.6	
Natural pasture with high tree density	0.0	0.9	1.8	2.7	2.7	
Fodder bank (monocrop)	0.0	0.5	0.5	0.5	0.5	
Scrub habitat	3.9	0.0	0.0	0.0	0.0	
Secondary forest	0.0	5.7	5.7	5.7	5.7	
Wire fences with trees	0.0	0.5	1.0	1.5	1.5	
Total points for the farm	12.6	15.4	16.2	17.0	17.0	
Baseline points	12.6					
Incremental points		2.8	3.6	4.4	4.4	
Income from environmental services (US$)	126	210	270	330	330	

Avoiding Perverse Incentives

The initial plan involved paying land users only for incremental improvements in land use practices. The extent to which land users had already adopted practices that conserved biodiversity or sequestered carbon prior to the project was to have been reflected in their baseline environmental service index, and only increments to this index were to be compensated. It soon became clear that this approach entailed a substantial risk of creating perverse incentives. "Bueno, corto todo," was a common reaction by land users when told they would not be compensated for pre-existing trees: "fine, I'll cut them all." It might have been possible to avoid this risk among project participants by imposing contractual restrictions on such actions, though this would certainly have required an increased monitoring effort, and thus increased costs.

But there was also a broader risk that non-participants in surrounding areas would postpone adopting silvopastoral practices that they might have been tempted to adopt, so that they might wait for a project to come and compensate them for doing so. As a result, the initial plan was modified to allow for a payment to be made for pre-existing environmental services. A one-time payment of US$10/point will be made for the baseline points, up to a maximum of US$500 per farm. This payment has the further benefit of helping to alleviate financing constraints to implementing silvopastoral practices. As part of the effort to avoid perverse incentives, the contract also specifies that burning in pastures is banned, and that the contract will be terminated if the participants cut down primary or secondary forest in their farms. A related problem is that of minimising 'leakage'-that is, avoiding environmentally-damaging activities being simply displaced, so that there is little net benefit. The RISEMP minimises this problem by computing the points on which payments are made over the entire farm and basing payments on the net points over the entire farm. If land users cut down trees in one plot even as they plant them in another, the negative points earned from the adverse change will offset those gained from the positive change.

MONITORING RESULTS

Monitoring is always important, as it allows the effectiveness of the project to be determined. It can also allow mid-course corrections to be made if they should prove necessary. The need for monitoring is particularly high in pilot projects, which are intended to serve as guides for future projects.

Changes in Land Use

For the purpose of monitoring compliance with the contract and computing payments owed, observation of participants would be sufficient. However, the RISEMP has the broader objective of piloting the application of PES systems to silvopastoral practices. This requires not only monitoring the behavior of participants, but determining that the project has been instrumental in affecting this behavior. Monitoring the degree to which the project is encouraging participants to undertake the desired changes in land use entails monitoring the changes in land use of the participants themselves, and of a control group. To address these issues, monitoring will be undertaken on three groups. In addition to the target group of participants, a control group of 30 households will be monitored in each country. This control group will be selected so as to have similar characteristics as the households in the target group, but will not receive any payments for environmental services. The target group itself will be partitioned into two groups, so that the impact of technical assistance provided by the project can also be tested. The main part of the group will receive both the payments and technical assistance, while a subgroup of 30 households will only receive the payments for environmental services, without the technical assistance. Each household in each of these groups will be monitored every other year. Their land use will be monitored, and a socio-economic survey will be conducted.

Impact of Land Use Change

To verify that the silvopastoral systems promoted under the project actually generate the expected environmental benefits, biodiversity and carbon sequestration will be monitored in all land use types in the three pilot areas. For biodiversity, counts of bird species will be the main indicator of biodiversity used, but they will be complemented by studies of butterflies, ants, and mollusks. Factors such as endemicity and rarity in the species observed will be taken into consideration. Water quality will be monitored only in the Colombian site, as funds did not permit a more general assessment of the contribution of these systems to improved water quality.

The changes will be compared to baseline measurements made at the start of the project. A study of the Matiguás-Río Blanco sites in Nicaragua, for example, found 131 bird species. The key test for the project will be whether biodiversity increases significantly compared to the baseline.

The results of the monitoring will also be used to revise and refine the biodiversity and carbon sequestration indices. These revised indices could be used to determine payments under any future project. They will not, however, affect payments under the current project, which will be made based on an ex ante estimates of the global environmental benefits of each land use, as expressed in the environmental service index.

Impact of the Project on Participating Households

In addition to the project's impact on the global environment, it is also important to understand its impact on household welfare: does welfare increase, and if so how much and in what way, and are there differences in how welfare increases across income groups? Payments for environmental services have been hypothesized as having the potential for improving the welfare of the poor in target areas, but there has been little empirical work on this topic to date. Data collected through the socioeconomic survey will help to address these questions. It will allow low-income households to be identified, and will provide a variety of measures of household welfare, including income-related measures and other indicators.

Assessing the Success of the Project

Will the desired land use changes be induced? The extent to which the desired land use changes are induced depends in part on the payment level being sufficient to 'tip the balance' between current and improved practices. The appropriate level of the payment was a subject of intense debate. The higher the payment offered, the greater adoption is likely to be. However, higher payments also militate against the cost-effectiveness of the approach. Moreover, a higher payment per point also increases the risk that participants will adopt otherwise un-profitable practices only temporarily, so as to receive the payment, with the intention of abandoning them later. The payment level was initially set at US$50 per environmental service index point, but was raised to US$75 after field staff reported that participants considered US$50/point insufficiently attractive to justify widespread adoption of silvopastoral practices. Given the novelty of the approach, there is also a potential credibility problem of the project's promise to pay for environmental services. The baseline payments probably played an important role in this sense, over and above their benefits in avoiding perverse incentives.

Assessing the Success of the Approach

How cost-effective is the PES approach in terms of biodiversity conservation? Both the cost and the effectiveness of this approach remain to be determined. The two are to some extent inversely related. Payments will be high if land users adopt practices with high point values, and low if they adopt practices with low point values. The transaction costs of implementing the project must also be considered. Some have argued that an incremental conservation dollar would be most effectively spent on other approaches, such as protected areas. Certainly protected areas are likely to have much lower transaction costs than the approach discussed here. But their implementation costs may well be higher. Establishing a protected area would require buying the land from its current owners. That is, it would require compensating them for the loss of the entire flow of benefits it might generate in its most profitable use. The PES approach only requires compensating them for the difference between the net benefits they obtain under the conservation use and the most profitable use. Moreover, buying the land outright requires paying for the entire present value of the future flow of benefits up-front. In contrast, a PES approach makes payments over time. PES will also attract lower-opportunity cost lands, while a protected area approach usually privileges conservation benefits and thus may include higher-opportunity cost land. The PES approach is likely to be particularly advantageous if, as may be the case in the RISEMP, a short-term payment is sufficient to result in sustainable adoption of the desired land uses. Finally, buying land outright may simply not be politically feasible, or may entail undesirable social consequences because of the need to relocate the landowners. In the case of the Costa Rica Ecomarkets Project, the PES approach was found to be much more cost-effective than establishing a protected area of the same size.

Replicating the Approach

How can transaction costs be reduced? The transaction costs involved in implementing a PES approach are a key determinant of its cost-effectiveness, its sustainability, and its replicability. They also play a critical role in the extent to which poorer land users can participate. Because of its pilot nature, the RISEMP has relatively high costs for detailed monitoring and other activities that would not necessarily be needed in a scaled-up project. The environmental service index used in the RISEMP allows a very fine-tuned targeting of payments to expected benefits, but it also imposes relatively high monitoring costs. There is a need to find proxy indicators that are highly correlated with biodiversity conservation but are easy and cheap to monitor, ideally using remote sensing. If forest cover provides an adequate proxy, for example, it would be relatively cheap to observe. The current environmental services index cannot be monitored solely from remote sensing, as it includes elements of the type and quality of vegetative cover. A crucial question which needs to be explored is that of the tradeoff between the precision of the index and the transaction costs involved in implementing it.

How can the approach be made sustainable? Where biodiversity services differ, however, is in the long-term sustainability of payments. Emerging lessons indicate that payments under PES programs usually have to be made on a long-term basis if the desired services are to be generated sustainably. The specific practices promoted by the RISEMP project may not require a long-term payment, but this is likely to be the exception rather than the rule. In the more general case in which the systems with the highest external benefits are not the most profitable to land users, the short-term payment approach used in the RISEMP is unlikely to result in sustainable adoption of the desired land uses. Rather, longer-term, probably indefinite payments will be needed. In turn, this means that sustainable long-term financing sources will be required. Even if only short-term payments are sufficient, substantial additional funding flows will be required if the approach is to be extended beyond the pilot areas. For water services, potential sources for such financing can readily be identified-although capturing them can be difficult. Moreover, funding streams for water services can in principle be very sustainable, as they are tied to services that will continue to be used indefinitely.

IMPACT OF CONVENTION ON BIOLOGICAL DIVERSITY ON RESEARCH

The ethical, commercial and policy context in which biodiversity prospecting, research and conservation take place has shifted significantly over the past ten years. The Convention on Biological Diversity (CBD) reflects and has bolstered a number of these shifts, which include the following.

- Biodiversity, previously considered the "common heritage of mankind", is now considered the "national patrimony" of host countries and is under their sovereignty.
- Genetic and species diversity have commercial potential not only as a source of material, or as commodities, but also as a source of information for product development.
- The rights of indigenous peoples and local communities to control and benefit from biodiversity research, prospecting and conservation are recognised, and increasingly

formalised in policy instruments as well as documents developed by indigenous peoples' groups.

— Benefits of biodiversity prospecting, research and conservation must be equitably shared.

The CBD objectives of biodiversity conservation, sustainable use and fair and equitable sharing of benefits (Article 1) reflect a balance between a range of diverse agendas and perspectives on what biodiversity is, and whom its conservation and use are intended to serve. Most of the world's biodiversity is found in inverse proportion to technological and industrial wealth. The CBD negotiation and implementation process has been an arena for expression of conflicts between North and South over financial and natural resources and differing conceptions of environmental problems and the meaning and value of nature. This process has made clear what was always the case in practice - that biodiversity and genetic resources are not only biological resources but political resources as well.

Simply put, high-biodiversity developing countries asked to set aside or manage large areas for conservation purposes argue that they should capture a fair share of the economic revenue generated by the genetic, species and ecosystem diversity that they conserve. Developed countries seeking the conservation of biodiversity have sought to maintain continued access to genetic resources for academic and commercial research. The CBD reflects the agreement reached by these various groups, and has been called the "grand bargain". The CBD clearly links environmental concerns with emerging human rights and trade issues; while it establishes general and qualified terms for trade in genetic resources, the CBD also acknowledges the value of non-marketed goods and services, and makes explicit ethical commitments to fairness and equity.

Biodiversity Prospecting

Biodiversity prospecting, or "bioprospecting", was first defined by Reid et al. as: "the exploration of biodiversity for commercially valuable genetic resources and biochemicals". It encompasses a wide range of commercial activities, including the pharmaceutical, biotechnology, seed, crop protection, horticulture, botanical medicine, cosmétic and personal care, and food and beverage sectors.

The most dramatic example of a forest tree species holding commercial potential, and the associated challenges to conservation and sustainable use, is the Pacific yew, *Taxus brevifolia,* whose bark yields the cancer drug taxol. Developed from collections made in 1962 as part of the United States National Cancer Institute's research programme, taxol, marketed by Bristol-Myers Squibb Co. under the brand name Paclitaxel, has been one of the top selling drugs in recent years; in 1998 worldwide sales were US$1.2 billion. Wild supplies in the tree's native range from northern California in the United States to British

Columbia in Canada proved insufficient to support commercial production. However, public concerns about the sustainability of *Taxus brevifolia* supply were counterbalanced by concerns that limiting wild harvest of *T. brevifolia* could restrict access to an important life-saving drug. Methods to convert the taxol precursor compound baccatin from Taxus baccata were developed in France, and today raw material sourcing for taxol is based in Europe.

There is considerable variation within and among the different industry sectors that use genetic resources. Differences include the following:

— *Size of industries and markets for products.* The global market for pharmaceuticals is more than US$300 billion a year, while sales in the botanical medicine industry are not much more than US$20 billion, and those of ornamental horticultural products lie between US$16 billion and $19 billion. The cosmetic and personal care industry has annual sales between US$50 billion and $75 billion a year, although the "natural" component of this industry is not more than US$3 billion.

— *Share of natural products in these markets.* In the pharmaceutical industry, natural products contribute somewhere between 25 and 50 percent of total sales of products on the market. Commercial botanical medicines, ornamental horticultural products and agricultural seeds are 100 percent natural products. "Natural" personal care and cosmetic products make up less than 10 percent of global sales in this sector today.

— *Relationship between commercial products and the genetic resources from which they are developed.* In the pharmaceutical and crop protection industries, for example, commercial products might be chemically identical to the pure natural product, might derive from chemical modification of a natural product, or might be synthesized to a design based on a parent structure that comes from nature.

— *Use of traditional knowledge in research and development.* Most commercial sectors involved in biodiversity prospecting today have their roots in traditional knowledge systems - that is, histories of traditional management and improvement of food and medicinal species dating back hundreds of years, and complex cultural relationships between people and the natural world. For example, of the approximately 120 pharmaceutical products derived from plants in 1985, 75 percent were discovered through the study of their traditional medical use. However, some sectors - the horticulture, seed and biotechnology industries, for example - appear to make little direct use of traditional knowledge in their research and development (R&D) programmes today; companies do not conduct field ethnobotanical collections and only rarely, if ever, use traditional knowledge gathered from second-hand sources such as literature. In the botanical medicine, cosmetics and personal care, pharmaceutical and crop protection sectors, a few ethnobotanical collecting programmes supply commercial companies, but the movement of traditional knowledge to the private sector takes place primarily through publication of academic research results in literature and databases

Provisions of CBD

The CBD deals with biodiversity research and prospecting issues in a number of ways. Article 15 establishes the sovereign rights of States over their natural resources and encourages Parties to the Convention to facilitate access to genetic resources. Access, where granted, must be on mutually agreed terms, subject to the prior informed consent of the Party supplying resources, with the aim of sharing in a fair and equitable way the results of R&D and the benefits arising from commercial and other utilisation of genetic resources. Article 16 encourages the transfer of technology to countries that provide access to genetic resources. Article 19 promotes the effective participation by providers of genetic resources in biotechnological research on the genetic resources they provide. Article 8j addresses traditional knowledge and promotes wider application of the knowledge, innovations and practices of indigenous and local communities, with their approval and involvement, and encourages the equitable sharing of benefits arising from this use or application.

The design of the CBD was influenced by a range of trends in the conservation, development and research communities, including recognition of links between sustainable development and conservation; awareness of the need to involve a range of stakeholders in resource management decision-making processes; and moves to ensure that indigenous peoples and local communities provide prior informed consent and receive benefits from these activities.

The development and implementation of the CBD is also influenced by trends outside the traditional domain of forest and biodiversity conservation, as described in the following paragraphs.

Scientific and Technological Advances

Scientific and technological advances in fields that make use of biodiversity (and in some cases traditional knowledge), such as biology, chemistry, genomics and information technology, are rapid and have changed the way natural products are used in R&D. At the same time, new technologies such as combinatorial chemistry, ultra-high throughput screening and laboratories on a chip have dramatically accelerated the pace of R&D, including that on natural products.

Increased Strategic Partnerships

Within and across commercial sectors, companies are entering into partnerships that allow them to participate in increasingly specialised research. As a result of the increased fragmentation and specialisation associated with biodiversity prospecting, no one individual or group tends to hold all the necessary technological infrastructure or expertise. Globally, complex networks of collaboration and partnership have become the norm. Partners to commercial companies are often academic research institutions, and one result of the trend towards collaboration and out-sourcing is increased blurring of the divide between academic and commercial research. At the same time, many companies are consolidated through mergers and acquisitions, and lines between sectors are blurring as companies seek cross-sector synergies to develop new knowledge and novel products. Large life science companies now often combine pharmaceutical, food, seed and chemical divisions under one umbrella. One result of increased consolidation is that already large companies become even larger, and corporate revenues dwarf the gross domestic product (GDP) of countries from which they seek to obtain genetic resources (Table 1). This has led to wariness on the part of high-biodiversity countries that cannot muster the same resources to negotiate and monitor partnerships.

Table 1. Gross domestic product (GDP) and annual pharmaceutical company sales, 1997(million US$)

Country (GDP ranking)/company	*GDP*	*Health care revenue*	*Pharmaceutical sales*
United States	8 083 400		
Japan4 706 877			
Germany	2 128 903		
China	962 389		
Brazil	808 147		
Australia	390 493		
South Africa	129 803		
Malaysia	97 240		
Bangladesh	31 359		
Merck & Co (USA)		23 637	13 282
Johnson & Johnson (USA)		22 629	7 696
Ecuador	19 428		
Novartis Group (Switzerland)		16 377	9 7327
Sri Lanka	15 139		
Bristol-Myers Squibb (USA)		14 996	9 932
American Home Products (USA)		14 485	11 076
Glaxo Wellcome Plc (UK)		13 087	13 087
SmithKline Beecham Plc (UK)		12 784	7 498
Pfizer Inc. (USA)		12 504	9 239
Costa Rica	12 067		
Côte d'Ivoire	10 453		
Cameroon	9 247		
Fiji	2 183		
Guyana	695		

Privatisation of Biological Resources

The patent system - paralleling scientific and technological advances - has undergone a process of regulatory globalisation and harmonisation, and the scope of what is regarded as patentable has "quietly expanded". In recent years, patent offices have begun to issue patents for the discovery of information already existing in the natural world, such as the genetic sequences of living organisms, and for plants, animals and microorganisms containing genes that have been modified in the laboratory. The Agreement on Trade-Related Intellectual Property Rights of the World Trade Organisation (WTO) has incorporated these tendencies. This trend has led to concerns associated with increased privatisation of the natural world, inequitable ownership of resources and knowledge, and appropriation of developing country resources and traditional knowledge systems by private corporations.

National Access and Benefit-sharing Measures

The CBD, as a framework convention, lays out legally binding objectives and overall obligations and rights of parties, but specific activities are regulated by protocols to the convention or other instruments such as national law. Access and benefit-sharing legislation developed to date can be grouped into five categories: environmental framework laws; sustainable development, nature conservation or biodiversity laws; dedicated stand-alone national laws and decrees on access to genetic resources; modification of existing laws and regulations; and regional measures.

Cases in which particular species showed commercial promise have helped to spur development of access and benefit-sharing measures in many countries. For example, random collections in the 1980s of the biologically diverse flora of the forest of Korup National Park in Cameroon produced a number of interesting leads, most strikingly the anti-HIV compound michellamine-B, derived from the forest liana *Ancistrocladus korupensis*. Although research on this compound is now stalled because of its toxicity, it showed tremendous promise in the United States National Cancer Institute's research programme for many years. The apparent commercial potential of *A. korupensis* influenced Cameroon's 1994 Law Regulating Forestry, Wildlife and Fisheries and the 1996 Framework Law on the Environment, which include provisions asserting national sovereignty over genetic resources and requiring prior informed consent and benefit-sharing from any commercial use of Cameroon's genetic resources. As a result of this case, a Prime Ministerial Committee was formed to address issues relating to medicinal plant commercialisation, and widespread public attention was paid to the benefits that Cameroon would receive as a result of research on its biodiversity.

Similarly, the adoption of the Sarawak Biodiversity Ordinance in 1998 arose partly in response to awareness raised by the case of *Calophyllum lanigerum*, collected from forests in Sarawak, Malaysia in 1987 by botanists from the Sarawak State Forestry

Department and a United States university. When this species yielded the anti-HIV compounds (+)-Calanolide A and (-)-Calanolide B, an agreement was signed between the Sarawak State Government and the United States National Cancer Institute. In 1996 the United States company Medichem Research and the Sarawak Government formed a joint venture to carry through the clinical development of these compounds, to facilitate investigation of other drug candidates from Sarawak's forests and to train Malaysian scientists. Although no commercial products have emerged from the joint venture to date, the case indirectly led to the development of the new ordinance, which established a new legal and administrative structure for access and benefit-sharing - including the creation of the Sarawak Biodiversity Council, which has the power to regulate access, collection, research, experiment, protection and utilisation of Sarawak's biodiversity, including the removal of any of the biodiversity from the state.

National access and benefit-sharing measures have advanced complex issues and trends, but have regularly encountered difficulties. A common problem has been a shortage of expertise and understanding within governments and NGOs of the industries they hope to regulate through these measures. In order to draft effective access and benefit-sharing measures, governments must have information available on the scientific, technological and marketing profile of each industry. They must understand how best to maximise benefits through partnerships. They must be familiar with biodiversity contracts and how they can reflect best practices to date, and with innovative ways of sharing financial benefits through mechanisms such as trust funds.

Addressing access and benefit-sharing issues requires an empirical and flexible approach. For example, the Philippines pioneered the first comprehensive access and benefit-sharing measure, and has subsequently adapted and adjusted the national regulations in light of practical experience. In May 1995, following a process of national consultation, Executive Order No. 247 "Prescribing a Regulatory Framework for the Prospecting of Biological and Genetic Resources, Their By-Products and Derivatives, for Scientific and Commercial Purposes, and for Other Purposes" was signed by President Fidel V. Ramos. In June 1996, Department Administrative Order No. 96-20, "Implementing Rules and Regulations on the Prospecting of Biological and Genetic Resources" was issued by the Department of Environment and Natural Resources. The Executive Order requires anyone seeking access to genetic resources to conclude either an Academic Research Agreement or a Commercial Research Agreement with the Government. Representatives of the pharmaceutical industry and many academic researchers found the Executive Order and Implementing Rules and Regulations to be overly bureaucratic and costly in practice, and companies criticised a requirement to license technology to the Philippines. In response, the government is redesigning administrative and other elements of the regulation to ensure it does not act as a direct disincentive to research.

The following are some of the central lessons offered by the experience of implementing an access and benefit-sharing measure in the Philippines.

— Stakeholder participation is essential in developing, enacting and implementing access and benefit-sharing policies, laws, rules and regulations.
— Defining the scope and coverage of a national access and benefit-sharing regulation is a priority concern.
— The potential impacts on scientific research activities must be carefully considered in designing and implementing national access and benefit-sharing measures.
— Creative approaches to obtaining consent from and sharing benefits with local communities, including indigenous peoples, need to be explored and developed.
— An efficient and effective institutional system should be put into place to implement the measure.
— Executive orders, rather than legislative acts, can be useful as a way of exploring and testing approaches to regulating access and benefit-sharing, but they are also inadequate.
— In regions where countries share genetic resources, national frameworks alone are inadequate and regional mechanisms may be required.

Reported changes in industry practice in response to the CBD and national access and benefit-sharing measures include: a decrease in collecting activities, consolidated in fewer countries where laws and procedures are clear and efficient, and an increased focus on domestic collection; greater recourse to material held in ex situ collections; an increased reliance on intermediaries as brokers of access and benefit-sharing relationships, as well as suppliers of samples; and the increased use of material transfer agreements to clarify terms of partnerships.

Protected Areas and Access and Benefit-sharing

Protected area managers are seeking innovative partnerships with companies and researchers to supplement declining or chronically inadequate budgets. For example, the National Institute of Biodiversity (INBio) in Costa Rica conducts collections in protected areas through commercial partnerships with companies such as Diversa and Bristol-Myers Squibb (United States), Givaudane Roure (Switzerland-United States), Indena (Italy), Analyticon (Germany), La Pacifica (Costa Rica) and the British Technology Group (United Kingdom) as a way to fund biodiversity research, conservation and support for the national system of Conservation Areas. INBio was established by the Ministry of Environment and Energy as a private non-profit organisation to help conserve, study and use the country's biological diversity. As laid out in a cooperation agreement between the Ministry and INBio, INBio will provide roughly 10 percent of the total annual budget of any project to the country's Conservation Areas, and 50 percent of any financial benefits from commercial product development resulting from collections in protected areas.

The CBD and national access and benefit-sharing measures regulate both academic and commercial research. Indeed, distinctions between the two are increasingly blurred. Academic researchers often undertake contracts for companies, and academic data flow to the private sector through publications and databases. As a result, it is important for both academic and commercial researchers to ensure that the manner and terms under which all research takes place are equitable for local groups. There remains a need to instill in the academic community - which sets the standards for most research - an appreciation of the new ethical and legal envelope within which research takes place, and of new demands that biodiversity research should contribute concretely to wider social and conservation objectives while furthering scientific understanding.

In addition to international and national- level policy and law, access and benefit-sharing policies are being developed for research institutions, and professional research groups are developing codes of ethics and research guidelines that incorporate the objectives of the CBD. Examples of institutional policies include the University of South Pacific Guidelines for biodiversity research and bioprospecting, the Limbe Botanic Garden in Cameroon's Policy on access and benefit-sharing, and the Common policy guidelines to assist in the preparation of institutional policies based on the "Principles on Access to Genetic Resources and Benefit-Sharing for Participating Institutions" developed by a consortium of botanic gardens and research institutions. Codes of ethics, research guidelines and other documents that address issues related to access and benefit-sharing and respect for traditional resource rights include those of the International Society of Ethnobiology, the Society of Economic Botany, the American Society of Pharmacognosy, the American Anthropological Association and the Asian Symposium on Medicinal Plants.

COMMUNITY-BASED BIODIVERSITY CONSERVATION

In the traditional approach to biodiversity conservation, local people and their economic activities were viewed as threats to the undisturbed functioning of natural ecosystems and were to be excluded from protected areas. However, it became evident that the social costs of exclusionary conservation projects were sometimes high, and that their success rate, even in biological terms, was disappointing. As a result, the classical approach has been replaced by integrated conservation and development projects (ICDPs) whose goal is to "enhance biodiversity conservation through approaches that attempt to address the needs, constraints and opportunities of local people" by involving local people as active partners. The success rate of ICDPs is also discouraging. However, decentralisation and devolution of management responsibilities continue to be so widely viewed as the only solution for maintaining ecosystems that it has become heresy to question them.

Classical vs Participatory Approaches

The preservation of natural ecosystems has long been on the agenda of institutions concerned with biodiversity. Representative samples of ecoregions have been set aside and put under strict protection. This "northern" vision of an untouched wilderness has permeated global policies and politics for decades. The conventional approach to conserving biodiversity requires the following:

- the identification of biodiversity loss as serious, indicating that conservation is urgently needed;
- the design of projects in which, if exclusion is not an option, the cooperation of local communities is sought;
- the implementation of plans through a combination of encouragement, persuasion and subtle threats, and sometimes by more coercive powers.

Local people are viewed as "the target population" or "beneficiaries" and are frequently excluded from the areas considered important for biodiversity conservation. In the conventional approach, biodiversity is seen to be at its optimum in undisturbed natural areas. The national government is viewed as the guardian of biodiversity and has sovereignty and nominal control over the areas required for conservation. In reality, however, effective control rests with any of a number of forest users and particularly with the population that lives in and around the forests.

Conservation projects and programmes that fail to consider the interests of local residents, that undermine existing indigenous management systems or that restrict local authorities' decision-making in resource management have often been seen to intensify the loss of biodiversity. Furthermore, they raise highly contentious debates about national and local interests and can lead to open protests and conflicts; for example, in many protected areas in India, restrictions on access to resources, harassment by forestry officials, exposure of crops and livestock to ravage by wild animals and invasion by noisy tourists - now freely roaming the same areas where villagers have been banned -have led to resentment on the part of village communities. Moreover, it appears neither politically feasible nor ethically justifiable to deny the use of natural resources to the poor and marginalised without providing them with alternative means of making a living or compensatory payments.

Recent consensus holds that local people do not pose a threat to biodiversity but rather are the victims of its loss and those most affected by forest degradation. When they have been implicated in forest destruction they are seen to have been obliged to overexploit resources because of inappropriate policies and legal systems. Hence, attention has shifted away from blaming rural people for deforestation and loss of biodiversity, and towards identifying the larger processes that are the primary cause of unsustainable land use practices and overexploitation of forest resources.

The recognition that the solutions to ecosystem management problems lie in social, cultural and economic systems stimulated the development of a new paradigm, which views local people as part of the solution and not as part of the problem. Thus, top-down exclusionary management has been replaced by forms of participation and devolution. According to this thinking, the conventional approach mainly failed because of a lack of coherence between interventions and local livelihood strategies and because of the exclusion of local people in project design, planning, implementation, monitoring and evaluation.

Yet, devolution or the participatory approach in many cases does not appear to be more effective than the conventional approach. One reason may be that genuine participation is still the exception rather than the rule: most organisations merely pay lip service to the goal of involving local communities. But a closer look at the assumptions underlying devolution and participatory approaches suggests other failings.

Challenging Assumptions

The concept of partnership in conservation is based on the following, often contested, assumptions:

- local populations are interested and skilled in sustainable forest resource use and conservation;
- contemporary rural communities are homogeneous and stable;
- local, community-based tenurial, knowledge and management systems are uniquely suitable for forest conservation.

Degree of Interest and Skills

The devolution of natural resource management and conservation activities is predicated on the assumption that local people have the necessary motivation and skills. Examples from the management of non-wood forest products (NWFPs), often held out as a pathway for the conservation of forests and natural ecosystems, challenge this assumption.

In the past, a number of social and environmental constraints held overharvesting of NWFPs and timber for local purposes in check. In low-use areas, most products are still used sustainably and traditional restrictions and regulations are heeded. However, empirical evidence suggests that few wild resources can sustain commercial exploitation, and trade will either result in local depletion or extinction, or initiate the domestication process. Destructive harvesting practices are used in certain places for many NWFPs (e.g. ironwood, wild honey, sandalwood and fruits, mushrooms, rattan and bird nests), seriously reducing the abundance of particular species. The conventional NWFP sector may be less harmful to the resource than the newer commercial sector, but this issue is complex, particularly because there is no clear distinction between subsistence and commercial uses.

The traditional collectors suffer most from the intensification of harvests, which partly explains why they are usually more concerned about the threats to forests posed by intrusions from outsiders than about ecological sustainability per se. Traditional collectors are often poorly organised, if at all. Their ability to take collective action in managing common or wild resources, and in controlling access, is limited. At the same time, market expansion for many products has led to greater competition among collectors and traders.

Many traditional societies have sensibly manipulated natural forests, even if they have not been overly concerned about conservation. On the island of Borneo, for example, many Dayaks have been manipulating old growth and secondary forests for several hundred years to raise their productivity. Even though the Dayaks exploit their forest-fruit gardens actively, such forests have a species diversity and a vegetation structure resembling natural forests. The impact of Dayak "forest agriculture" on forest cover -although not necessarily on plant composition and biodiversity - is minor as long as population densities remain low and no other stakeholders appear on the scene, but this is rarely the case and is becoming increasingly uncommon. Colfer and Soedjito point to an area that, until the early 1960s, was a lowland primary dipterocarp forest but, today, is covered by a significantly different forest affected by fires, extensive large-scale logging and small-scale agroforestry activities.

In other instances, for example in Mali, archaeological evidence suggests that, historically, resource management has been marked by cyclical periods of overexploitation, out-migration, environmental recovery and subsequent in-migration. Local groups therefore may not have developed the knowledge, skills and techniques for long-term sustainable management in an increasingly confined environment where out-migration is no longer an option.

Local communities' interest in forest conservation depends at least to some degree on how much they are still part of the ecosystem and how much their management of resources directly affects their own survival. Cultural mechanisms that have been developed as adaptations to the forest environment over hundreds of years may be easily cast aside when trade and new technologies free people from traditional ecological constraints. Thus, traditional resource use patterns are only sustainable under specific circumstances, usually characterised by low population densities, land abundance, use of simple technologies and limited involvement in the market economy. On the other hand, even under such conditions people can be exceedingly wasteful of resources, although the magnitude may not be enough to have a significant impact.

Human relationships with the landscape are dynamic and may change from generation to generation. Younger people who are alienated from traditional culture, oblivious to past natural resource management practices and impatient for development may be a threat to biodiversity conservation.

It is naïve to assume that there are always ways to improve local incomes without depleting biodiversity, that people are interested in conserving biodiversity and that they prefer to hang on to traditional practices and knowledge. Local people may desire to obtain material benefits and a better standard of living from community-based forest management, and this goal may not be consistent with sustainability.

Role of Local Communities

Inherent in the premise that local communities should be given a central role in reaching conservation objectives is the dilemma of defining and understanding "local community", "indigenous knowledge" and "traditional culture". The idea of a unified community is a social construction that is perhaps comforting to policy-makers and foreign donors, who base on it assumptions about local management of resources. Advocacy for stronger legal rights and government recognition for community-based systems sometimes presents an idealised description of local communities. They may be portrayed as custodial and non-materialistic in their attitudes to land and natural resources, when in fact local communities in the forests and at the forest margins are often anxious for development and increasingly affected by rapid marketisation and modernisation processes. The idealised image is not only misleading, but defeats the purpose of constructive consensus building and frustrates those who view the empowerment of local communities as a precondition for successful biodiversity conservation.

Although small groups such as neighbourhoods or extended families - which can be considered "interest groups" -may share resources and livelihoods, whole village communities in fact tend to be heterogeneous, factional and stratified. Villagers are often politically fractured and socially differentiated along gender, wealth, class, age or ethnic lines. Perceptions and definitions of biodiversity, as well as the implications of biodiversity loss and the costs of conserving biodiversity, are similarly differentiated. Furthermore, the divisions within a community are changeable; communities may present a unified front to a perceived threat from outside, or may be divided and redivided by internal struggles over land and resources. A hierarchical society can hinder the performance of participatory projects and grassroots development, as Rigg has shown in Thailand.

The most daunting problem is that the different interest groups subsumed in the category "community" interact with the local environment and its resources in different ways. These interactions are constantly changing and depend as much on the type of prevailing agro-ecosystem as on the local economy and influences of external forces.

Appropriateness of Community-based Tenure

Community-based tenurial systems are rarely acknowledged by national governments or logging operators in any meaningful way. While changing forest ownership and the transfer of authority over forests to local communities are not viewed as a panacea to the problem of resource degradation, they are usually viewed as prerequisites for biodiversity conservation. Increased tenure security has been linked to sustainable farming practices and is assumed also to apply to forest management. The assumption is that people are only willing to invest their scarce resources in conservation if they know that, ultimately, they will reap the benefits.

Much uncertainty remains regarding the implications of tenure change and devolution for resource conservation. In Bolivia, for example, both positive and negative examples are seen. Some local governments in indigenous areas patrol their areas to avoid encroachment from logging companies, ranchers or agricultural colonists, while others have suffered from petty corruption and sold their timber resources to logging companies with little concern for sustainable production. On balance, giving indigenous communities greater control over their natural resources by strengthening both their land tenure security and their local governments seems to have assisted resource conservation, although most groups are still more concerned with access to existing resources and short-term income than with long-term sustainable development.

Although necessary, tenure security is not a sufficient condition for sustainable forest management and conserving biodiversity. Many communities plan for their perceived immediate needs rather than for their future welfare necessities, ignoring or not understanding the medium- to long-term social, economic and environmental consequences of their current land use practices. In Papua New Guinea, for example, many forest-edge communities have been campaigning for years to attract extractive development and mining to their communally held areas and are opting to sell harvest rights to timber companies. The reasons include the desire for rent capture, consolidation of power and the perception that this constitutes "development". Conservation, because it yields future, diffuse and often intangible benefits, some of which have no direct monetary value, tends to be undervalued in this context and is seen to conflict with community aspirations.

Resource and property rights are changing with the transformation of agrarian societies and livelihood strategies as well as the influx of migrants. For example, in parts of Kalimantan, several of the forests managed by Dayak families but communally shared for many products are being rapidly privatised. In particular, where roads are diminishing the relative isolation of formerly remote areas and markets for land are developing, it is not uncommon to find parce ls of forest that, de jure, belong to the state but, de facto, are traded in the market. The interest in communally held resources or common property is thus diminishing.

The relationship between devolving responsibility and community-based tenure, on the one hand, and biodiversity conservation, on the other, defies broad generalisations. However, if biodiversity conservation requires some management, the contested issue of land tenure should figure in the discussion of viable options — but it should not be viewed as the solution for all of the problems.

REFERENCES

Pagiola, S., and G. Platais. Forthcoming. *Payments for Environmental Services: From Theory to Practice.* Washington: World Bank.

Schneider, R. 1994. "Government and the Economy on the Amazon Frontier." LAC Regional Studies Program Report No.34. Washington: World Bank.

Thiollay, J.-M. 1995. "The Role of Traditional Agroforests in the Conservation of Rain Forest Bird Biodiversity in Sumatra." *Conservation Biology*, 9, pp.335-353.

Tomich, T.P., M. Van Noordwijk, A. Budidarsono, A. Gillison, T. Kusumanto, D. Murdiyarso, F. Stolle, and A.M. Fagi (eds). 1998. *Alternatives to Slash-and-Burn in Indonesia: Summary Report and Synthesis of Phase II*. Bogor: ICRAF.

Zelek, C.A., and G.E. Shively. 2003. "Measuring the Opportunity Cost of Carbon Sequestration in Tropical Agriculture." *Land Economics*, 79:3, pp.342-354.

12

Biodiversity Informatics

Biodiversity Informatics is the application of computer informatics techniques to biodiversity information for improved management, presentation, exploration and analysis. It typically builds on a foundation of taxonomic, biogeographic, or ecological information stored in digital form, which with the application of modern computer techniques can yield new ways to view and analyse existing information, as well as predictive models for information that does not yet exist. Biodiversity informatics is a relatively young discipline but has hundreds of practitioners worldwide, including the numerous persons involved with the design and construction of taxonomic databases. The term "Biodiversity Informatics" is generally used in the broad sense to apply to computerized handling of any biodiversity information; the apparently broader term "bioinformatics" is often used synonymously with the computerized handling of data in the specialized area of molecular biology.

Biodiversity informatics has been defined as "the creation, integration, analysis, and understanding of information regarding biological diversity", and "[the] field that brings information science and technologies to bear on the data and information generated by the study of organisms, their genes, and their interactions"

Broadly speaking, it seeks to draw upon and integrate information held in various taxonomic databases and other digital sources to answer biodiversity questions at scales ranging from global to local. Such questions might range from "How many described species exist in the world?" to "Predict the effects of a global temperature rise of X degrees C. on the geographic range of species Y", a question which involves not only biodiversity in the basic sense but related domains of ecology, geographic distributions of environmental parameters, global climate models, and more. In addition to handling formally named taxa, biodiversity informatics may also have to cope with managing information from unnamed taxa such as that produced by environmental sampling and sequencing of mixed field samples. The term biodiversity informatics is also used to cover the computational problems specific to the names of biological entities, such as the development of algorithms to cope with variant representations of identifiers such as species names and authorities, and the multiple classification schemes within which these entities may reside according to the preferences of different workers in the field, as well as the syntax and semantics by which the content in taxonomic databases can be made machine queryable and interoperable for biodiversity informatics purposes.

According to correspondence reproduced by Walter Berendsohn, the term "Biodiversity Informatics" was coined by John Whiting in 1992 to cover the activities of an entity known as the Canadian Biodiversity Informatics Consortium, a group involved with fusing basic biodiversity information with environmental economics and geospatial information in the form of GPS and GIS. Subsequently it appears to have lost any obligate connection with the GPS/GIS world and be associated with the computerized management of any aspects of biodiversity information.

CURRENT BIODIVERSITY INFORMATICS ISSUES

Global List of All Species

One major issue for biodiversity informatics at a global scale is the present absence of a machine queryable (or even non-digital) master list of currently recognised species of the world, although this is an aim of the Catalogue of Life project which has been quoted as aiming to achieve this goal (for extant species only) by 2012; in its 2009 Annual Checklist edition a total of 1.16 million valid species names and 0.76 million synonyms were included, out of an estimated target 1.8 million extant described species. A similar effort for fossil taxa, the Paleobiology Database documents some 100,000+ names for fossil species, out of an unknown total number.

Problems with Genus and Species Scientific Names

Application of the Linnaean system of binomial nomenclature for species, and uninomials for genera and higher ranks, has led to many advantages but also problems with homonyms (the same name being used for multiple taxa, either inadvertently or legitimately across multiple kingdoms), synonyms (multiple names for the same taxon), as well as variant representations of the same name due to orthographic differences, minor spelling errors, variation in the manner of citation of author names and dates, and more. In addition, names can change through time on account of changing taxonomic opinions (for example, the correct generic placement of a species, or the elevation of a subspecies to species rank or vice versa), and also the circumscription of a taxon can change according to different authors' taxonomic concepts. One proposed solution to this problem is the usage of Life Science Identifiers (LSIDs) for machine-machine communication purposes, although there are both proponents and opponents of this approach.

Achieving a Consensus Classification of Organisms

Organisms can be classified in a multitude of ways which can create design problems for Biodiversity Informatics systems aimed at incorporating either a single or multiple classification to suit the needs of users, or to guide them towards a single "preferred" system. Whether a single consensus classification system can ever be achieved is probably an open question, however in an attempt to provide at least a degree of consensus, the Catalogue of Life project has recently released a document that attempts to list some of the issues in this area, and may lead to a more coherent classification that can be promoted via that project's future products at least.

PRIMARY BIODIVERSITY INFORMATION

"Primary" biodiversity information can be considered the basic data on the occurrence and diversity of species (or indeed, any recognizable taxa), commonly in association with information regarding their distribution in either space, time, or both. Such information may be in the form of retained specimens and associated information, for example as assembled in the natural history collections of museums and herbaria, or as observational records, for example either from formal faunal or floristic surveys undertaken by professional biologists and students, or as amateur and other planned or unplanned observations including those increasingly coming under the scope of citizen science. Providing online, coherent digital access to this vast collection of disparate primary data is a core Biodiversity Informatics function that is at the heart of regional and global biodiversity data networks, examples of the latter including OBIS and GBIF.

As a secondary source of biodiversity data, relevant scientific literature can be parsed either by humans or (potentially) by specialized information retrieval algorithms to extract the relevant primary biodiversity information that is reported therein, sometimes in aggregated / summary form but frequently as primary observations in narrative or tabular form. Elements of such activity (such as extracting key taxonomic identifiers, keywording / index terms, etc.) have been practiced for many years at a higher level by selected academic databases and search engines.

However, for the maximum Biodiversity Informatics value, the actual primary occurrence data should ideally be retrieved and then made available in a standardized form or forms; for example both the Plazi and INOTAXA projects are transforming taxonomic literature into XML formats that can then be read by client applications, the former using TaxonX-XML and the latter using the taXMLit format. The Biodiversity Heritage Library is also making significant progress in its aim to digitize substantial portions of the out-of-copyright taxonomic literature, which is then subjected to OCR (optical character recognition) so as to be amenable to further processing using Biodiversity Informatics tools.

BIODIVERSITY INFORMATICS STANDARDS AND PROTOCOLS

In common with other data-related disciplines, Biodiversity Informatics benefits from the adoption of appropriate standards and protocols in order to support machine-machine transmission and interoperability of information within its particular domain. Examples of relevant standards include the Darwin Core XML schema for specimen- and observation-based biodiversity data developed from 1998 onwards, plus extensions of the same, Taxonomic Concept Transfer Schema, plus standards for Structured Descriptive Data and Access to Biological Collection Data (ABCD); while data retrieval and transfer protocols include DiGIR (now mostly superseded) and TAPIR (TDWG Access Protocol for Information Retrieval). Many of these standards and protocols are currently maintained, and their development overseen, by the Taxonomic Databases Working Group (TDWG).

Darwin Core

Darwin Core is a Biodiversity informatics data standard that consists of a vocabulary of terms to facilitate the discovery, retrieval, and integration of information about organisms, their spatiotemporal occurrence, and the supporting evidence housed in biological collections. Often abbreviated as DwC it was originally created as a Z39.50 profile by the ZBIG (Z39.50 Biology Implementers Group) supported by funding from a USA National Science Foundation award . The name "Darwin Core" was first coined by Allen Allison at the first meeting of the ZBIG held at the University of Kansas in 1998 during discussions about elements required in the Z39.50 natural history collections profile (prototypical Darwin Core) and the conceptual similarity with the Dublin Core. The Darwin Core profile was later expressed as an XML Schema document for use by the DiGIR (Distributed Generic Information Retrieval) protocol. A TDWG task group is currently engaged in revising the standard.

The latest version of Darwin Core is based on the ideas of Dublin Core and decouples the definition and semantics of individual terms from application of these terms in different technologies like XML, RDF or simple CSV text files. Darwin Core defines separate guidelines on how to encode the DwC terms as XML or text files .

Although never ratified as an accepted TDWG standard, DwC has seen several version in production usage. Different versions tried to define their own namespace, but some flavours of Darwin Core are around that collide and use the same namespace although they define different terms. Please refer to the history for details.

Darwin Core Terms

The Darwin Core is body of standards. It includes a glossary of terms (in other contexts these might be called properties, elements, fields, columns, attributes, or concepts) intended to facilitate the sharing of information about biological diversity by providing reference definitions, examples, and commentaries. The Darwin Core is primarily based on taxa, their occurrence in nature as documented by observations, specimens, and samples, and related information. Included are documents describing how these terms are managed, how the set of terms can be extended for new purposes, and how the terms can be used. The Simple Darwin Core is a specification for one particular way to use the terms - to share data about taxa and their occurrences in a simply structured way - and is probably what is meant if someone suggests to "format your data according to the Darwin Core".

Motivation

The Darwin Core standard was originally conceived to facilitate the discovery, retrieval, and integration of information about modern biological specimens, their spatiotemporal occurrence, and their supporting evidence housed in collections (physical or digital). The Darwin Core today is broader in scope. It is meant to provide a stable standard reference for sharing information on biological diversity. As a glossary of terms, the Darwin Core is meant to provide stable semantic definitions with the goal of being maximally reusable in a variety of contexts.

Rationale

The Darwin Core is based on the standards developed by the Dublin Core Metadata Initiative and should be viewed as an extension of the Dublin Core for biodiversity information. The purpose of these terms is to facilitate data sharing by providing a well-defined standard core vocabulary in a flexible framework to minimize the barriers to adoption and to maximize reusability. The terms described in this standard are a part of a larger set of vocabularies and technical specifications under development and maintained by Biodiversity Information Standards TDWG.

Guiding Principles

Each term has a definition and commentaries that are meant to promote the consistent use of the terms across applications and disciplines. Evolving commentaries that discuss, refine, expand, or translate the definitions and examples are referred to through links in the Comments attribute of each term. This means of documentation allows the standard to adapt to new purposes without disrupting existing applications. There is meant to be a clear separation between the terms defined in this standard and the applications that make use of them. For example, though the data types and constraints are not provided in the term definitions, recommendations are made about how to restrict the values where appropriate.

Taxonomic Concept Transfer Schema

The availability, exchange and interpretation of taxonomic information (such as species check lists, distribution and identification data) is of critical importance to taxonomists, ecologists and other biologists and legislators, amongst others. This information is provided by a number of global and local taxonomic database services. These databases hold records, often based on valid scientific names for species, according to their own models of what constitutes a taxonomic 'entity' or concept (i.e. a species or higher level taxon). Databases typically model a single view of taxonomy, whilst making some attempt to relate their concepts to synonymous names or concepts.

The rules and history of taxonomic nomenclature mean that scientific names do not unambiguously identify taxonomic concepts as represented according to different models of taxonomy. Therefore the same name will apply to different taxon concepts according to different taxonomic viewpoints (conversely, very similar concepts according to alternative taxonomies may sometimes be given different names). The problems inherent in using names as identifiers are recognized and might be addressed by representing richer taxon concepts, based on a name plus a reference to the definition of the concept or at least by capturing the usage of that name according to some original taxonomic source.

The need for a common mechanism for the providers of taxonomic information to exchange information with other providers and users of varying expertise in taxonomy was recognized at TDWG Lisbon 2003. Such a mechanism must adequately represent the data as modelled by the owners of the data, whist facilitating integration with data provided according to different models of taxonomy.

The development of an abstract model for a taxonomic concept, which can capture the various models represented and understood by the various data providers, is central to this project. This model is presented as an XML schema document that is proposed as a standard to allow exchange of data between different data models. It aims to capture data as understood by the data owners without distortion, and facilitate the query of different data resources according to the common schema model.

The TCS schema was conceived to allow the representation of taxonomic concepts as defined in published taxonomic classifications, revisions and databases.

As such, it specifies the structure for XML documents to be used for the transfer of defined concepts. Valid transfer documents may either explicitly detail the defining components of taxon concepts, transfer GUIDs referring to defined taxon concepts (if and when these are available) or a mixture of the two.

The TCS schema is not designed to facilitate the exchange or documentation of information about Taxon Concepts where this information is not part of a taxonomic revision creating new concepts. The amount and variety of (additional) information that can be potentially assigned to concepts is outside the scope of a taxonomic concept transfer schema, but we would encourage the development of domain specfic models that use or extend this schema. XML supports this flexibility by allowing the use of different name spaces.

For example, whilst a TCS taxon concept definition may include details of specimen circumscription (i.e. list specimens that are asserted to define the taxon concept) datasets that merely include observations identifying specimens as being examples of a taxon concept would reference a defined taxon concept, not constitute a new or modified concept. That is, TCS documents are for transferring the definitions of taxon concepts, not for detailing observations of these defined concepts.

Examples of observational datasets that could refer to defined taxon concepts might include

— ecological field records, which record abundance data on a variety taxa
— descriptive datasets that include descriptions of representative specimens or taxa
— specimen records for biological collections

The observations recorded in these would not constitute definitions of the concepts, but document instances of the concept. (However, observation datasets might themeselves be included as part of a taxon concept definition, e.g. as specimen or character circumscriptions - if this is what the 'creator' of the concept intended).

Once concept definitions are globally available and have been assigned GUIDs, referencing them in observation datasets etc. is simplified. In the absence of GUIDs some form of user key could be recorded that would identify a preexisting concept. This might consist of the scientific name and concept author (AccordingTo).

Where there is no appropriate, prexisting globally defined concept to reference, datasets could provide TCS definitions of the concept being described/observed/ identified in TCS format. These would minimally consist of a name, but optimally include reference to the origin of the concept definition (The AccordingTo element in TCS). Thus field data that records presence of Aus bus, Aus cus and Aus dus, as identified in the field guide of should include this information in the data description/mark-up. It then will be possible to resolve these incompletely defined concepts with other concept records represented in TCS format.

Access to Biological Collection Data

The Access to Biological Collections Data (ABCD) Schema is an evolving comprehensive standard for the access to and exchange of data about specimens and observations (a.k.a. primary biodiversity data).

The ABCD Schema attempts to be comprehensive and highly structured, supporting data from a wide variety of databases. It is compatible with several existing data standards. Parallel structures exist so that either (or both) atomised data and free-text can be accommodated. Version 1.2 is currently in use with the GBIF (Global Biodiversity Information Facility) and BioCASE (Biological Collection Access Service for Europe) networks.

Apart from the GBIF and BioCASE networks, the potential for the application of ABCD extends to internal networks, or in-house legacy data access (e.g. datasets from external sources that shall not be converted and integrated into an institution?s own data, but be kept separately, though easily accessible). By defining relations between terms, ABCD is a step towards an ontology for biological collections.

Structured Descriptive Data

In taxonomy, descriptive data takes a number of very different forms. Natural-language descriptions are semi-structured, semi-formalised descriptions of a taxon (or occasionally of an individual specimen). They may be simple, short and written in plain language (if used for a popular field guide), or long, highly formal and using specialised terminology when used in a taxonomic monograph or other treatment.

Dichotomous keys are specialised identification tools comprising fragments of descriptive data arranged in couplets forming a branching tree. Each fragment (lead) comprises a small (occasionally verbose) natural-language description.

Raw data descriptions usually comprise repeated measurements of parts of individual specimens, and are the basis from which the more abstracted descriptions in natural language and coded descriptions are derived. Few taxonomists consistently record and archive their raw data in a standardised format.

The goal of the SDD standard is to allow capture, transport, caching and archiving of descriptive data in all the forms shown above, using a platform- and application-independent, international standard. Such a standard is crucial to enabling lossless porting of data between existing and future software platforms including identification, data-mining and analysis tools, and federated databases.

CURRENT PRIORITIES IN BIODIVERSITY INFORMATICS

The term "Biodiversity Informatics" was coined to circumscribe the application of IT tools and technology to biodiversity information, principally at the organismic level. It thus deals with information capture, storage, provision, retrieval, and analysis, focused on individual organisms, populations, and taxa, and their interaction. It covers the information generated by the fields of systematics (including molecular systematics), evolutionary biology, population biology, behavioural sciences, and synecological fields ranging from pollination biology to parasitism and phytosociology.

Biodiversity Informatics is considered a part of biological informatics sandwiched between - and strongly overlapping with - environmental informatics and molecular bioinformatics. It will provide the skeleton for a generalized scientific information infrastructure in biology.

Presently, the overriding objective of research and other activities in Biodiversity Informatics is to provide a sound information management infrastructure for biodiversity and Global Change research. The basic objectives are:

- Mobilizing existing information resources
- Increasing research efficiency by timely provision of fundamental data for a steadily increasing number of problems
- Providing the information base and the tools for diverse biodiversity modelling tasks.

These are general problems, common to many scientific disciplines, so why is there a need for a new sub-discipline? The reason lies in the inherent complexity not only of the problem domain itself but also of its basic data. In meteorology, for example, the basic data are essentially location-referenced numeric data plus information on the methods used to obtain these measurements. This is not to negate that the interpretation of the data, e.g. those obtained with remote sensing techniques, may be highly complex. However, in biodiversity research and data capture we are dealing with concept-related data at the very basic level. To cite but one example: the presence or absence of a certain species at a specific place is not only a question of recording the name, location, and time, but it also relates to the definition of the species - what we find is the organism, what we record is its classification. The matter becomes even more complex in palaeontology, where even the location (stratum) where the species is found is the designation of a concept.

There are also a number of specific sociological and political obstacles to general access to biodiversity information. For example, primary information, which has been made accessible for research purposes, may be used for commercial aims (bioprospecting). According to the Convention on Biological Diversity and to many national laws, the rights of stakeholders such as nations or indigenous people must be respected.

Data capture is proceeding in most institutions in a more or less organized form, ranging from the individual scientist to entire organizations and thematic networks, from collections of text documents or spreadsheets to sophisticated databases. However, as mentioned before, the basic information we are dealing with is complex, and so structural features (e.g. in database design) and content definitions (e.g. in controlled vocabularies, i.e., lists of applicable terms) vary widely. Biodiversity Informatics has to meet this challenge by providing consensus reference systems.

On the structural side, much progress has been made in the last decade, when several groups published standard formats or even comprehensive information models for subject areas such as natural history collections and biological collections in general, palaeontological collections, and taxon names.

On the content side progress is also notable, but less rapid. The main reference system for biodiversity information, the classification of animals, plants, fungi, and micro-organisms into groups called taxa is essentially a classification of concepts, and consequently no "standard" system can be devised without impeding progress in systematics and evolutionary research. Taxon based information systems (or systems using taxon names) must find ways to map individual taxon concepts reliably.

However, taxon concepts are but one of the problem areas designers of biodiversity information systems face. Most biodiversity research is associated with organisms in the field, which must be properly identified. Practical species identification tools can and must be delivered for field and laboratory research. Up to now, this is still largely a process of formal scientific description, and the terminology used and ontology needed for the description of organisms is very extensive. Striving for universal coverage of descriptive terms is clearly not the way to go. The current challenge for biodiversity informatics lies in the creation of descriptive systems (and identification tools based on these), which are useful also for the non-specialist.

In scientific research, results must be falsifiable, so in biodiversity research the possibility to re-examine the organism itself should be maintained wherever possible. This is and has been done by depositing voucher specimens, suitably conserved organisms or part of them, in natural history collections. These institutions not only provide a solid base for modern biodiversity research, they also serve as archives of biodiversity. The specimens deposited there over several hundreds of years represent samples of biological space-time. They document the existence of an organism at a certain time at a defined place. They are samples of biological material, which can be analysed for example to measure environmental parameters or genetic variation and document their change over time. Consequently, biological collections are a cornerstone of biodiversity research, and the electronic representation of their content is a priority task for Biodiversity Informatics.

One of the principal reference systems for biodiversity research actually lies outside biology: the spatial reference system, including palaeontological space. Biodiversity information systems need sophisticated spatial reference systems, which are able to unite several layers of information related to a geographic reference point, which are able to scale horizontally and vertically from very large features right down to the small (e.g. strata in a rain forest) and the very small (e.g. soil communities inside a dung heap). Since many of the historic references found on specimens are not exact (referencing historic locality names and not co-ordinates), sophisticated gazetteers have to be incorporated which include the historical component of place names. In contrast to the domain of taxon names, the problem of language representation becomes very important in geographic names.

Information structures and data references

Research in information modelling and standardization of biological databases has been an important (though sadly under-funded) issue in the past decade. The IUBS Commission for Taxonomic Databases has been a driving force and an important forum for information exchange. However, most of the experience was gained in the course of the implementation of information systems in individual biological institutions, large and small, with a heavy rate of failure due to sub-estimation of the complexity of tasks and lack of expertise and liaison.

With respect to taxon names and biological collection data, detailed information models exist which can serve as a base for implementation projects. Some of the hot issues in that realm include concept mapping and data quality and automatic linkage evaluation tools. On the data side, we have extensive data collections, covering the names of higher plants, mosses, fungi, and zoological names. Apart from these name lists, promising attempts exist to unite information on the species level. However, a global federation of species databases, as envisioned by the Species 2000 project is still lacking the coordinated funding effort, which is needed mainly to create the underlying group-specific information systems. The FishBase information system provides a particularly fine example for what several years of consistent and well-managed funding can achieve in terms of international research collaboration and data provision. For collections, attempts are made to create metadata based systems which are thought as an intermediate step on the way to specimen level information systems. The research potential of specimen level information networks has been demonstrated by the Australian Environmental Resources Network, the NSF funded Species Analyst project , the European Natural History Collection Information Network, the Comisión Nacional para el Conocimiento y Uso de la Biodiversidad and other systems on national level.

Investigating user interests

Two EU financed projects have recently examined the interests of (potential) users and data providers for an electronic information system on biological collections: ENHSIN - the ongoing European Natural History Specimen Information Network, a research infrastructure network initiated by CETAF, the Consortium of European large-scale Taxonomic Facilities and BioCISE, a concerted action project which finished in December 1999 and published its findings (Berendsohn 2000). Since collection data include many of the aforementioned aspects of biodiversity information, their results may here be used in a broader context.

BioCISE differentiated four major groups of users of collection information (after Felinks & al. 2000):

- Users working in the environmental sector (monitoring and impact studies, development of guidelines, administration) e.g. the environmentalist compiling information on past changes in the flora and fauna of a particular area to assess the possible impact of building activities or invasive species (but also: the customs officer seeking information on the legal status of plant or animal specimens to be imported).
- Users from biological systematics and environmental research (basic research) e.g. the zoologist looking for type specimens within a taxonomic group to finish her revision of the group, the ecologist investigating communities and populations, the plant pathologist looking at host-parasite relationships.
- Users from the fields of biotechnological research and industry (applied research and production), e.g. the biotechnology company looking for a particular bacterial strain or for a plant producing a specific natural substance.
- The general public (education and entertainment), e.g. the teacher inquiring for the whereabouts of oak species in the Botanic Garden to prepare a visit with his second grade, or the child interested in large dinosaur skeletons exhibited in a particular city.

This grouping is essentially valid for biodiversity information in general, although basic biodiversity research includes wider questions such as the actual basis for species richness, evolutionary mechanisms, etc.

BioCISE also concluded that the primary data access points are geographical areas and taxon names. Users want user-group specific, visual, interactive, intuitive user interfaces, with scalable, standardized, high-quality and documented output.

Information Provider's View

Information providers face a different set of challenges and requirements. Efficient data capture software is needed, which is flexible enough to be adapted to ever changing research needs. With respect to the networking of information sources, the question of the protection of intellectual property rights, ensuring the proper crediting of data sources, maintenance of control over the use of their information, and (sometimes) commercialisation of the information are issues ardently discussed among, e.g., the big natural history museums. Long term archiving of primary data from research projects, which are not directly connected to specimens, is an unsolved problem. Here, cooperation with the library community must be sought.

Technical realization

Biodiversity data are mostly international in scope and interdisciplinary in content, so their acquisition and distribution asks for wide ranging collaboration. The World Wide Web has had a tremendous levelling effect on information distribution techniques, from early attempts (Gopher etc.) to HTTP and now XML standards. Web browsers offer universal common user interface functionality independent of hardware and of the location of the user. Information access through a network of distributed and largely independent nodes is now feasible. A prerequisite for this is the provision of standardized interfaces to data resources on the Internet, and international working groups are tackling this aim (e.g. the recently instituted CODATA working group on access to biological collection data). However, data retrieval is but one side of the equation. Biodiversity informatics will also be in the forefront of the development of user interfaces adequately representing highly complex data structures in distributed systems.

Global Biodiversity Information Facility

In the 1990ies, in the wake of the Convention on Biological Diversity, governments increasingly realized the scope of the needed informatics infrastructures related to biodiversity research. Consequently a (limited) dedication of research funding to took place, spearheaded by Australia and the United States, but also e.g. by Canada, Great Britain, and Mexico. In the course of tackling the problem on the national level, the global scope and magnitude of the problem became apparent. In January 1996, the U.S. submitted a proposal to the Organization for Economic Cooperation and Development's (OECD) Megascience Forum to create a Working Group on Biological Informatics, with a subgroup on Biodiversity Informatics (Edwards, pers. comm.). Over the following 3 years, this international expert panel elaborated a proposal for the creation of a Global Biodiversity Information Facility (GBIF, Edwards 1999). With the signing of an intergovernmental Memorandum of Understanding by 14 States the GBIF finally came into existence on the first of March, 2001.

The creation of the GBIF, and the process leading to it, is of twofold importance. On the one hand, the Secretariat will provide funding bodies, institutions, and individuals with the much needed guidance on the allocation of resources and the availability of solutions. Current priorities for coordination and (to a limited amount) direct action address the problems specified above. Topmost items on the agenda are access to collection information and the creation of a global checklist of species names. On the other hand, the discussion process has alerted governmental funding bodies to the existence of the problem domain, and much discussion has be set in motion on the national level. The recent creation of specific programmes for Biodiversity Informatics in, e.g. Germany, Belgium and the creation of a European Biodiversity Information Network with dedicated funding can be considered results of that process. Apart from the Secretariat, the GBIF is to be created by national research funding; so further programmes will have to follow. A first step to be taken by signatory states is to create national GBIF nodes, which are to organize and coordinate the contributions of member states. This process poses a challenge to existing structures many of which up to now work in splendid isolation, to co-ordinate their efforts and to clearly identify priorities.

USE OF BIODIVERSITY INFORMATION SYSTEMS

The realm of biodiversity informatics are information systems which use taxa, specimens, and (species) observation records as their reference and index systems. They are able to link geographic, climatic, and environmental information with data on the molecular or physiological aspect of organisms, with functional data, bioindicative value, human usage, natural substances, etc. Specimen and observation data are parts of a global environmental monitoring process, they document for example changes in the biotic spectrum of an area, including the occurrence and expansion of biological invasions, the distribution of indicator species (for example those mirroring air quality), the distribution of pests and diseases, changes in the behaviour of species, to name but a few.

Biodiversity information systems also form part of obligations taken by governments e.g. in the context of the Convention on Biological Diversity, Bern Convention, Bonn Convention, and EU directives.

Knowledge discovery and discovery of gaps in knowledge will be the direct result of the creation of a Global Biodiversity Information Facility. The role of biodiversity informatics in this context lies currently primarily in the documentation of biodiversity as a resource and of its change, the networking of inhomogeneous information resources, the development of adequate user interfaces for data capture and retrieval, the standardization of information exchange and content creation methods, the develpment of methods for quality control, and archiving as well as securing the recycling of research results.

REFERENCES

Baud, R. H., Lovis, C., Rassinoux, A.-M. & Scherrer, J.-R. 1998: Alternative ways for knowledge collection, indexing and robust language retrieval. *Meth. Inform. Med.* 37: 315-326.

Berendsohn, W. G., Costello, M. J., Emblow, C., Güntsch, A., Hahn, A., Koenemann, J., Thomas, C., Thomson, N. & White, R. 2000: Concepts for a European Portal to Biological Collections. Pp. 59-70 in: Berendsohn, W. G. (ed.), *Resource Identification for a Biological Collection Information Service in Europe* (BioCISE). - Berlin.

Bisby, F.A. 1998: Putting names to things and keeping track: the Species 2000 programme for a coordinated catalogue of life. - Pp. 59-68 in: Bridge, P., Jeffries, P., Morse, D. R. & Scott, P. R. (ed.), *Information technology, plant pathology & biodiversity*. - Oxon, New York.

Duckworth, W. D., Genoways, H. H. & Rose, C. L. 1993: *Preserving natural science collections: Chronicle of our environmental heritage*. - Washington.

Edwards, J. L. 1999: "The Global Biodiversity Information Facility: An international network of interoperable biodiversity Databases". *ASC Newsletter* June/August 1999:6-7

Felinks, B., Hahn, A., Olsvig-Whittaker, L. & Los, W.: Users and uses of biological collections. Pp. 19-32 in: Berendsohn, W. G. (ed.), *Resource Identification for a Biological Collection Information Service in Europe* (BioCISE). - Berlin.

Bibliography

Ahituv, Niv & Newmann, Seev (1990). *Principles of information systems for management*. Third edition. Won C Brown Publishers, 2460 Kerper Boulevard, Dubugue 145 2001

Anderson, A. J., Tari, P. H., and Tepper, C. S. 1988. Genetic studies on the role of an agglutinin in root colonization by *Pseudomonas putida*. Appl. Environ. Microbiol. 54:375-380.

Bannerman, R.H.1982. Traditional medicine in modern health care. *World Health Forum*3(1): 8-13.

Banpot Napompeth, 1994. *Leucaena psyllid in the Asia-Pacific Region: Implications for its management in Africa.* FAO, Regional Office for Asia and the Pacific, Bangkok, Thailand, RAPA Publication: 1994.

Baud, R. H., Lovis, C., Rassinoux, A.-M. & Scherrer, J.-R. 1998: Alternative ways for knowledge collection, indexing and robust language retrieval. *Meth. Inform. Med.* 37: 315-326.

Campbell, C. L. and L. V. Madden. 1990. *Introduction to Plant Disease Epidemiology*. Wiley, New York.

Cunningham, A.B.1993.*African medicinal plants. Setting priorities at the interface between conservation and primary healthcare.* - Paris, UNESCO (People and Plant Working Paper 1).

Dallmeier, F. 1992. *Long-Term Monitoring of Biological Diversity in Tropical Forest Areas.* MAB Digest, UNESCO.

Dickinson, M . *Molecular Plant Pathology*. BIOS Scientific Publishers. 2003.

Doane, C.C. & McManus, M.L. 1981. *The gypsy moth: Research toward integrated pest management.* USDA Forest Service, Science and Education Agency and Animal and Plant Health Inspection Service, Technical Bulletin 1584, 757 pp.

Duckworth, W. D., Genoways, H. H. & Rose, C. L. 1993: *Preserving natural science collections: Chronicle of our environmental heritage.* - Washington.

FAO1995.*Non-wood forest products for rural income and sustainable development.* - Rome (Non-wood Forest Products 7).

Francl, L. F. and D. A. Neher (eds.) 1997. *Exercises in Plant Disease Epidemiology*. American Phytopathological Society, APS Press.

Furniss, R.L. & Carolin, V.M. 1977. *Western forest insects.* USDA Forest Service, Miscellaneous Publication 1339, 654 pp.

Gehue, Hazen L. (Hazen Leslie) *GPS integrated systems for precision farming.* Geomatics Engineering, University of Calgary. 1994

George N. Agrios. *Plant Pathology,* Academic Press. New York. 1997.

Gilbert, J.M. & Miller, L.W. 1952. An outbreak of *Sirex noctilio* (F.) in Tasmania. *Australian Forestry,* 16: 63-69.

Gilligan, C.A. 2002. An epidemiological framework for disease management. Advances in Botanical Research 38:1-64

Groombridge, B. & M. Jenkins1994.*Biodiversity data sourcebook.* - Cambridge, UK, World Conservation Press (WCMC Biodiversity Series 1).

Hennessy, Nik & Lawlor, Damian 2000. *Computer software requirements in forest management.* National Council for Forest Research and Development (COMFORD)

Iqbal, M.1993.*International trade in non-wood forest products. An overview.* - Rome, FAO.

Jones, D. G. (ed.) 1998. *The Epidemiology of Plant Diseases.* Kluwer Academic Publishers. Dordrecht, Boston.

Lucas, Henry C. (1986). *Information systems concepts for management.* McGraw-Hill series in management information system. McGraw-Hill Inc.

Maciariello, Joseph A. (1984). *Management control systems.* Prentice-Hall International, Inc., Englewood Cliffs, New Jersey 07632.

MacIver, D.C., 1998. *Atmospheric Change and Biodiversity. Environmental Monitoring and Assessment* 49:177-189.

Maloy, O. C. 1993. Plant Disease Control: Principles and Practice. John Wiley and Sons, Inc., New York.

Manion, P.D. 1991. *Tree disease concepts.* Prentice Hall, Engelwood Cliffs, NJ (USA), 402 pp.

Mark S. Boguski, "Bioinformatics—A New Era," *Trends Guide to Bioinformatics* (Elsevier Science, 1998).

McDowell, J.M. and Woffenden, B.J. "Plant disease resistance genes: recent insights and potential applications". *Trends Biotechnol.* 21, 178–183. 2003.

Murdick, Robert G. (1986). *Management information systems concept and design.* Prentice-Hall International Inc., Englewood Cliffs, New Jersey 07632

Newman, William H. (1975). *Constructive control: design and use of control systems.* Prentice-Hall International Inc., Englewood Cliffs, New Jersey.

Ohmart, C.P. & Edwards, P.B. 1991. Insect herbivory on *Eucalyptus. Annual Review of Entomology* 36: 637-657.

Ordentlich, A., Elad, Y., Chet, I. 1988. The role of chitinase of *Serratia marcescens* in the biocontrol of *Sclerotium rolfsii.* Phytopathology 78:84-88.

Pagiola, S., and G. Platais. Forthcoming. *Payments for Environmental Services: From Theory to Practice.* Washington: World Bank.

Rafea, A. 1998. Agriculture. In J. Liebwitz (ed) *Handbook of applied expert systems.* CRC Press, Boca Raton.

Smit, B. I. Burton and R.J.T. Klein. 1999. The Science of Adaptation: A Framework for Assessment. Mitigation and Adaptation Strategies to Global Change 4.

Speight, M.R. & Wainhouse, D. 1989. *Ecology and management of forest insects.* Clarendon Press, Oxford, UK, 374 pp.

Stankey, G.H. 2001. Too Early to Tell, or Too Late to Rescue? Adaptive management under Scrutiny. US Forest Service, PNW, Science Findings, April 2001.

Stuiver, M.H. and Custers, J.H.H.V. "Engineering disease resistance in plants". *Nature* 411, 865–868. 2001.

Swortzel, Kirk A, "Role of Information Technology in Agriculture". *Agricultural Education Magazine,* v73 n6 p26-27 May-Jun 2001.

Thiollay, J.-M. 1995. "The Role of Traditional Agroforests in the Conservation of Rain Forest Bird Biodiversity in Sumatra." *Conservation Biology,* 9, pp.335-353.

Vanderplank, J. E. 1963. *Plant Diseases: Epidemics and Control.* Academic Press, New York.

Whetzel, H. H. 1929. The terminology of plant pathology. Proc. Int. Cong. Plant Science, Ithaca, NY, 1926:1204-1215.

Wormald, T.J., *Mixed and pure forest plantations in the tropics and subtropics.* FAO, Rome, Italy, Forestry Paper 103, 152 pp. 1992.

Zadoks, J. C. and Schein, R. D. 1979. *Epidemiology and Plant Disease Management.* Oxford University Press, London and New York.

Zelek, C.A., and G.E. Shively. 2003. "Measuring the Opportunity Cost of Carbon Sequestration in Tropical Agriculture." *Land Economics,* 79:3, pp.342-354.

Index

Agro-ecosystems 119
Alternative strategy 125
Alternative tree species 119
Audio recording industry 5

Behavioural sciences 294
Better stand management 122
Biodiversity research 296
Bioinformatics 10
Biological collections data 293
Biological collections 293
Biological control methods 122
Biological databases 297
Biotechnological research 298
Bioterrorism 21

Cellular subsystems 15
Clinical image analysis 15
Collaborative software 9
Communication devices 1
Comparative genomics 14
Computational biomodeling 17
Computational evolutionary biology 11
Computational genomics 17
Computer based technology 2
Computer related technology 2
Computer simulations model 12
Computer technology 2
Computerized management 288
Cultural tactics 127

Damage assessment technologies 129
Darwin core 290
Decision process 126
Digital archiving 4
Digital capabilities 1
Digital environment 1
Digital revolution 1

E-business applications 7
Ecosystem dynamics 119
Ecosystem stability 124
Effective management 1
Electronic information management 8
E-mail address 6
Enterprise architecture 8
Environmental parameters 287
Evolutionary biology 294
Evolutionary mechanisms 298
Exotic forest plantations 122
Exotic pests 121
Expressed cDNA sequence tag 12

Forest plantation programmes 120

Genome annotation 11
Genome-wide association studies 10
Genomic revolution 10
Geographic information systems 127
Geospatial information 288
Global biodiversity data networks 289
Global climate models 287

High-throughput image analysis 15
Human resource information systems 7

Information infrastructure 295
Information modelling 297
Information superhighway 2
Information systems 2
Information technology 1, 5, 17
Integrated pest management 126
Intellectual property rights 298
Intellectual property rights 3
Interaction overload 1
Interoperable Informatics Infrastructure Consortiu 20

Invasive plants 119

Landline telephony 6
Landsat satellite imagery 129
Life Science Identifiers (LSIDs) 288
Longhorn wood boring beetles 124

Machine-machine communication purposes 288
Machine-machine transmission 290
Mass spectrometry 13
Massively parallel signature sequencing 12
Mixed-species ecosystems 124
Modeling biological systems 15
Modern computer techniques 287
Morphometrics 15
Mycoplasma-like organism 124

National Archives and Records Administration (NARA 4
Nuclear magnetic resonance spectroscopy 16

Palaeontology 295
Parasitism 294
Personal digital assistant 6
Pest management tactics 129
Phytosociology 294
Pollination biology 294
Population biology 294
Protein microarrays 13
Protein-protein docking 16
Protein-protein interactions 10

Regulatory tactics 127

Single species plantations 125
Sophisticated statistical software 3
Species richness 298
Spruce budworm 120
Synecological fields 294

Taxon based information systems 296
Taxonomic databases 287
Taxonomic information 292
Technical realization 298
Transforming information exchange 5
Tree breeding programmes 128
Tropical rain forest 119

Verification technologies 1

X-ray crystallography 16